PRAISE FOR *OVERHEATED*

"A compelling indictment of capitalism. . . . Well documented and necessarily provocative."

—*Kirkus*, starred review

"Kate Aronoff is so sharp, witty, and relentlessly on target that reading her fills me with hope. *Overheated* is a blistering account of the many varieties of denial that have prepared the ground for climate catastrophe—and a thrilling tour of the kind of visionary politics and policies that could put the future back in our hands. Please: read this book."

—Naomi Klein, author of
This Changes Everything: Capitalism vs. the Climate

"As this masterful volume makes clear, Kate Aronoff is one of the most important writers ever to take on the climate crisis. She's hard-headed in her assessment of how neoliberalism put us on the brink of civilizational collapse, but she's not hard-hearted: she offers a persuasive case for how, with lots of solidarity, we could still escape the worst of this mess. This book is careful, comprehensive, and compelling, and it should be widely read, since it offers a baseline understanding for thinking through the greatest challenge humans have ever faced."

—Bill McKibben, author of *Falter: Has the
Human Game Begun to Play Itself Out?*

"In this deep and vital analysis, Kate Aronoff subjects every flimsy pretext and prevarication for postponing the climate emergency to a forensic analysis. She uncovers the genealogy of ideas, and the flows of money behind them. And then, with a razor-sharp intellect, she eviscerates the climate deniers, one lie at a time. If we are to win the battle for a livable planet, we'll only do so with the moral clarity and intelligence that *Overheated* has in abundance."

—Raj Patel, author of *Stuffed and Starved:
The Hidden Battle for the World Food System*

"The Bible observes that 'the heart is deceitful, who can understand it?' The greed at the heart of climate denialism isn't new, but understanding how it has warped our imagination for what is possible is critical work for our time. I'm grateful to sister Kate Aronoff for doing this work and for making plain exactly what sort of bold action is needed if we are not only to preserve a livable planet but also revive the heart of our democracy."

—WILLIAM J. BARBER II, cochair of the Poor People's Campaign and author of *We Are Called to Be a Movement*

"Climate-driven apartheid that sacrifices the many for the few, or reformed democracy that will work for the vast majority? That is the choice, explains Kate Aronoff in this bracing call to save ourselves by saving our planet from fossil fuel corporations and their enablers."

—NANCY MACLEAN, author of *Democracy in Chains: The Deep History of the Radical Right's Stealth Plan for America*

OVERHEATED

OVERHEATED

How Capitalism Broke the Planet—
and How We Fight Back

KATE ARONOFF

BOLD TYPE BOOKS
NEW YORK

Bold Type Books
116 East 16th Street, 8th Floor, New York, NY 10003
www.boldtypebooks.org
@BoldTypeBooks

Printed in the United States of America

First Edition: April 2021

Published by Bold Type Books, an imprint of Perseus Books, LLC, a subsidiary of Hachette Book Group, Inc. Bold Type Books is a co-publishing venture of the Type Media Center and Perseus Books.

The Hachette Speakers Bureau provides a wide range of authors for speaking events. To find out more, go to www.hachettespeakersbureau.com or call (866) 376-6591.

The publisher is not responsible for websites (or their content) that are not owned by the publisher.

Print book interior design Jeff Williams

Library of Congress Cataloging-in-Publication Data

Names: Aronoff, Kate, author.

Title: Overheated : how capitalism broke the planet--and how we fight back / Kate Aronoff.

Description: First edition. | New York : Bold Type Books, 2021. | Includes bibliographical references and index.

Identifiers: LCCN 2020049065 | ISBN 9781568589473 (hardcover) | ISBN 9781568589961 (ebook)

Subjects: LCSH: Environmental policy—Economic aspects—United States. | Capitalism—Environmental aspects—United States. | Climatic changes—Government policy—United States. | Climate change mitigation—United States. | Energy industries—Environmental aspects—United States. | Energy policy—Economic aspects—United States. | Business and politics—United States.

Classification: LCC HC110.E5 A785 2021 | DDC 363.738/745610973—dc23

LC record available at https://lccn.loc.gov/2020049065 ISBNs: 978-1-56858-947-3 (hardcover), 978-1-6528-0046-8 (international), 978-1-56858-996-1 (ebook)

LSC-C

Printing 1, 2021

For Trent, Gavin, and Madison

CONTENTS

INTRODUCTION

From Great Acceleration to Great Transformation

THE EARTH IS an unforgiving scorekeeper and an ethically neutral one. The stratigraphic record has silently catalogued key moments in history, taking careful note as we leave behind little sediment markings on layers of rock. At some point, humans ceased to be only one of the billions of species participating in this planet's dynamic systems and became its defining feature—a geological epoch that since the early 2000s has been known informally as the Anthropocene.

When exactly we became the stars of the show is still the subject of some debate. In 2017, a cross-disciplinary group of scientists in the Anthropocene Working Group (AWG) reached the preliminary conclusion that plutonium fallout from Cold War–era nuclear arms tests marked the start of the Anthropocene around the year 1950.[1] The decades that followed were a Great Acceleration of human influence, helped along significantly by the unprecedented burning and unearthing of the long-dead creatures stuffed down into the layers of rock below us: fossil fuels. Three-quarters of the greenhouse gas emissions ever produced on earth have been created during that period by a global economy that grew fifteenfold, in the process depositing the remains of dinosaurs into the atmosphere as heat-trapping gases that help to warm everything below. So far, global average temperatures have risen by about 1 degree above preindustrial levels. Roughly 1 percent of the earth's surface is currently so hot as to be uninhabitable by humans. By 2070, a fifth of the planet could fit the same definition.[2]

Like rock formations or atmospheric concentrations of carbon, the causes of that Great Acceleration took time to accumulate. Rejected theories on the origins of our current epoch—Anthropocene precursors, as the AWG calls them—offer clues. Some have suggested a start date of 1784, when James Watt refined the design of the steam engine that, powered by coal, would fuel Britain's Industrial Revolution.[3] Still, others have argued that the Anthropocene began even earlier, and that the Industrial Revolution owes its existence to developments a century and a half before it. "The arrival of Europeans in the Caribbean in 1492," geographers Simon Lewis and Mark Maslin write, kicked off "a swift, ongoing, radical reorganization of life on Earth without geological precedent." The Columbian Exchange saw valuable species like maize and tobacco for the first time cross the Atlantic Ocean to be introduced to vast trading networks that already spanned Asia, Africa, and Europe. To European settlers escaping their disease-riddled and war-torn continent, the New World promised unlimited expansion and the freedom of a fresh start. Freedom and limitless growth for some meant annihilation for others, and within decades, the new arrivals had killed between 85 and 90 percent of North America's existing inhabitants—an estimated fifty million people—through disease, enslavement, famine, and murder. By 1610, the end of farming, fire, and other land use practices that resulted from that genocide led to the regeneration of over fifty million hectares of forest, woody savannah, and grassland across the so-called New World. That newly unfettered plant growth sucked up between 5 and 40 gigatonnes of carbon dioxide, enough to contribute to a Little Ice Age. From the Latin word for world, Lewis and Maslin call this extraordinary moment the Orbis spike, implying that "colonialism, global trade and coal brought about the Anthropocene." Annexing the Americas made industrialization possible, creating new flows of cheap resources and new markets for manufactured goods, subsidized by slaughter and slavery. The United States' original sin took an enormous amount of carbon out of the atmosphere. Not all emissions upticks and reductions are created equal.[4]

The sweeping scientific terms used to describe the climate crisis— of epochs and degrees and parts per million—obscure the sheer amount of human suffering that has fueled it. The largest single cause of climate change is particular humans moving about this earth and doing particular things to it and to one another, including pouring prodigious amounts of greenhouse gases into the sky. Some humans deem some lives more valuable than others in the hunt for profit. An imperfect shorthand for these processes is capitalism.

Capitalism hasn't tended to be a popular protagonist in stories about the climate crisis. Often, it's said to be a matter of faulty psychobiology: we humans are hopelessly greedy, hardwired not to deal with the earth-shattering consequences of our wasteful ways. "We have trained ourselves, whether culturally or evolutionarily, to obsess over the present," journalist Nathaniel Rich wrote in his blockbuster *New York Times Magazine* story about a failed early round of climate talks. We "worry about the medium term and cast the long term out of our minds, as we might spit out a poison."[5] Facing undoubtedly long odds, it's tempting to turn inward, too, seeking personal absolution by lowering your carbon footprint: have fewer kids, take fewer flights, and turn off the lights when you leave the room.

Yet, not long after Watt first fine-tuned his steam engine, just ninety corporations—almost all of them fossil fuel producers—have been responsible for two-thirds of all greenhouse gas emissions.[6] Since 1965, just twenty shareholder and state-owned fossil fuel producers have spewed out 35 percent of the world's energy-related carbon dioxide and methane emissions.[7] The richest 5 percent of the world's population, by and large those most insulated from the effects of the climate crisis, consume more energy than the poorest 50 percent.[8] In 2015, then ExxonMobil CEO Rex Tillerson's $145 million worth of shares in his company made the future secretary of state responsible for over 52,000 metric tons of carbon dioxide emissions that year alone—well over 3,200 times that of the average American, 6,400 times of the average resident of China, and 38,400 times of the average Indian.[9] Unlike ExxonMobil and other fossil-fuel firms, the average residents of those countries had not funded elaborate disinformation campaigns to spread doubt about whether those emissions were a problem, lobbied governments for their rights to continue extracting, nor spent vast sums painting themselves as the solution to problems they have continued to fuel. Diffusing responsibility for this crisis to the masses—or chalking it up to innate human greed—is a convenient narrative for Tillerson and his ilk. It also happens not to be true.

This book is about various kinds of climate denial and how to overcome them in time to salvage a livable future. Among them is the idea that we humans have collectively dug ourselves into this mess, a fable that distracts attention away from the mountains of evidence that particular humans, industries, and ideologies are still quite proudly holding shovels. Like other myths about the climate crisis, those about collective personal responsibility were mostly cooked up by the fossil fuel industry.[10] British Petroleum first popularized the concept in the mid-2000s, telling anyone who navigated to its carbon footprint calculator at the time that it was "time to go on a low-carbon diet."[11] This meshed well, of course, with the warped moral philosophy that had by that point captured political common sense on both sides of the Atlantic. "Too many people," the Tory British prime minister Margaret Thatcher famously said, "have been given to understand that 'I have a problem, it is the Government's job to cope with it!' . . . they are casting their problems on society and who is society? There is no such thing! There are individual men and women and there are families and no government can do anything except through people and people look to themselves first."[12]

How capitalism has developed as an economic and belief system shapes not just the carbon content of the atmosphere but world governments' continued inability to respond to this crisis with the requisite scale and speed. The way it developed in the United States is an especially important part of that story given its outsized footprint, and the primary focus of this book. Without a major course correction, capitalism will define, for the worse, how the US deals with the consequences of having waited so long, from wildfires to mass migration. Like everything else this country does, those consequences will reverberate well beyond its own borders.

Until recently, the Herculean level of effort required to curb runaway warming has seemed unfathomable. A pandemic put that and a number of other things in perspective. A preliminary analysis by the Global Carbon Project found that global shutdowns to contain the spread of COVID-19 reduced 2020 emissions by 8.8 percent during the first half of 2020, the largest single drop-off since formal record-keeping of that sort began in 1900.[13] Much of that, researchers projected, would bounce back as economies restarted, even as many pursued so-called green recoveries. In 2019, the UN Environment Program called for sustained global emissions reductions of 7.6 percent each year between 2020 and 2030 to cap warming at the "well below 2 degrees" threshold outlined in the Paris Agreement, reductions that ought to happen fastest in the parts of the world—like the US—where years of carbon-intensive plunder have helped build national fortunes that allow them to transition out of the fossil age more quickly.[14] If the novel coronavirus was a test run for the scale of transformation demanded by the climate crisis, the US has failed it.

IT'S TEMPTING TO lay blame for each of these catastrophic failures solely at the feet of Donald Trump and the GOP. History's sediments have, however, accumulated in our politics and on both sides of the aisle. Though the question of global warming itself has been polarized along partisan lines in the United States, there has been plenty of bipartisan consensus around the things that have already made navigating this century's crises more difficult. Since the 1990s, at least, Democrats have been eager collaborators with Republicans in casting doubt on the ability of government to get big, good things done, shifting far to the right on everything from financial reform to trade to immigration in the name of bipartisanship. That's not to draw some false equivalence between the GOP and the Democrats on either climate or the coronavirus but to point out that the kind of widespread governmental failure that has characterized the US response to the latter took decades, if not centuries, to evolve. That same government now has to respond to a climate crisis that, aside from making pandemics like COVID-19 or worse diseases more likely, could make the casualties the virus racked up look modest by comparison. By 2030, annual climate-related deaths are expected to reach at least 250,000 worldwide, according to the WHO.[15]

For years, though, this country's bizarre climate politics have revolved around a single question that has precluded any meaningful conversation about what it will take to stem that destruction: Do you believe the climate is changing or not? Brewed in C-suites and think tanks and universities, climate denial has always been a peculiar American invention, managing to distill so much about this country's reactionary past and present into one neat, crank-filled spirit: one part paranoid anticommunism, one part corporate capture of politics, and a healthy dose of manifest destiny, funded by the executives with the most to lose in the transition to a low-carbon world.

It's also waning. Trump has been a shot in the arm to the old-school climate deniers spouting off junk science, but theories about sunspots and global greening aren't long for this world. Today, the term *climate denier* is mostly used to set whoever lobs that charge apart from the person it's being lobbed at, that is: *I'm* not like *them,* those uncultured brutes. And yet an imagined enlightened future free of climate deniers could easily be shaped by oil companies, which help write the climate policies that should constrain them, and the right, which responds by strewing the world with fences and cages to keep out refugees from the climate crisis it increasingly agrees is happening. Like Trump himself, climate denial—so often understood as a strange, frustrating sideshow—is a conventional, even predictable product of US politics.

The nominal opposition to these forces has been sorely lacking. By now mostly devoid of the really committed free-market ideologues of the Bill Clinton–era New Democrats, the Democratic Party establishment today operates according to a set of aesthetic sensibilities about what constitutes reasonable policy. They express a quasi-religious belief in climate science, contra Trump and the GOP. At the same time, they laugh off the transformative measures many climate scientists urge are needed as unrealistic fantasies. What drives these politicians forward isn't an ideology or vision for a better world so much as an all-encompassing urge to defend shrinking turf against challenges on their right and left. They've gotten both the politics and the policy wrong. During the same period in which Democrats tailor-made climate measures to win over the GOP, they lost over a thousand seats in statehouses around the country and control over every branch of government. Aside from their general uneasiness about big, egalitarian government, politicians in both parties have avoided a head-on confrontation with a fossil fuel industry whose business model is incompatible with a livable future. Meanwhile, lawmakers who have taken millions of dollars in campaign contributions from them—mostly Republicans—still liken anything tinted green to Stalinist five-year plans.

The same interests and institutions that propagated climate denial—a small if phenomenally successful piece of their overall project—have constrained imaginations about what climate action could look like, too. In order to starve the parts of the state now most needed to avert climate chaos, the right mainstreamed radical ideas about how economies and states work that have placed the most straightforward means of curbing emissions on the fringe. So long as that broader consensus remains intact, the fact that more people than ever believe climate change is real is basically irrelevant to whether it's dealt with successfully.

If Beltway climate politics can feel a little hopeless, these last few years should also serve as a warning to anyone looking for a billionaire or a silver bullet new technology to save the world—or even just the US—from rising tides. To assume the 1 percent has some master plan for a greener future, even a grossly unequal one, gives too much credit to this demographic's planning abilities. Smash-and-grab profiteering can extend onward forever until there's no one left to scam. Burn all the fossil fuels because there's money to be made off them now. Whenever that stream runs dry, buy up the distressed assets of flailing electric utilities and oil and gas companies for pennies on the dollar and litigate the hell out of any restructuring deals. Or fashion a new speculative financial product out of the risks of them not paying back their exorbitant debts. If all else fails, bet on how bad the weather will get. Then bet on those bets. Invest the profits in a plan to inject reflective aerosols into the atmosphere to block out the sun and cool the planet. Have a twenty-three-year-old McKinsey analyst making six figures design the rollout plan. There's no secret long-term vision for what the world will look like in thirty or three hundred years, just a series of mostly disconnected schemes for how to make as much money as possible at any particular point in the stratigraphic record. The vultures will come, and they will leave richer than they came, whether the planet happens to be running on fossil fuels or not. Extraction predated the age of fossil fuels and may well outlive it.

Neither is there any accumulation of feel-good corporate and personal climate epiphanies that will convert all of the country's cooktops and space heating systems to run on electricity or string the thousands of miles of transmission lines necessary to support that transition. Voluntary pledges will not rewire the grid to accept electrons from millions of people rather than distributing them from fossil-fueled power plants, or change the antiquated laws governing the power sector that have made that so difficult, even as rooftop solar has proliferated. Personal responsibility will not erect electrified mass transit systems, nor will it transform the toxic and unwieldy supply chains that comprise the methane-spewing disease vector that is the world's food system. And it will not, by 2050, wind down global coal, oil, and natural gas usage by 97, 87, and 74 percent, respectively—the levels of reduction the Intergovernmental Panel on Climate Change (IPCC), a clearinghouse for climate science from around the world, suggests are needed to cap warming at the 1.5 degree Celsius outlined by the Paris Agreement, a threshold low-lying island nations and other climate vulnerable countries have long urged is critical to their survival.[16] Sufficiently motivated individuals will not keep that carbon buried. Leaving these changes up to the planning prowess of the market is a good way to make sure they never happen. If there is to be such a thing as a low-carbon society, it will be the government's job to build it.

In the US, at least, that will take a very different kind of government than the one we've got. Thanks to persistent pressure from social movements around the world, fighting against fossil fuel infrastructure and for a Green New Deal, among other things, there is now broad agreement among Democrats that government standards and investment need to play a leading role in curbing emissions. Where Barack Obama's top advisers balked at the notion of a more than $1 trillion stimulus to prevent the United States from slipping into a depression after the 2008 financial crisis, Joe Biden—who ran his primary campaign as a climate moderate—pledged to spend $2 trillion on climate priorities alone: to decarbonize the power sector by 2035, zero out emissions from buildings five years before that, install 500 million solar panels within four years, create a Climate Conservation Corps, and allocate 40 percent of all clean energy and infrastructure investments to communities living on the front lines of climate change and fossil fuel development.[17] Though modest compared to Bernie Sanders's $16 trillion plan for a Green New Deal, it was the most sweeping and ambitious suite of climate measures ever championed by a Democratic presidential candidate.

It's not nearly enough. As old-school climate denial loses ground, a new denialism that's every bit as dangerous is taking its place: that building more of the stuff that's needed to create a low-carbon future will replace the stuff that's killing us. Today's deniers don't spread misinformation about the reality of the crisis so much as what's needed to curb it. Painting decarbonization as a hopelessly complicated, indefinite undertaking—one that only today's polluters are fit to lead—isn't so far removed from questioning whether it needs to happen at all. My argument in this book is not that capitalism has to end before the world can deal with the climate crisis. Dismantling a centuries-old system of production and distribution and building a carbon-neutral and worker-owned alternative is almost certainly not going to happen within the small window of time the world has to avert runaway disaster. The private sector will be a major part of the transition off fossil fuels. Some people will get rich, and some unseemly actors will be involved. Capitalist production will build solar panels, wind turbines, and electric trains. But whether we deal with climate change or not can't be held hostage to executives' ability to turn a profit. To handle this crisis, capitalism will have to be replaced as society's operating system—setting out goals other than the boundless accumulation of private wealth.

The trouble is that growing private wealth is what the US government does best. Prison abolitionists Craig Gilmore and Ruth Wilson Gilmore have summed this dynamic up aptly: "The history of the United States is, in large part, the history of capitalists figuring out how to develop and use large-scale complex governmental institutions to secure their ability to get rich."[18] Despite all the right's cloying rhetoric about the virtues of small government, its leaders have creatively expanded the functions of the state to suit their own ends. Bipartisan statecraft has helped birth supranational institutions like investor-state dispute settlement systems that allow corporations to sue sovereign governments should they dare to infringe on profits—provisions used most frequently to protect investments in energy.[19] Coal, oil, and gas companies collected an estimated $5.2 trillion worth of direct and indirect subsidies in 2017 and were generously showered with support from the Federal Reserve during the most recent downturn even as local and state governments plunged into fiscal crises.[20]

There's a reason why the US government is so good at helping corporations extract fossil fuels, starting wars, and locking people up and so bad at providing health care and restricting carbon emissions, to name a few. As safety nets were starved and regulations rolled back, the right poured money into prisons and police departments throughout the country as part of wars on crime and drugs, which have treated the residents of predominantly working-class Black and brown communities around the country as if they were enemy combatants. In the last forty years, the number of people warehoused in state and federal prisons has risen by 500 percent.[21] Though it's home to just 5 percent of the world's population, the US is home to 25 percent of its total prison population; African American adults are nearly six times as likely to be incarcerated as whites. Liberal cities like Minneapolis and Oakland spend as much as 50 percent of their budgets on police departments.[22] The US has spent $6.4 trillion on wars in the Middle East and Central Asia since 2001, complemented by an elaborate build-out of domestic surveillance operations. Over the same period, the US has spared no expense to militarize the southern border with Mexico and erect new federal agencies devoted to extending the border into communities around the country, conducting raids aimed at detaining and deporting migrants. Government hasn't gotten smaller in the past several decades. It's gotten meaner, keeping capital free and people contained.

In this context, a lower-carbon society—should such a thing take shape—won't automatically be a more decent one. Driving these skewed investment priorities is an ugly, antidemocratic throughline in American politics. Similar logic stretches from the murder and dispossession of indigenous people to slavery; from the Redeemer governments of the post–Reconstruction South to Jim Crow apartheid to mass incarceration; and from the public choice theory economics so popular with fossil fuel billionaires like the Koch brothers to the Trump administration. The Founding Fathers themselves warned of "mobs" and "majorities" and wrote a Constitution intended to keep them at bay. All these forces have, through various means, looked to enshrine the minority rule of white property owners against persistent attempts to transform the United States into an egalitarian, multiracial democracy and squash movements pursuing similar aims abroad.

The 1 percent's most effective tactic for solidifying their power has been to divide the 99 percent against itself within and beyond our borders, deploying racism to justify the enormous inequalities baked into an economic system built around the extraction of labor and land. The public policy products of these brutally successful efforts—sclerotic safety nets, voter suppression, defanged regulations, and punishing trade deals, among others—stand directly at odds with the relatively narrow goal of decarbonization, much less building an enjoyable or sustainable twenty-first century. Bringing down emissions means declaring trillions of dollars worth of fossil fuel assets—all those reserves that can't be safely burned—worthless. If carried out, this would represent the single largest evaporation of private wealth since the Emancipation Proclamation. There are many more forces standing in the way of that than the Republican Party, and many more things wrong with capitalism than the fact that it runs on fossil fuels.

COVID-19 AND THE protests that emerged during the summer of 2020 laid bare the consequences of this country's radically misguided, typically bipartisan priorities. Nurses crowdfunded for protective equipment as tanks rolled through the streets of US cities on the prowl for unarmed protesters. Oil and gas companies were bailed out for months as they fired employees who had no safety net to catch them when the shaky foundations of fracking finally began to buckle under the weight of the industry's massive debt overhang. As tens of millions of people lost their jobs, they lost their health care, too. Hospitals told by management consultants to operate on razor-thin margins were overwhelmed, as cities considered digging mass graves to intern the dead. Around the world, more functional economies gradually eased back to something like normal, having implemented large-scale lockdowns and contact tracing systems. For the US, at least, normal was the problem.

When the Green New Deal first emerged with its goal of equitably decarbonizing the US economy within ten years, critics on the right and the center-left whinged about its pledges to "counteract systemic injustices" and its commitment to "repairing the historic oppression of indigenous peoples, communities of color, migrant communities, deindustrialized communities, depopulated rural communities, the poor, low-income workers, women, the elderly, the unhoused, people with disabilities, and youth."[23] Sure, they conceded, climate change is a big problem. And justice is nice. But isn't all this talk of a federal job guarantee and Medicare for All a bit *distracting*?

"Alexandria Ocasio-Cortez's Green New Deal appears to take every big spending idea that has emerged on the political left in recent years and combine them into one large package deal, with little notion of how to pay for them all," Bloomberg columnist Noah Smith wrote in one such take after Ocasio-Cortez introduced H.R. 109, a nonbinding resolution laying out the principles of a Green New Deal. Smith complained that "although a big push for renewable energy is needed, the Green New Deal's vast program for economic egalitarianism could make it unworkable."[24] He wasn't alone. Conventional wisdom in Washington had to that point wagered that the best route to climate policy was to sneak it in, hiding provisions in omnibus bills and lathering up proposals with enough bureaucratic jargon to make them virtually incomprehensible. A carbon price here. Some clean energy tax credits there. Eventually, the thinking went, it would all add up to cap warming at 2 degrees. Just don't shake the boat too much.

The problem was assuming the boat didn't already have holes. They were obvious to the people forced to live with polluting fossil fuel infrastructure in their backyards, some of whom have spent decades organizing in climate and environmental justice groups to push for an end to fossil fuel extraction and infrastructure build-outs. Energy Transfer Partners' $3.8 billion, 1,172-mile Dakota Access Pipeline had originally been slated to run through Bismarck, North Dakota, which is 90 percent white. After residents raised a fuss, the US Army Corps of Engineers rerouted it farther south, to cross just upstream of the Missouri River and under the main water source of an 84 percent Native residential area, through the unceded territory of the 1868 Fort Laramie Treaty.[25] In response, Native residents from several tribes constructed a network of camps along the proposed path to block construction; the largest of these—the Oceti Sakowin encampment—at one point housed as many as fifteen thousand people. The sites would become a flashpoint for indigenous sovereignty and environmental justice both, attracting Native and non-Native visitors eager to stop the pipeline and media coverage from around the world.

Water protectors faced down police water cannons, snarling guard dogs, and hired company thugs through the harsh North Dakota winter.[26] Eventually, they won a major victory in having the pipeline rerouted, though many had hoped it would be scrapped entirely; in 2020, it was emptied of oil pending a full environmental review.[27]

It's not some great historical accident that some of the places worst hit by the pandemic and police violence are also the fossil fuel industry's sacrifice zones. Organizers in Indian country and other communities on the front lines of climate change and extraction have for years pushed predominantly white-led green groups to take fights from the Unist'ot'en Camp to Protect Mauna Kea seriously, in both their opposition to extraction and vision for what can come after it. As scholar, organizer, and Lower Brule Sioux Tribe citizen Nick Estes explains, these varied encampments "rise *against* colonial and corporate extractive projects. But what's often downplayed is the revolutionary potency of what Indigenous resistance stands *for*: caretaking and creating just relations between human and nonhuman worlds on a planet thoroughly devastated by capitalism."[28]

That the Green New Deal reflects long-held demands for justice is a credit to that intramovement organizing. Its limitations—and ambivalence about fossil fuel phaseouts, in particular—are a testament to the fact that there is more work to be done. Encouragingly, it is a living and open source document. Rather than a suite of preordained policies, the Green New Deal is a framework for reimagining the fractured social contract upon which this country was built. As I draw out in the second half of this book, there are more lessons and cautionary tales to be drawn from its namesake in the New Deal and the domestic mobilization around World War II, invoked frequently as a model for the scale of industrial policy and administrative coordination needed to build a clean energy future. There are less common reference points, too—in the democratic experiments during Radical Reconstruction and the Freedom Budget's ambitious plan for a full employment peacetime economy, crafted by leaders in Black Freedom struggles seeking to extend the gains of the civil rights movement into the realm of economic democracy. Yet if it's to live up to its lofty promise, the Green New Deal won't be limited to history.

The past decade has seen a resurgence of social movements teeming with visions and policy proposals about what a better world can look like, from Occupy Wall Street to waves of teachers strikes, to Native-led up-risings against fossil infrastructure, and the ongoing movement for Black lives. An expansive coalition including fisherfolk, racial justice groups, labor unions, environmentalists, tribes, and college activists came to-gether in 2019 to launch the Gulf South for a Green New Deal Policy Platform. Born out of facilitated sessions across Texas, Louisiana, Ala-bama, Mississippi, and Florida, it includes demands for reparations and land redistribution, the repeal of regressive "Right to Work" and "At Will" employment laws, and the cleanup of toxic lands polluted by the area's extractive industries. "Relocation processes must be self-determined by communities and must assure the social, cultural, and economic require-ments for a transitioning community to survive and thrive," the platform states; Louisiana is home to the United States' first officially internal cli-mate migrants, whose home on Isle de Jean Charles is being rapidly in-undated.[29] The Red New Deal, drafted by Red Nation—the indigenous resistance organization Estes cofounded—similarly outlines a plan for restoring indigenous land to indigenous people, keeping fossil fuels in the ground and defunding the police and military to build a caretaking economy. There's resonance here, too, in the Movement for Black Lives Policy Platform, developed over more than a year of thorough collabora-tion among M4BL network members. Released in 2017, its invest-divest plank demands "investments in Black communities, determined by Black communities, and divestment from exploitative forces including prisons, fossil fuels, police, surveillance and exploitative corporations."[30]

Through the fall and early winter of 2020, the Sunrise Movement and Iowa Citizens for Community Improvement brought together Iowans in cities and rural areas to map out what their state could look like in 2030, focused largely around farm policy and regenerative agriculture.[31] Aside from adapting the Green New Deal framework to local contexts, painting a vision for what a world with it can look like helps people connect with and imagine a different reality. Shawn Sebastian, from Ames, Iowa, who worked to convene the visioning sessions in his home state, likened that translation challenge to a box of brownies. "You don't put the ingredients on the front of the box. When you look at brownie mix in the grocery store, you don't see a list of ingredients," he told me. "What you're seeing is a completed, luscious, moist brownie. I think as policy people we tend to list out the ingredients and say, 'We need x and y and z.' But we don't put the whole picture together of what we're fighting for."[32]

ACROSS THE POLITICAL spectrum, mitigating and adapting to climate change has been framed as a problem of scarcity. The right fearmongers about the Green New Deal coming to take your airplanes and hamburgers. On the left, scarcity arguments can occasionally take on a Puritanical bent: the earth can only hold so many people, and so those of us on it need to forgo our earthly pleasures in service of planetary salvation. Economists translate this as a trade-off between ecological preservation and economic prosperity, wherein any efforts to curb those inherently valuable activities that produce emissions now represent a tax on the earnings of generations down the line.

There are things those of us living comfortably in the Global North have come to expect and that won't have a place in a sustainable future: Amazon Prime's two-hour delivery, perhaps, or lawns in desert suburbs and diets rich in factory-farmed meats. Still, it's worth asking who today is really prospering, even just in the US. Between 1973 and 2013, productivity rose by 74 percent as wages stayed roughly flat.[33] A staggering racial wealth gap means that the typical Black household owns just ten cents for every dollar owned by whites.[34] The annual incomes of the top 0.01 percent have grown by 343.2 percent since 1979. And all this has happened as greenhouse gases have been poured into the atmosphere with abandon. Whatever this economy and its prolific carbon emissions are making, it is not a more prosperous life for most people.

The most meaningful trade-offs for building a low-carbon future, I'd wager, aren't the ones economists generally have in mind. The business model of the fossil fuel industry can't continue to exist in one that's tackling the climate challenge head-on. A world where Jeff Bezos can accumulate $13 billion in a single day isn't one that's compatible with valuing the work of teachers, nurses, and other essential, low-carbon workers. A US attempt to jingoistically dominate clean energy export markets and have its companies hoard intellectual property isn't consistent with a rapid global energy transition. The economists might also do well to remember the risks of failing to mitigate emissions, the real costs of which could easily creep into the range of hundreds of trillions of dollars. The climate debate is less an issue of how to distribute the planet's scarce resources and more of how to share its abundance more equitably. That also means reassessing what sorts of activities are really valuable in a just and sustainable society. *We*, broadly speaking, can have nice things—including a habitable planet.

I use the word *we* here and in the subtitle of this book advisedly. We did not somehow land ourselves in this mess: they did, though its effects are distributed unevenly. As dirty a word as populism has become, a low-carbon populism—defining an encompassing we to go up against the polluting elites—may be our best shot at a decent future. One of the scarier concepts in the science of global warming has to do with feedback loops: disasters that feed on and exacerbate one another, like California's wildfires in 2020 unleashing thirty million more tons of carbon dioxide that year than the state's power sector.[35] We can harness a different kind of feedback loop: by prioritizing climate policies that make people's lives better in the short run and grow the power of democratic institutions like labor unions, a Green New Deal can swell the multiracial, working-class coalition invested in designing and fighting to expand those programs as they scale back emissions and build up a fairer, cleaner economy. And it can create durable electoral majorities that ensure those changes stick for decades to come.

What critics of the Green New Deal have tended to miss is that its policy ambitions are one and the same with its political strategy. It's an opportunity to plan out a vision for a future that isn't either some parched Mad Max–style dystopia or a techno-optimist fantasy, where fleets of Tesla EVs ferry between sprawling single-family suburban smart homes. A Green New Deal won't stop climate change; at this point, nothing can. Yet even as those effects of the climate crisis already coming our way play out, a low-carbon future can be a more leisurely, abundant, and democratic one. Shorter workweeks can make more time for trips on electrified trains to beaches along remediated coastlines, spent sipping wine grown by vintners paid to sequester carbon. Those transitioning from work in carbon-intensive sectors like coal or oil can take advantage of free college and five years of guaranteed wages to retrain in a new field, or make just as much as they were plugging up orphaned wells, reclaiming abandoned mines for nature preserves, or weatherizing old housing stock. Single-payer health care and full employment can give workers the freedom to leave jobs and bosses they don't like and participate in the energy transition, whether as a solar engineer or a preschool teacher or a playwright, one of the many living wage, low-carbon jobs on offer through a federal job guarantee. Instead of new prisons, the government will invest in millions of new units of energy efficient schools and affordable housing in cities, suburbs, and rural areas.

As they trend away from old-school denial, right-wing politicians are likely to keep making their favorite jabs: that Democrats' climate policies are an unwieldy lefty wish list and a Trojan horse for big entitlement programs. Why not give them and their fossil fuel industry donors something worth being scared of?

PART 1

NEW DENIAL, OLD IDEAS

CHAPTER 1

CLIMATE DENIAL IS DEAD

STANDING BEFORE THE hundred and fifty-odd people gathered in a Dusseldorf hotel conference room on a November afternoon in 2017, Wolfgang Müller, the general secretary of the European Institute for Climate and Energy, asked attendees to return from lunch a few minutes early so they could take a group photo. About one hundred of them did. As they gathered to take the shot, Müller walked around, distributing stemware and pouring champagne. On camera, they toasted: "To Donald Trump pulling out of the Paris Agreement!"

They clinked their glasses at the 11th annual International Conference on Climate Change, cohosted by the Müller's organization (EIKE, in the German abbreviation), the Competitive Enterprise Institute (CEI, an American outfit), and a handful of smaller groups of self-identified climate skeptics. Billed as a contra-COP23, it was about an hour's train ride from the twenty-third annual Conference of the Parties (COP23) of the UN Framework Convention on Climate Change (UNFCCC) in Bonn, the first such gathering since Donald Trump took office and pledged to pull out of the Paris Agreement the preceding summer.

There's more than one way to be a climate denier, and just about every kind was floating around Western Germany that November. Müller and company fit the stereotype: cranks poking holes in the scientific consensus, railing against the pointy-headed academics—often, though not in Müller's case, with generous industry funding. His kind are a dying breed. Explaining that EIKE couldn't provide translation into languages other than English and German, Müller quipped that "the check from Exxon-Mobil keeps getting lost in the mail."

It's easy to see why EIKE sits on the margins. In one presentation, a historical building preservationist argued that medieval construction—castles with two-foot-thick stone walls—was better suited to insulate heat than Germany's apparently tyrannical energy efficiency standards, in a talk that included an extended, only half-joking anecdote involving sex and boar skins. A session on renewables pleaded sympathy for wildlife; literature handed out by the earnest young Swiss presenter featured a picture of a dead bird at the foot of a wind turbine. The sole caption, in German, asks: "Bird shredder?"

It's all pretty pathetic. The US is one of the few countries on earth where climate deniers making similar claims have enjoyed access to the reins of power. Given its status as the world's largest economy and its second-largest polluter, that's not something to be taken lightly. Former EPA administrators, including Christie Todd Whitman and Gina McCarthy, estimate that the damage wrought by their Trump-era successors could take three decades to repair.[1] Clearing the way for new fossil fuel exploration and infrastructure was a key priority for the Trump administration, serving not just to legitimize the kinds of climate denial that had been on the outs before him but also to lock in toxic coal, oil, and gas projects and—via regulatory rollbacks—make them tougher to shutter down the road.

Yet an hour's train ride from the EIKE confab, at a sprawling UN campus along the Rhine River, was a preview for the kinds of climate politics that could soon dominate in the US. Unfortunately for the rest of us, they are only marginally more in touch with scientific reality than our German revelers.

The EIKE diehards' conclusions may not be empirically grounded, but they go down easy: relax, everything will be OK. Another version of that message was being marketed across COP23. As climate scientists called for a dramatic transformation of the world's economy, a different set of deniers started to coalesce around plans to tackle the climate crisis that acknowledge its urgency but concretely offer only market tweaks, technological quick fixes, and hopeful messaging as an alternative. These plans, in other words, may well still not avert disaster. Unlike Müller, these softer deniers have been at the center of the climate policymaking debates the world over. Exhibition halls at COP23 in 2017, as in the years before and since, were dotted with stalls sponsored by fossil fuel companies proselytizing carbon capture and storage technology; international investment banks eager to discuss the central role of private finance in driving the new green revolution; and polluter-backed researchers exploring the necessity of spraying particulates into the air to block out the sun.

That climate change exists, is man-made, and presents one of humanity's most pressing challenges is common wisdom for the attendees of UNFCCC talks, including some of the world's most right-wing heads of state. The relevant question for them isn't whether the earth is heating up but what we intend to do about it: to reduce emissions domestically, push other countries to do so, close off borders to climate migrants, or some combination of all three. That's a radically different conversation about climate change than the one that's been going on in the United States for most of my lifetime and an approach only just starting to creep into the national debate. Here, decades of propaganda from the fossil fuel industry and the think tanks they support have forced the debate to orbit around whether there's a problem at all, prying open the Overton window of acceptable policies to accommodate conspiracy theorists and Nobel Prize winners alike.

As Naomi Oreskes and Eric Conway document in *Merchants of Doubt*, there were several staging grounds that built the playbook for climate denial in the US: defending cigarettes, acid rain, and more against regulation. Many of the scientists recruited by industry to be early soldiers in these fights started their careers researching weapons technology for the Pentagon, meant to keep the world safe from the Soviet Union. "When the Cold War ended," Oreskes and Conway write, "these men looked for a new great threat. They found it in environmentalism."[2] That their funders in organizations like the American Enterprise Institute (AEI) and the Competitive Enterprise Institute (CEI) did too isn't incidental. As Naomi Klein has noted, these free-market propagandists rightly saw in the climate crisis a problem that could only be dealt with through the sorts of big government interventions they hated most.[3] Republican congressman Tom McClintock summarized the sentiment during his keynote address to the Heartland Institute's 2019 conference: "If the earth truly hangs in the balance, well then no measure is too extreme. No cost is too great. No governmental excess is too oppressive to enact their agenda," he told the crowd. "How much of a sacrifice is it if the alternative is a dead planet?"

In what Klein has called an "epic case of bad timing," public awareness about climate change coincided with the zenith of a wide-ranging political project known as neoliberalism, which lays out an often loosely defined set of rules for and beliefs about how capitalist economies should function.[4] The advocates who spearheaded the neoliberal revolution on either side of the Atlantic Ocean after the oil and inflation crises of the 1970s—familiar names include Milton Friedman and Friedrich Hayek—didn't sell lower corporate tax rates or rollbacks of labor protections on their own merits; they sold freedom and small, responsible government, realized through markets. In this country, American business leaders offered sage guidance and cold hard cash, eager, with a little coaxing, to get behind a program that was looking out for their profits. Rather than seeking a wholesale scale back of government, neoliberalism's various strains coalesced around theories of state design, seeking to deploy laws and supranational institutions like the World Trade Organization in closing markets off from alleged threats—including and often democracy.[5] Central to the success of that project was breaking the power of unions, whose national policy priorities and power in the workplace posed a threat to unmitigated corporate profits. In the US, antidemocratic militants were especially aggrieved by the claims of nonwhite people to ballots and public budgets, fearing that a majoritarian democracy might vote to distribute the hoarded wealth of white plutocrats to multiracial masses. To get the word out about such dangers, neoliberal stalwarts in the US took over economics departments at places like the University of Chicago and George Mason University, convened closed-door confabs, built institutions like AEI, CEI, and the Heritage Foundation, and participated in global forums like those hosted by the Mont Pèlerin Society, an international organization of neoliberal thinkers founded in 1947. They cultivated promising talent for higher office, including Barry Goldwater and Ronald Reagan; wrote and published books, like *The Limits of Liberty* and *The Road to Serfdom*; and founded magazines and radio shows, like *National Review* and *The Manion Forum*, aimed at spreading the gospel.

Fossil fuel fortunes, in particular, would be generous backers of these policy entrepreneurs. Young, fiercely ideological, and newly at the helm of his family's fossil fuel empire, Charles Koch took a particular liking to James M. Buchanan, whose public choice theory—as described in rich detail by Nancy MacLean's *Democracy in Chains*—offered a sort of cleaned-up continuity with the racist John Birch Society conspiracy theories his father had imbued in him from a young age. One of the earliest fights Buchanan took on from his post at UVA's Thomas Jefferson Center in Charlottesville was against the Supreme Court's decision in *Brown v. Board of Education*. He hoped to preserve the apartheid order of the Old South against the prospect of integration and the slippery slope it presented to a multiracial democracy—and he'd dismantle the state's system of public education in order to do so. Destroying such a cornerstone public good was more than a means to an end, though. For Buchanan, MacLean writes, "venal self-interest was at the core of human motivation." Public choice theory took the ensuing position that a government run democratically by more humans could only hope to reflect the venal self-interest of the politicians they put in charge to—as he saw it—provide for voters so as to win reelection. "Why," Buchanan wondered in *The Limits of Liberty*, "must the rich be made to suffer?" He asked what set "simple majority voting" that might raise taxes on wealthy men apart from "the thug who takes his wallet in Central Park?"[6] Democracy, in his estimation, bore a fatal flaw: "How can the rich man (or the libertarian philosopher) expect the poor man to accept any new constitutional order that severely restricts the scope for fiscal transfers among groups?" To adequately safeguard property rights, he argued, there needed to be a "generalized rewriting of the social contract," keeping the few in charge at the expense of the many.

Whereas Koch considered the likes of Milton Friedman and Alan Greenspan "sellouts," Buchanan—a fellow member of the Mont Pèlerin Society—was a true believer. They would join forces sometime after the latter's failed crusade against *Brown v. Board of Education* and pursue similar ventures that ranged from school privatization to climate denial, which

Koch's Cato Institute would play a foundational role in seeding.[7] Other Koch-sprouted outfits—including the American Legislative Exchange Council, State Policy Network, and Americans for Prosperity—would work diligently in the decades to come, turning Charles Koch's most hardline dreams of market supremacy and minority rule into reality.

Beyond enthusiastic efforts by the US Chamber of Commerce and the National Association of Manufacturers to spread doubt about the existence and causes of global warming, the Olin Foundation—seeded by a chemical and munitions fortune—was instrumental through the 1970s and '80s in spreading a like-minded law and economics movement, "an intellectual enterprise that approached law using the tools of neoclassical economics."[8] As legal scholars Jedediah Britton-Purdy, David Singh Grewal, Amy Kapczynski, and Sabeel Rahman write, law and economics "simultaneously recognizes and embraces the fact that law makes markets, while demanding that the satisfaction of markets becomes the aim of politics," at the exclusion of, say, environmental protections. Neither this ecosystem of right-wing of thinkers, think tanks, and academic departments that took root after the New Deal nor their funders were a monolith, but in the US especially, they pushed forward a few core ideas. If they didn't oppose democracy on philosophical principle, they certainly opposed the version of it in which the civil rights movement expanded the electorate and the environmental movement demanded more safeguards against polluters' economic liberty. An insurgent right created institutions to do battle with both and more at every level imaginable. In many cases, the bodies that translated white reaction into public policy were those that most aggressively pushed a neoliberal agenda and climate denial. Frequently, it has been difficult to draw clear lines between those projects, helmed as they were by the same institutions and even the same people.

When it came to climate denial, the right's media strategy was especially effective. That the cranks and credentialed scientists spent years cohabitating on the same cable news panels meant that the climate debate has played out on deniers' terms. And once broad swaths of the American public had become convinced that market forces are society's best problem solvers while state action is an only occasionally necessary evil, any discussion of reasonable, large-scale climate solutions—stringent

regulation, massive public investment, an economy planned around reducing emissions—was off the table. As the US political landscape shifted
more dramatically to the right once Reagan took office, these two phenomena produced a parade of utopian market-based solutions for everything from health care to climate policy, dreamed up by the likes of
Milton Friedman and eventually embraced by left and right alike. In the
climate debate, neoliberals have successfully insisted that the fossil fuel
industry is a good-faith partner in policy formation rather than an actor
who must be brought to heel. The same disinformation campaigners that
created a debate over the reality of climate change have hedged their bets
and have now staked a claim to solving a problem they tried to convince
the world didn't exist.

In the spring of 2018, Royal Dutch Shell—Europe's biggest oil company—released a pathway to meeting the low-bar commitment laid out
in the Paris Agreement to cap warming at 2 degrees Celsius above preindustrial levels. The plan, to reach net-zero emissions by 2070, is by most
accounts hugely ambitious. It's also premised on two big fantasies: that
fossil fuel production and consumption can continue at roughly similar
levels for the next several decades and that at some point between now
and then we will have figured out how to suck massive amounts of carbon from the atmosphere with so-called negative emissions technologies,
which remain unproven and uneconomical at scale after decades and billions of dollars in research investments.

The wishful thinking baked into Shell's decarbonization scenarios,
though, also plagues the research compiled by the UN's IPCC, where the
"least-cost" pathways to decarbonization are also those anointed to keep
warming below catastrophic levels. It's an admittedly tough line to walk
between the right's decades of attacks on scientific authority and the reality that scientific knowledge is itself the product of heated debates and
rigorous process—all of which is vulnerable to being taken out of context by bad-faith actors. I appeal on occasion in this book to scientific
research and also recognize that climate scientists can agree with one another about the broad strokes of a problem—for instance, the existence
of climate change and the urgency of decarbonization—while disagree-
ing about which particular policy prescriptions work best, for reasons
that may or may not directly relate to their academic research. Scientists

are humans, and like all fields, the various disciplines included under the broad umbrella of climate science have their own internal politics and microdynamics that naturally find their way into bodies like the IPCC, shaped as they are by the political debates playing out around them. It's my view that climate science helps to define the boundaries within which policy should happen, but that policy will necessarily be the product of political choices and democratic processes that look to meet any number of other criteria. Political scientists, sociologists, historians, economists, and members of other academic disciplines have a lot to offer those conversations, as do grassroots organizers, storytellers, investigative journalists, and the many other experts on various subjects who devote their lives to understanding this vast problem.

The fossil fuel industry, I argue, can only play a destructive role in the climate policymaking process—and are actively angling for it to be a major one. A study by the London-based watchdog InfluenceMap found the world's five largest oil companies have spent $1 billion rebranding themselves as "green" since the Paris Agreement, all the while pushing aggressively to access new supplies of oil and undermine climate rules and regulations at various levels of government.[9] Shell alone has sent 111 representatives to UN climate talks in the last several years, where there is still no formal conflict of interest policy.

As carbon-guzzling multinational corporations stand ready to play hardball, the debate over what to do about climate change is much harder to win than the one over whether it's happening. Major polluters, keen to have that climate debate play out on their terms, are coming to the table at international climate talks with ready-made plans. Forty years into the neoliberals' long march through our institutions, even well-intentioned policymakers have had trouble coming up with alternatives. With precious few years left to course-correct away from catastrophe, political theorist Fredric Jameson's most famous, probably apocryphal quote has taken on a more literal meaning: "It's easier to imagine the end of the world than the end of capitalism."

TITANS OF INDUSTRY are nothing if not good materialists, ready to adapt when they sense a change in the political weather. That they're now changing their tune on climate is also why the likes of EIKE are starting to seem more like living anachronisms on the world stage. Perhaps more than anywhere else, the vein of outright climate denial on display in Dusseldorf that winter has long been a marginal force in Germany. While hardly without its flaws, that country's state-led Energiewende, or energy transition, has been lauded as a model for other industrialized countries looking to get off fossil fuels. As of 2020, they had passed a rule to entirely phase out coal production by 2038. It's all part of a larger plan for the country to reduce its carbon emissions by 40 percent below 1990 levels by 2020, even if that is a goal it will almost certainly miss. The tenor around warming is different in Germany, where even conservative political parties don't tend to question either the existence of climate change or the fact that something needs to be done about it. Among EIKE's biggest concerns is what they see as a blackout in the German media around climate skeptic points of view. "They act as if we don't exist," Müller lamented. The notable exception to Germany's climate groupthink—and a rare friend to EIKE's top brass—is Alternative for Deutschland (AfD), the upstart far-right party that, as of my writing, remains the biggest opposition party in the Bundestag, Germany's national parliament. Founded in 2013 by a handful of academic economists, the AfD rose to prominence and electoral success largely by filling the gap in right and center-right politics created by German chancellor Angela Merkel and her ruling coalition's drift leftward on issues such as immigration, LGBTQ rights, and climate change. "Carbon dioxide . . . is not a harmful substance but part and parcel of life," the AfD manifesto asserts, before laying out a handful of common denier talking points. "The IPCC and the German government," the party contends, "conveniently omit the positive influence of CO_2 on plant growth and world nutrition. The more CO_2 there is in the air, the more plant growth will be."[10]

The AfD's ascendance coincided with the revival of France's National Rally, the nationalist Austrian Freedom Party's short-lived entry into a governing coalition, Viktor Orbán's Nazi-curious prime ministership in Hungary, and Poland's ruling Law and Order Party, as well as right-wing strongmen outside the European Union like Jair Bolsonaro, Vladimir Putin, Rodrigo Duterte, and Narendra Modi. For most of these parties, the climate crisis has been an afterthought. Rather than denying global warming outright, they filter their response to it through their other reigning beliefs. Similarly, in the US, climate denial is at this point much less central to the right's agenda than its racism and xenophobia. But climate denial is still nonetheless at home in the Republican coalition, where its conspiratorial mind-set fits neatly alongside John Birchers and QAnon rantings about child sex trafficking rings. An American keynote speaker at the EIKE conference, Mark Morano, made headlines at COP22 in Morocco for walking around the convention center in a Make America Great Again hat.

While Morano didn't have the ear of the Trump administration, so far as we know, plenty of people saying similar things certainly did. Climate skepticism under Trump was the ruling party line in the US, aped by congresspeople and the heads of key regulatory agencies. The director of CEI's Center for Energy and Environment, Myron Ebell, was tapped by Trump to head the transition effort at the Environmental Protection Agency (EPA). William Happer—a physicist who got his start in denial casting doubt on the ozone layer in the 1990s—briefly served as Trump's senior director on the National Security Council. Trump's EPA and Department of Interior have opened the door to extraction and rolled back regulations on mining and drilling at breakneck speed, fulfilling in a matter of months many of the priorities industry groups have advocated for years. HuffPost's Alexander Kaufman reported that top Interior officials cited the work of prominent deniers in official communiques.[11] Beyond chiding Obama's supposed "war on coal," Trump himself has been known to flirt with denialist talking points. Visiting California amid the West Coast's destructive 2020 wildfire season, he said it's "going to get cooler," when presented with evidence about the links between the blazes and rising temperatures. When a state official disagreed, he said, "I don't think science knows, actually."

In the middle of what would seem to be an organizational high point, the vibe at the Heartland Institute's 2019 conference—held, symbolically, in the gaudy ballroom of Washington's Trump Hotel—wasn't much more optimistic than the one at EIKE's conference in Dusseldorf. The median age of the crowd at each was about sixty-five, though a few young staffers dutifully manned information desks. It's hard at places like the EIKE and Heartland Institute conferences to feel like these shrinking groups of old men are much of a threat, commanding smaller audiences than when cable news networks regularly trotted them out to give contrarian takes on global warming. Previous conferences had brought in senators helming powerful committees. This year's big draw was Tom McClintock, a House backbencher. Before it was postponed by COVID-19, the 2020 Heartland lineup didn't feature any sitting politicians to speak of. With beleaguered funding and clout, what's driving deniers now is a variation on what's driving their opponents: preserving their way of life. The denier network has become a kind of community for people shocked at the possibility of a threat so big it might threaten even them, and for the network's funders, it provides a way for aging cranks to pay the bills. Some believe what they're saying, and some don't, and there's no point trying to find any coherence in it. What they all firmly believe, though, is that—whether from climate "alarmists" or #MeToo or the Black Lives Matter movement—the world they have known is under attack. For this one day, the Trump International Hotel is their safe space, a category that, for them and few others, used to envelop most of the country and, thanks to various colonial adventures, the world.

I asked Pat Michaels, a conference circuit veteran who had just been ousted by the Cato Institute and one of the few credentialed climatologists in the bunch, whether the age of the crowd made him worried that there wasn't much fresh blood being injected into the so-called climate realist movement. Puffing out his chest a bit, Michaels made a point of informing me that he "hangs out with quite a few young people" because "my wife is quite a bit younger than me." And yes, he conceded, there is a generational difference when it comes to climate. But the moment we're in is unique since the earth, as opposed to a decade ago, is actually warming. A little. But, he insisted, it won't last! Once we're through with this particular and very normal cyclical variation in temperatures, everyone will calm down, and he and his friends will be vindicated by the truth. Like the tobacco industry's manufactured insistence that the link between smoking and cancer is spurious—a crusade Michaels joined in on—the main function of climate denial has never been to convince the public of any particular point of view, just to cast doubt on reality, positioning the deniers as brave truthtellers willing to buck dogma.[12]

By the time we spoke, Michaels had spent decades going against the grain of established science. In doing so, he was nurtured by the same institutions that had sponsored other reactionary ideas. It was in his capacity as a professor at the University of Virginia—where he taught environmental studies for nearly thirty years—that he attended a 1991 Cato Institute conference entitled Global Environmental Crises: Science or Politics?, one of many such events to follow. In a 2013 blog post railing against public funding for scientific research, Michaels lamented that Buchanan had been too optimistic in seeing science as insulated from the same public sector rot that had infected other knowledge fields. "In reality, public choice influences on science are pervasive and enforced through the massive and entrenched bureaucracies of higher education," Michaels countered.

While Buchanan's minoritarian ideas have flourished these last few years, Michaels and his cohorts' contrarian science has withered in the public eye. Partway through our conversation, we got interrupted as an old buddy of Michaels's walked up and started to chat, reminiscing about old times. Until 2015, Fred Palmer, now in his late seventies, had served as the senior vice president of Government Relations at Peabody Energy and before that in various other positions at coal industry lobby groups and trade associations, following a stint in government. As we spoke, he lit up describing a new project of his that he called "propeople, proenvironment, pro-CO_2." It's not all coming up roses, though. "My frustration level has been very high, seeing everybody in the industry just pull back," he told Michaels. "People have gone to ground because of the stigma associated with resisting," referring to climate science. "It's finished." When I asked Palmer to explain, he paraphrased his former colleagues in the coal industry: "While we're not going to get out of the fossil fuel business, we're just going to be invisible. And we will make do with what we can and try to get things put in place that are good for us in the meantime and let this pass, or not. But not be activist against it." Funds that once flowed to denier groups from the fossil fuel interests, including coal companies, have dried up, he said.[13]

Circa 2010, Michaels estimated that roughly 40 percent of his funding came from the oil industry.[14] By 2015, the George C. Marshall Institute—a legacy denier think tank whose papers George W. Bush's administration used to justify more lax climate policies—folded into the Center for Strategic and International Studies, which still collects generous checks from the oil industry but has gotten out of the denial business. "You can forget about asking money from Exxon; they send all their money to Stanford [University] or to Princeton [University] for greenwashing," Happer, former chair of the Marshall Institute's board, told E&E News when asked about the merger.[15]

"Coal has pulled back," Palmer said. "Coal now talks about 'reducing our emissions,'" through methods like carbon capture and storage or sequestration. "But the game has changed . . . You cannot find any public advocacy on behalf of coal like you used to. I was at Peabody . . . I haven't changed. I happen to think more CO_2 in the air is good and not bad. You cannot find one executive at a coal company that will say that." Asked how all this has impacted his work, Palmer was circumspect. "Well, we're still here. But the funding . . . you've got to be more clever in terms of hustling," adding that he hoped to raise money for his new venture through crowdfunding and Facebook.

Stephen Milloy—a blogger at JunkScience.com and another member of the small crew that frequents these gatherings—voiced similar gripes to me about oil funding drying up, railing against ExxonMobil for having abandoned the fight when it stopped funding Heartland more than a decade ago.[16] This set apparently couldn't count on the White House, either. Happer ended up leaving the National Security Council about a year after taking his post, reportedly facing internal dissent over his attempts to conduct an "adversarial" review of climate science, in which he has no formal training.[17] Onstage, Ebell fretted that their side was losing the battle inside the administration. Trump's inner circle was allegedly full of people who are "squishy" on climate questions.[18]

In the months after the conference, things got even more dire for Heartland. Though the dark money group Donors Trust stepped in with support for denier groups after polluting industries began to walk away, Heartland in March 2020 laid off several staff as part of a "reorganization." The shedding followed both the loss of funds and repeated sexual harassment scandals involving senior staff. As one longtime staffer put it bluntly in a leaked text message, "Heartland is broke." The smart money on climate denial has moved on.

THOUGH THE DENIERS are struggling, their footprint lives on. Among the most pernicious effects of just how toxic the climate debate in the US has become is the dangerously low bar it has set for what constitutes progress. Simply *believing* well-established science has been enough for Republicans to garner breathless news coverage, and a steady trickle of them have started talking in word salads about innovation and tree planting. Notorious GOP pollster and strategist Frank Luntz—who helped the Koch brothers build their fossil fuel empire and counseled George W. Bush's administration to rebrand global warming as climate change—issued a mea culpa in the summer of 2019, stating that "I was wrong" and pointing to rising support among young Republicans for something called climate action.[19] "That was a lifetime ago," he pleaded. "I've changed." Mitch McConnell has said bravely that he does "believe in human-caused climate change," and a trickle of Republican-sponsored bills addressing climate change has appeared, attempting to stake out some ground. Congressman Matt Gaetz—who at one point called to dismantle the EPA—put out the vague Green Real Deal in 2019, aimed at creating "market-driven clean energy solutions."[20]

If the fossil fuel industry's recent history offers any indication, Republican climate plans will become more common in the coming years, not less, as Republicans look to distance themselves from the old-school denialism quickly falling out of fashion and follow the lead of corporate donors, who recognize the need to say the right things about climate change to investors and voters increasingly alarmed about the risks of rising temperatures. Economic historian Philip Mirowski, whose work has focused on the network and institutions he calls the neoliberal thought collective (NTC), suspects that conventional denialism has always been more of a useful distraction than a belief system for the right. "I don't think most of these people really believe in denialism," Mirowski told me. "The left can feel all noble fighting them because they're fighting ignorance. But I think denial is just a feint to absorb all that energy while they push forward the stuff they really believe." He suggests climate denial is—more than anything—a short-term strategy to buy time while industry-aligned lawmakers and think tanks work out a longer-term plan. Fossil fuel companies' performative shift toward caring about the climate would seem to bear out his theory.

The climate crisis wouldn't be the first blow the NTC has improbably weathered. Mirowski's book, *Never Let a Serious Crisis Go to Waste*, tracks the rise of neoliberalism and how its biggest ideas—the primacy of the market and the veiled but active role of the state—survived what should by all accounts have proven its death knell: the 2008 financial crisis. Mirowski argues that what made neoliberal doctrines so resilient was the ideology's decades-long project of institution building and eventually statecraft, bit by bit shifting the terms of economic common sense. Though it is the water economics departments and public policy debates swim in today, so-called free-market ideas were once fringe, supported by a minority of economists when men like Friedman and Hayek first started trying to spread their doctrine via the Mont Pèlerin Society, comprised of economists, philosophers, and scientists. Drowned out by the Keynesian post–World War II consensus, they and their collaborators around the world began developing new ideas, and working through internal ideological divisions and developing what would become a multifaceted worldview, to be unveiled on the world stage by—among others—the administrations of Margaret Thatcher in the UK and Reagan in the US. All those decades of work paid off. As Reagan settled into the Oval Office in 1980, he handed every member of his prospective cabinet a copy of a 1,100-page Heritage Foundation document that detailed some two thousand conservative policy priorities. The document, which would later be published as the Mandate for Leadership series, served as the cornerstone of right-wing leadership; in his first year alone, Reagan would take up nearly two-thirds of its proposals.

"One of the reasons that the neoliberals have come to triumph over all their ideological rivals in recent decades is that they have managed to venture beyond any simplistic notion of a single 'fix' for any given problem," Mirowski writes, "but always strive instead to invent and deploy a broad spectrum of different policies," argued for and rolled out by a revolving door of industry groups, think tanks, and lawmakers."[21] Neoliberals have always operated within a world of contradictions, arguing in public that the market offers freedom and that it needs to be insulated from democracy; painting the market as a natural part of human existence while pushing to pass policies that keep it functional. Just like climate denial, there are some dupes who truly believe in *laissez faire* and the invisible hand, but the real movers and shakers have always been more pragmatic than dogmatic.

The Cato Institute, Heritage Foundation, and AEI have all, at one point or another, held strong ties to NTC members and accepted vast sums of money from the coal, oil, and gas industries to curry favor for proextraction policies. Several of those same organizations, however, also devote resources to drafting climate policy, including some—carbon taxes, carbon trading markets, and carbon capture and storage—that have been championed by progressive Democrats. The AEI regularly publishes papers fleshing out plans for levying fees on pollution, even while its staff attend Heartland Institute gatherings.[22]

Politically savvy oil companies have done the same. Like Shell, most multinational petrol firms have dropped old-school denialism altogether. Even the US-based producers that have tended to have less progressive messaging on the climate have changed their tune. ExxonMobil has for years factored some level of carbon pricing into its long-term projections and in recent years has been vocal about its support for such a policy. In the lead-up to the Paris climate talks in 2015—the ones that resulted in the Paris Climate Agreement—BP, Eni, Royal Dutch Shell, Statoil, and Total all called on the UNFCCC and world governments to adopt a carbon-pricing scheme, the shortcomings of which will be explored in depth in the chapters to come. The American Petroleum Institute (API), the lobbying arm of the oil and gas industry, followed suit. In line with API member oil and gas companies, the group's president and CEO, Jack Gerard, in 2015 invited world leaders to develop a "market-driven blueprint that achieves emissions reductions without sacrificing jobs, economic growth and energy production."[23] Since then, API has created a Climate Change Task Force. European oil producers, especially, have gone so far as to talk about the need for an energy transition—however far off. If they're listening to their donors, savvy Republicans aren't likely to keep bucking their fossil fuel industry donors by spouting outright denial for too much longer. Before long, more of them will start writing climate policy.

If carbon-pricing and trading schemes really are part of neoliberals' medium-term plan for pretending to address climate change, progressive lawmakers have been more than happy to play along. Former California governor Jerry Brown spent much of his time at COP23 championing his state's freshly renewed cap-and-trade program—drafted in part by the state's oil and gas lobby—as a model for state-level action in the Trump era, despite what many experts have described as its negligible role in reducing the state's emissions relative to standards and regulations.[24] A carbon tax proposal first introduced by climate hawks Democrat Senators Sheldon Whitehouse of Rhode Island and Brian Schatz of Hawaii in 2017 mirrored several conservative proposals for the same thing, citing as a positive its similarity to a plan endorsed by ExxonMobil and former Bush and Reagan cabinet members. Fittingly, it was unveiled at the AEI's headquarters in Washington, DC. Even within UNFCCC processes, talk of market-based solutions to climate change—unleashing private financing for renewables, letting the market itself phase out carbon-intensive coal plants, rolling out emissions trading schemes—is now dominant, with a growing buzz around prospects for carbon removal schemes and geoengineering.[25]

Lawmakers' imaginations aren't only constrained when it comes to reducing emissions. For about as long as I've been alive, a belief that markets composed of rational actors are the best tool for solving the world's most pressing issues has been at the heart of Democratic and Republican agendas. Big government spending packages on infrastructure and entitlements and stringent industry regulation were, until fairly recently, standard operating procedures for Western politicians across the political spectrum. But if, per Thatcher, there is "no such thing as society," then what obligation could the government possibly have to it and its myriad problems? Perhaps the most serious constraint on ambitious government action has been fearmongering about the size of the federal deficit. Although it'd be virtually impossible for the US government to default, the right has cast big spending programs as an intolerable burden on future generations, despite its presidents' penchants for driving up big deficits while in power through both wars, tax breaks for the wealthy and corporations, and the realization that big budget items like Social Security are actually popular. As with climate denial, trying to find any coherence in deficit hawk rhetoric is a fool's errand.

Though it sprouted on the right, Republicans and Democrats alike echoed these sentiments during tense budget debates in the nineties, when Luntz advised lawmakers to evoke images of "parents sitting around the kitchen table going over bills" in their efforts to curb spending.[26] If hardworking Americans have to balance their budgets, after all, then why shouldn't their government? Of course, Mom and Dad don't have access to central bank monetary policymaking. That fact hasn't stopped Democrats from adopting similarly wrongheaded lines about budgets. Before retaking the reins as Speaker, Nancy Pelosi wrote a so-called pay-go provision into the rules of the 116th Congress after Democrats retook the House in 2018, pledging a commitment to not enact policies that would add to the federal deficit. "We all have responsibility for reducing the debt for our children," she had urged. "Democrats believe that you must pay as you go. Whatever you want to invest in, you must offset."[27]

Neoliberalism's best trick may not have been convincing a few useful idiots in high places that climate change isn't a problem but convincing both sides of the political spectrum that an all-powerful market is the best way to deal with the crisis it created—and that a big, active government is bound to do more harm than good.

At a dinner party in 2002, Thatcher was allegedly asked by a guest what she saw as the greatest achievement of her political career. Her answer, an attendee of that event reported, was "Tony Blair and New Labour. We forced our opponents to change their minds."[28] Shortly before his election in 1997, Thatcher was similarly rosy about the future Labour leader's prospects, saying the UK had "nothing to fear" from a Labour government ready to enforce spending cuts and mirror the Tories' supply-side economics. By the early 2000s, Reagan's mind had largely succumbed to Alzheimer's. Had he been cogent, his answer to the same question might have been similar: Bill Clinton and the modern Democratic Party. That the same party is now tasked with decarbonizing the economy should make us all nervous.

THE CLIMATE CRISIS has come along at the worst possible moment: amid starved public spheres and anemic economic thinking in the Global North and an ascendant, xenophobic far-right. It would be naïve to blame global warming on neoliberalism, but still more so to say it hasn't crippled our ability to deal with it in a way that's anything other than dystopian.

The upside is that reasonable solutions to climate change lend themselves well to today's populist times, and these kinds of redistributive policies could help stem the rise of the far-right. Transforming the electric grid, fortifying coastlines against sea-level rise, and manufacturing solar panels on a large scale could form the backbone of the biggest jobs program America has ever seen and set millions of people up with well-paid, fulfilling work. On the left and the right, populism is built on pitting *us* against *them*; climate change makes those sides all too clear, and they will only become clearer as storms and droughts continue to batter poor communities worst. That people like Michael Bloomberg were until recently some of the most visible faces of the climate fight—offering their own support for piecemeal market-based solutions—does little to diminish the idea that only elites have the luxury of caring about climate change. Billionaires jet-setting around to UN climate talks and the World Economic Forum make it almost too easy for right-wingers to call out their hypocrisy: Why do they get to travel the world while asking us to give up our jobs, vacations, and hamburgers?

The Sunrise Movement and politicians like Alexandria Ocasio-Cortez have begun to articulate a kind of left populist climate politics here in the US, building largely on longtime demands from climate and environmental justice groups. They understand that the scale of changes climate science demands will require doing something the neoliberals early on recognized as crucial: taking state power, then using it to radically rethink the relationship between the state and the economy—in this case, toward building a more equitable, low-carbon world.

But the right has its own rising climate populism, pushed forward by fresher faces than the EIKE set. In the lead-up to the 2019 European elections, AfD leaders had doubled down on climate denial, mounting a bizarre campaign against Swedish teenager Greta Thunberg, who inspired the youth-led climate strikes that had helped turn the climate crisis into one of the election's top issues.[29] The party grew modestly, garnering 10.8 percent of the vote, but fared poorly compared to the Green Party's surge to second place there with over 20 percent. In a furious open letter to party leadership, Young Alternative Berlin chair David Eckert urged higher-ups to "refrain from the difficult to understand statement that mankind does not influence the climate," warning that the party risks losing touch with younger voters and that climate issues move "more people than we thought."[30] During the same election, France's National Rally (Rassemblement National in French, or RN) took a different tack than AfD, unveiling a climate change policy platform in advance of the European election.[31] "Borders are the environment's greatest ally," twenty-three-year-old RN spokesperson Jordan Bardella told a right-wing paper. "It is through them that we will save the planet." Marine Le Pen herself has argued that concern for the climate is inherently nationalist. Those who are "nomadic," she's said, "do not care about the environment; they have no homeland." In 2020, Austria's right-wing People's Party formed a coalition government with the country's Greens.[32] Chancellor Sebastian Kurz, a millennial, called their deal a "breakthrough," boasting that it "is possible to slash taxes and make environment-friendly tax policies. It is possible to protect the environment and protect the borders."[33]

This exclusionary logic has infected some center-left parties, too. In running against that country's far-right People's Party, Denmark's Social Democrats adopted a kind of green-tinged xenophobia, promising a "sustainable future" alongside harsher immigration restrictions. Before taking office, charismatic forty-one-year-old party leader Prime Minister Mette Frederiksen embraced legislation hardening rules around the official "ghettos" housing predominantly Muslim migrants, including harsher sentencing for crimes committed within them.[34] And she has linked her stance on migration explicitly to climate change: "Denmark and the world are facing a genuinely difficult situation. A new situation. Record numbers of refugees are on the move," she wrote. "Climate change will force more people to relocate. And add to that the fact that the population of Africa is expected to double by about 2050."[35]

If the end of climate denial and partisan polarization on climate sounds like welcome news, consider what might succeed it. There's no reason to think a GOP shift on climate will cause it to abandon the party's overarching commitments; it is simply more committed to white supremacy than it is to climate denial, and any postdenial turn can be expected to reflect that. Through travel bans and hardened borders ready to halt refugees fleeing heat, drought, and disaster, the GOP has already been writing big government climate policy. Should it start to openly embrace something like climate action, the party would almost certainly retain the racism and xenophobia that has animated it for decades.

Before exploring that possibility, though, it's worth asking the question: As undesirable as a market-oriented, neoliberal climate agenda might be, could it actually get the job done and bring down emissions? There's plenty not to like about the right-wing ideologues preaching market gospel, of course. Still, faced with an existential threat, the fact that there's some plan on offer to bring down emissions is certainly better than nothing—even if it might leave the wrong people in charge. Having witnessed the Trump administration in action, it's understandable for those rightly concerned about the future of life on earth to welcome any move away from denial in the United States as one in the right direction. Unfortunately, the climate policies on offer from a postdenial business-as-usual stance aren't likely to leave the planet much better off.

CHAPTER 2

LONG LIVE CLIMATE DENIAL!

There is no alternative.

—Herbert Spencer, apocryphally; Margaret Thatcher, popularly

There is basically no alternative to the market solution.

—William Nordhaus[1]

WHAT DOES A world warmed by 6 degrees Celsius look like?

Just 2 degrees of warming—the ambitious goal inscribed in the Paris Climate Agreement—could see hundreds of millions more people die of climate-related causes than in a world warmed by 1.5 degrees.[2] Coastal cities and whole nations will be swallowed by rising oceans. Heat waves could make areas around the equator uninhabitable, as an estimated 400 million people live without regular access to water. Three degrees would bring six times as many wildfires to the United States than it currently experiences and droughts persistent enough to cripple the world's food supply. Kevin Anderson, climate scientist and deputy director of the Tyndall Center for Climate Change Research, has called 4 degrees of warming "incompatible with an organized global community," although the United Nations estimates that continuing on with business as usual will heat the earth by at least 4.5 degrees Celsius come 2100. The last time the earth warmed by 5 degrees was 252 million years ago, when a feedback loop of increasing carbon concentrations triggered a sudden burst of methane emissions and caused the fifth mass extinction. According to science writer Mark Lynas, in his 2007 book on the subject, 6 degrees of warming would see most of the earth's surface become "functionally uninhabitable. . . . It's pretty much the equivalent of a meteorite striking the planet, in terms of the overall impacts."[3]

Economists tend to see things differently and in different terms altogether. As recently as 2017, Yale economist William Nordhaus's widely used climate and economy model—detailing the effect of climate policies and impacts on GDP growth—predicted that "damages are 2.1% of global income at a 3 °C warming, and 8.5% of income at a 6 °C warming." To put those numbers in perspective, GDP in the US fell 6.4 percent in 1931 and 12.9 percent in 1932. According to an earlier version of the same model, warming of 19 degrees Celsius would cut global GDP in half.[4] There is no life in a world warmed by 19 degrees Celsius, though perhaps the few remaining phytoplankton will continue compiling our national accounts.

Widely credited with having created the 2 degree target, Nordhaus now contends that a more optimal level of warming sits somewhere around 3.5 degrees, to be achieved through a $35 per ton price on carbon dioxide that nudges polluters and consumers toward lower carbon behaviors. Unveiling its report on what it would take to cap warming at 1.5 degrees in October 2018, the IPCC suggested a carbon price between $135 and $6,005 per ton by 2030.[5] During a press conference about the report in South Korea, its lead authors laughed at one reporter's question about whether a carbon tax alone would be enough to stay within the 1.5 degree threshold.[6]

The next day, William Nordhaus won the Nobel Prize in economics. Most economists don't deny that climate change exists or that humans are causing it. That doesn't make recommendations like Nordhaus's any more reassuring. Sadly, they've been empowered to make an outsized share of the decisions about how to deal with it.

THAT NORDHAUS AND his ilk are neither on the dole of the fossil fuel industry nor committed right-wing ideologues shows the success of the decades-long project to see all problems through the eyes of the market. Accordingly, their red line vis-à-vis climate policy isn't a certain amount of death or destruction wrought by rising temperatures. It's a slowdown in GDP growth.

Why is that number so all important to economists and just about every policymaker on earth? Until fairly recently, they turned to more intuitive metrics like unemployment to judge the health of the economy. But when the mobilization around World War II looked to churn out bullets and clothes to arm soldiers, the focus of national accounts shifted from measures of well-being to measures of productivity. And the new array of economic data spawned by wartime planning efforts helped create what we today know as an indicator called gross domestic product (GDP): household consumption plus business investment, government spending and imports minus exports, with the value of each reflected by prices.[7] Something without a price has no value, and the price of something reflects its value, as well as whether it gets included in national accounts.

Until the 1970s, the GDP more or less reflected the flow of real goods and services. As the financial sector grew in the 1970s alongside deregulation, this extraordinarily rapid growth of new economic activity—previously considered unproductive, intermediary transactions—wasn't being reflected in national accounts. Then, almost overnight, it was. Now, the rest of the economy had to keep pace. Unlike in the financial sector, which profited off ceiling-less fees tacked onto various banking activities, the exponential growth of the real economy was premised on using real resources like coal, food, and steel. Ecological economists like Herman Daly suggested there might be less destructive goals to chase than exponential growth, but they were outliers in their field. Those in the profession's mainstream cast these notes of caution as misguided Malthusianism. The idea of boundless growth and resource use meshed well, after all, with much older ideas in the West about nature as an external force to be mastered and exploited rather than the foundation on which all life depends. Even those neoclassical economists who bothered to think about climate change (there weren't many of them) saw exponential GDP growth as an engine of emissions reduction rather than a leading driver of it. Peak oil—a fixation of early growth critics—was not the crisis it was thought to be in the 1970s; if anything, there was too much oil. But treating nonhuman nature as little more than a productive input to an exponentially expanding economy has birthed catastrophic consequences.

More so than GDP growth, it was prices that took on a mystical quality for many members of the neoliberal thought collective. Perhaps the main prize of their broad-based revolution was to crown markets as the world's ultimate arbiter of knowledge and truth, deeming them uniquely capable of harnessing the collective wisdom of the masses to guide decision-making in ways that governments, by this tale, are simply unable to accomplish.

That's a compelling idea: we only vote every so often and—in most places—not everyone does it or can. But we shop constantly. Every transaction made on the market feeds it another piece of data with which it can make decisions about everything from sourcing to wages, as companies produce supply to meet demand in striving toward equilibrium. From the sum of these interactions emerge prices, which will ultimately reflect the value of whatever is being produced—from goods to services to wages. It's an awe-inspiring premise to expand to the scale of a whole country, let alone humanity. And the whole process seems more rational than trusting some disconnected group of bureaucrats to determine what people want. These bureaucrats, they argued—the economic planners, socialists, and New Dealers the neoliberals wanted to defeat—mistakenly assigned value to things with regulations and price fixing rather than letting a neutral accumulation of market forces pick winners and losers, thus distorting prices and throwing the whole system out of whack.

If a problem arises within the matrix of market transactions, then, the neoliberals asserted, the market has been fed bad information. The solution that flows from that problem—any so-called *market failure*—is simple: correct the misinformation. That's where the government should step in: not with burdensome regulations but to tweak or in extreme cases shape markets so that the prices of what is flowing through them are more accurate and they can start running efficiently again. The state's job is simply to set the rules within which the market can function, insulating it from unhelpful distortions. An ever-growing GDP would reflect the success of market actors unencumbered by any such barriers to their profits, to be gained through the rising price of the goods and services they produce.

Economists more generally describe one major source of bad information as externalities—costs or benefits not reflected in prices. Pollution—in the form of, say, a cancer-causing refinery—is among the most famous examples. Factories and mining operations, for instance, traditionally haven't paid for the chemicals they unload into the air and water, meaning the prices of their products, from cars to truckloads of coal, are inaccurate. As these costs become visible, it's up to the state to step in and correct for that with what's called a Pigouvian tax.[8]

As the concept's namesake Arthur Pigou put it in his influential 1920 book *The Economics of Welfare*, companies are always looking to make as high a profit as possible (what he called marginal private net product), a goal that doesn't always align with what's good for society as a whole (marginal social net product). In one direction or another, that ultimately drives down the "national dividend," the rough equivalent of what we now know as aggregate demand and that is still a major indicator of an economy's health.[9] "In these cases," he wrote, "certain specific acts of interference with normal economic processes may be expected, not to diminish, but to increase the dividend."[10]

These include instances when one party—an industrialist or a landowner, for example—performs services or disservices that either help or hurt some other party, from whom the provider can't either easily extract a fee for the service rendered or compensate the offended party for a disservice. For our purposes, a disservice might be Chevron not paying the health care bills of the people who live near its refinery in Richmond, California, and suffer disproportionately high rates of asthma, cancer, and heart disease.[11] For services, Pigou gives the example of afforestation (e.g., planting trees) on private land, "since the beneficial effect on climate often extends beyond the borders of the estates owned by the person responsible for the forest."[12,13]

The concept of Pigouvian taxes has lived many lives since across the political spectrum. The most common example when it comes to the climate is the carbon tax. Today, much of the right still maintains that it is a leftist idea, even a Marxist one; Heritage Foundation analysts Brian Cosby and Katie Tubb have called a carbon tax a plot for "centrally-planned tax-ation and wealth redistribution."[14] They might be surprised to learn that one of the earliest backers of pricing pollution in the US was none other than Milton Friedman. Asked during a televised forum in 1977 about how to deal with pollution, he told a shaggy, flannel-clad audience mem-ber that he would "like to tax those activities that create pollution, but we're going about it in a very unwise fashion. We're going about it by try-ing to regulate the equipment which people use. . . . Far better to impose [a tax] and then leave it to the ingenuity of people to minimize the cost.

"The only reason we have so much more attention to ecology and pollution today than we did fifty or seventy-five years ago is that we're rich enough to be able to afford the luxury," he added with a smirk. "The water is cleaner today and safer to drink than it was one hundred years ago, not only in the United States but all over the world. The air is cleaner today . . . don't think that the answer is always another governmental law, which will restrict the freedom of people to use their own resources."[15]

It was an odd time to make that argument. Friedman was speaking just a few years after the passage of the Clean Air Act (CAA) in 1970 and the Clean Water Act (CWA) two years later. Among the more mytholo-gized reasons for the latter's passage was the fact that Ohio's Cuyahoga River caught fire in 1969 for the thirteenth time, having been the dump-ing ground for waste from Cleveland manufacturers. Beyond establishing standards for limiting particulate pollution and cleaning up industrial waste, the two bills also set up a nationwide regulatory apparatus that equipped cash-strapped state governments with the necessary funds and administrative assistance to conduct cleanup efforts and enforce federal regulations. With the audacious goal of making all waters "fishable and swimmable" by 1985, the CWA further empowered the Environmental Protection Agency (EPA) to prosecute polluters.

Though pilloried by the right as an example of big government ex-cess, the CAA has delivered a thirty to one return on the government's investment, according to a 2011 study by the EPA, with most savings coming from the prevention of some 230,000 premature deaths caused by degraded air quality.[16] Another study from the George W. Bush–era Office of Management and Budget called "The Cost of Government Reg-ulation," analyzing 107 different regulations across departments, found that CAA rules provided the largest payback of any federal rules, with the EPA as a whole providing the highest returns of any agency.[17] As a result of a 2007 Supreme Court ruling in *Massachusetts v. EPA*, that agency now also has a mandate to regulate carbon dioxide and other greenhouse gases as pollutants—a provision under reliable assault from the right, which recognizes how powerful a lever it is for constraining corporate excess.[18]

Starting in 1995, the EPA established limited market-based mechanisms under the CAA—the Clean Air Markets Division—to regulate certain types of pollutants, most notably acid-rain-causing particulates. Yet the most effective policy responses to environmental harms have come via bans and firm limits. Such regulations are both straightforward and elegant—and are also derided as command-and-control methods by their critics. But their effectiveness and life-saving potential goes to show that if some substance represents a threat to life on earth, indirectly incentivizing corporations to gradually wean off those substances along a path of their choosing just won't cut it.

COMMAND-AND-CONTROL METHODS ARE a time-tested idea in environmental thinking that's well worth returning to. In the foundational text of modern environmentalism, *Silent Spring*, Rachel Carson skewered the byzantine rules that then governed the various synthetic pesticides she spends the bulk of the book describing. The scarcely staffed Food and Drug Administration used a system of tolerances, designed to encourage corporations to use fewer amounts of harmful substances in their products or, as Carson wrote, "allowing a sprinkling of poisons on our food—a little on this, a little on that." The idea was that the human body can safely ingest small amounts of chemicals like DDT. "Even if 7 parts per million of DDT on the lettuce in his luncheon salad were 'safe,'" Carson countered, "the meal includes other foods, each with allowable residues, and the pesticides on his food are . . . only a part, and possibly a small part, of his total exposure. This piling up of chemicals from many different sources creates a total exposure that cannot be measured. It is meaningless, therefore, to talk about the 'safety' of any specific amount of residue."[19]

Synthetic pesticides, she explained, were the by-product of a wartime economy: as part of the home-front mobilization, petrochemical companies had ramped up production of chemicals to protect soldiers from malaria, typhus, and other bug-borne ailments. Once the war ended, these companies sought out consumer markets to replace lucrative federal contracts. It wasn't the chemicals themselves that Carson opposed but the companies that kept producing them despite the availability of more effective, less harmful biological methods. They did so in large part, she writes, by "pouring money into the universities to support research on insecticides," creating "attractive fellowships for graduate students and attractive staff positions," not dissimilar from fossil fuel companies' lush funding of academic departments today. Setting arbitrary limits on harmful products, Carson contended, authorized contamination so that "the farmer and the processor may enjoy the benefit of cheaper production," at the expense of public health. To regulate these poisons properly, she wrote, "would cost money beyond any legislator's courage to appropriate."

Her proposed solution was both simpler and more efficient: "zero tolerance" and a robust regulatory apparatus capable of enforcing it. Already dying of cancer by the time *Silent Spring* was published, Carson would spend the last few years of her life fighting for exactly that approach, all while under vicious attack from the chemical companies. For some of those same firms, countering Carson would be a testing ground for the disinformation campaign they'd wage against climate science years later. In an era when social movements and labor militancy had put lawmakers and corporations both on the defensive, her writing and activism are credited with having helped usher the Clean Air and Water Acts into existence—among the last gasps of the New Deal order.

It would be impossible, of course, to issue an overnight, blanket ban on fossil fuels or any number of greenhouse gases, as eventually happened with the banning of DDT in 1972; our economy has never revolved around noxious pesticides. But like Carson's poisons, the source of the excess emissions wrecking our planet is no mystery: if we want to keep warming below catastrophic levels, 80 percent of known coal, oil, and gas reserves will need to stay buried, which doesn't account for the new stores that fossil fuel producers are constantly hunting for.[20] The target of climate action isn't molecules, which can't themselves be bound by new fees or regulations. It's the companies that spew them out into the atmosphere. Continuing to use some amount of fossil fuels as we transition off them doesn't mean we can't snuff out fossil fuel producers' quest for unlimited expansion and their ability to distort policymaking.

We can't get rid of fossil fuels overnight, and ensuring that transition happens rapidly and equitably will be the defining fight of the twenty-first century. Decarbonization will involve a massive uphill battle against some of the most powerful interests on earth and a level of governmental coordination not seen since at least World War II. It'll mean ensuring workers in the extractive sector aren't thrown under the bus and that no-carbon energy can meet society's needs. In short, an energy transition requires designing a world that isn't based root to branch around fossil fuels—a future for which the last three hundred or so years of history doesn't provide a great blueprint. Even given those massive planning and political hurdles, the actual mechanics of getting off coal, oil, and gas are, by contrast, pretty simple. With alternatives at the ready, our answer doesn't need to be much more complicated than Carson's: zero tolerance for the business model of fossil fuel companies, premised as it is on finding and burning as much of their main product as possible.

That this option seems inconceivable is one more testament to the neoliberal thought collective's success. Less than a month after Ronald Reagan's inauguration, his administration made finding "optimal" policy the law of the land with Executive Order (EO) 12291, establishing that "regulatory action shall not be undertaken unless the potential benefits to society from the regulation outweigh the potential cost to society." Though cost-benefit analyses (CBA) had been employed to some degree by administrations throughout the twentieth century, EO 12291 made it the law of the land.

A response from the right to the perceived inefficiency of the Clean Air Act and part of a broader attack on federal regulations, the order required any "economically significant" new policy to undergo a rigorous cost-benefit analysis, reviewed by the Office of Management and Budget. Environmentalists protested, claiming it wasn't possible to put a price on the value of life or natural beauty, for instance. "They are trying to put into numbers something that doesn't fit into numbers, like the value of clean air to our grandchildren," NRDC economist Richard Ayers told the *New York Times* after Reagan signed the order into law. "Cost benefit analysis discounts the future." Steven Kelman, of Harvard's Kennedy

School, offered a similar take: "The very process of placing a monetary value on such things as human life and pristine wilderness devalues those things. . . . It will discount the great value we place on saving life, and we will be doing less and less of it."[21]

Unsurprisingly, the push for EO 12291 didn't emerge out of thin air. Industry-funded wonks at the American Enterprise and Brookings Institutes had been pushing to scale back regulations for years and cheered on its signing. The National Association of Manufacturers joined them, and the right-wing Hoover Institute would later rank it among their top one hundred conservative victories of the 1980s—a crowded field, to be sure.[22] Though a product of the right, the logic behind regulatory reform has enjoyed bipartisan support in every administration since. The Clinton administration repealed the order in 1993, only to replace it with a largely similar measure requiring the government to intervene in the event of a clear market failure and evaluate the costs, too, of inaction. Other changes were made under both the Bush and Obama administrations, but the spirit of EO 12291 has remained intact.

EO 12291 was just a part of the broader war on regulation the right waged through the Reagan White House. By the time NASA's Dr. James Hansen gave his testimony before Congress about climate change in 1988—thrusting the issue into the national spotlight—the kinds of policy approaches that would seem to make the most sense for dealing with the problem had already been purged from conventional wisdom in Washington if not legality, thanks to the right's crusade through the courts. EO 12291's legacy meant that any response to the problem had to attempt to justify itself through a cost-benefit analysis.

Today's economic common sense, then, is running up against scientific reality. Few would argue a carbon tax is a bad idea for curbing emissions, in theory. But few outside the economics profession would argue that it's anywhere near enough. Atmospheric chemist Will Steffen has suggested a "wartime footing" toward decarbonization, implying something a good deal stronger than a tax incentive. The "obvious thing we have to do is to get greenhouse gas emissions down as fast as we can," he told me. "That means that has to be the *primary target of policy and economics*. You have got to get away from the so-called neoliberal economics," reimagining virtually every sector of the economy at a rapid clip. Keeping warming below 1.5 degrees Celsius, the IPCC urges, will "require rapid, far-reaching and unprecedented changes in all aspects of society."

That's a far cry from a Pigouvian tax. A higher price of carbon could be passed down to consumers without prompting much change at the corporate level, at least not to the degree needed. That might kill off coal, but it would have to be far higher than the taxes currently in place or being considered to prompt a rapid shift off fossil fuels writ large. Assuming decarbonization can flow from prices, nearly all the planning for that transition is left up to individual industries that have little interest in meeting such a goal. Backers of a pricing-first approach are explicit about this. Recommending a flat $75 per ton global carbon tax in 2019, the International Monetary Fund (IMF) wrote that such policies "are the most powerful and efficient, because they allow firms and households to find the lowest-cost ways of reducing energy use and shifting toward cleaner alternatives."

What if they don't? What if oil companies facing a $75 per ton carbon tax pass it all down to drivers, and households with no reliable alternatives at the ready stop being able to afford their commutes? What if the low prices that can be put in place by lawmakers barely make a dent in ExxonMobil's ability to profit off the continued production of fossil fuels that built our economic system and that still meet the vast majority of the world's energy needs? Simply put, climate breakdown isn't a market failure or consumer choice problem to be weeded out so the rest of the economy can keep humming along. Nor is it solely the fault of bad actors in the fossil fuel industry, although there are plenty of them. "The form of capitalism that dominates the discourse at the moment is clearly not compatible with dealing with climate change," climate scientist Kevin Anderson told me at COP23 in Bonn in 2017. "I'm absolutely categorical in my views on that. A lot of the modifications of this form of capitalism are incremental in their tweaks. That's not compatible either.

"We will need some root and branch changes to what we might call capitalism if it's ever going to deal with climate change," he continued. "And even then whether you can say it looks and sounds like capitalism I don't know."

FOR NOW, THOUGH, most climate policy takes business as usual as a given and avoids challenging insatiable mandates of a capitalist economy. Weary of regulation and keen on optimization, William Nordhaus would produce work that nestled comfortably into the constrained and corporate-friendly policymaking landscape already in place when the climate crisis rose to national and international prominence. After getting his PhD at MIT, he began researching climate issues at the International Institute for Applied Systems Analysis (IIASA) outside Vienna in 1975. He called then for a carbon tax, but at IIASA also began developing a related idea that would define his work for decades to come.[23] Nordhaus proposed a basic trade-off: since the activities driving pollution are valuable and drive growth, any efforts to scale them back will pose a cost to society in the short term. Yet continued economic growth will mean people in the future will be richer, so they will value each dollar less than we do now and have a far greater capacity for technological innovation. There's more than a few grains of truth to this: wealthier countries really are better positioned to make a costly, intensive shift to cleaner fuel sources. The theoretical underpinning of this is what's known as a discount rate, used to estimate what the relative price of different goods and services will be at different points in time. With a high discount rate—that is, discounting prices paid in the years to come—climate action in the short term can be relatively modest since our wealthy and inventive descendants (or even our richer future selves) will be able to pay more and do more. Following from this, he suggests a "climate-policy ramp," explaining in a later paper that

> in a world where capital is productive, the highest-return investments today are primarily in tangible, technological, and human capital, including research and development on low-carbon technologies. In the coming decades, damages are predicted to rise relative to output. As that occurs, it becomes efficient to shift investments toward more intensive emissions reductions.

When Nicholas Stern proposed using a lower discount rate than other environmental economists in 2007, Nordhaus accused him of a "radical revision of the economics of climate change." Hitting back against Stern's ensuing recommendation for considerable mitigation investments in the short term, he concludes that "the central questions about global warming policy—how much, how fast, and how costly—remain open." By the time this debate was happening, a year after the release of Al Gore's *An Inconvenient Truth*, this plainly wasn't true. Nordhaus's insistence on a low discount rate and modest emissions reductions has continued to be used as ammo against Stern and other "alarmists."[24]

Accordingly, Nordhaus's work has focused largely on figuring out how to balance emissions reduction with GDP growth and ensure the former doesn't cut into the latter. As he once framed it, "Good policies must lie somewhere between wrecking the economy and wrecking the world."[25] In this respect, accurately pricing carbon is to him an essential government intervention—and far preferable to bans or regulations. Carbon-intensive behaviors are discouraged as they become more expensive, allowing market actors—people and firms alike—to make low-carbon investment decisions, be that shutting down a coal-fired power plant or trading in a gas-guzzling SUV for a Tesla. A price sends a market signal to both firms and consumers, allowing the collective wisdom of the market to harness solutions.

All this is standard fare in the field and even vaguely left of center in that it considers the environment at all. Nordhaus's and Paul Samuelson's widely used economics textbook—railed against by their more conservative colleagues—lays the point out concisely. The nineteenth edition, released in 2010, notes that since the climate threat is a distant one, any government intervention beyond a simple market correction should be avoided.[26] "While it is possible that the regulator might choose a combination of pollution-control edicts that guarantees economic efficiency, in practice that is not very likely. Indeed," they write, "much pollution control suffers from extensive inefficiencies," bemoaning the fact that pollution rules sometimes fail to apply a rigorous cost-benefit analysis. More ideal are "market solutions," "the enhanced incentives" that "allow the ambitious targets to be met at a much lower cost than would be paid under traditional command-and-control regulation."

Nordhaus isn't single-mindedly focused on carbon pricing as the sole key to averting climate catastrophe. For him, the fact that carbon-intensive products are undervalued is simply a barrier to the real solution: economic growth.[27] Thanks to the work of Nordhaus and other environmental economists—including Robert Stavins, Richard Tol, and Robert Mendelsohn—carbon pricing and other market incentives are now more or less the consensus solutions among economists broadly, who mostly defer to the relatively tiny environmental economics sub-field. According to a report from the Institute for Policy Integrity at New York University's School of Law, 95 percent of economists with expertise on climate agree on the need to price carbon, with 81 percent saying it's the most efficient policy possible.[28] Conservative Harvard economist Greg Mankiw has urged his colleagues to join the "Pigou Club," which includes Reaganomics guru Arthur Laffer; the chairman of Trump's Council of Economic Advisors, Kevin Hassett; and Nixon-era treasury secretary George Shultz, who went on to serve as secretary of state under Reagan.[29] Shortly after the Green New Deal began making headlines in late 2018, forty-five of the country's most prominent economists signed on in support of a carbon tax—backed by ExxonMobil and BP, and building in amnesty from lawsuits and EPA regulations—as an explicit alternative. Contrasting carbon pricing with a Green New Deal, former Fed chair Janet Yellen called a price "much more efficient and less costly."[30]

Efficiency and cost are in the eye of the beholder, though, and economists can take a blinkered view of both. In a blistering takedown of his field's standard bearers on climate, economist Steve Keen points out that Nordhaus's own survey of climate "experts" found that natural scientists' predictions for how much damage rising temperatures would cause was as much as twenty to thirty times greater than those made by mainstream economists. One scientist refused to answer his question outright: "I marvel that economists are willing to make quantitative estimates of economic consequences of climate change where the only measures available are estimates of global surface average increases in temperature. As [one] who has spent his career worrying about the vagaries of the dynamics of the atmosphere, I marvel that [economists] can translate a single global number, an extremely poor surrogate for a description of the climatic conditions, into quantitative estimates of impacts of global economic conditions."[31]

And yet the sort of climate and economy modeling Nordhaus pioneered is commonly used by the IPCC to determine pathways for reducing emissions and calibrate the integrated assessment models (IAMs) that help create them; economists proliferate in Working Group III of the IPCC, dealing with mitigation. Some of the more outlandish assumptions he used to develop his widely utilized Dynamic Integrated Climate-Economy (DICE) model have gone mostly unquestioned. In 1994, for instance, Nordhaus assumed that 87 percent of economic activity would be "negligibly affected by climate change," since most of it happens indoors—nevermind the vulnerability of ports and global supply chains to sea level rise and brutally hot, inclement weather. All told, Nordhaus has found, agriculture, forestry, and fishing account for only about 4 percent of GDP, so the kind of widespread loss of cropland that warming may well bring about—thrusting millions into food insecurity and starvation—will register as only a tiny blip.[32] As Keen notes, the IPCC repeated similar claims as recently as 2014. "Economic activities such as agriculture, forestry, fisheries, and mining are exposed to the weather and thus vulnerable to climate change," the *IPCC Fifth Assessment Report*, released that year, states. But "for most economic sectors," the authors add, "the impact of climate change will be small relative to the impacts of other drivers."[33]

More troubling is how environmental economists have viewed the so-called fat tail risks of climate change. Climate economy models typically imagine climate impacts as playing out along a linear trajectory as temperatures rise, with no sudden jumps. The late economist Martin Weitzman repeatedly raised alarms about the potential for catastrophic events and the need to prepare accordingly. Though scientists have urged that the 3.5 degrees of warming that Nordhaus has characterized as "optimal" before 2100 risks triggering disastrous, hardly linear tipping points, he casts this prediction as too far off and unlikely to be worth worrying about. Yet even if humanity should face fiercer floods or fires, environmental economists tend to see our species as endlessly adaptive to catastrophe, engaged as they are in a collective quest toward equilibrium. In a Twitter exchange where he was questioned about the impact of a 10 degree Celsius rise in temperatures—a level of warming likely incompatible with human civilization—environmental economist Richard Tol asserted that we'd simply "move indoors, much like the Saudis have." And because poorer countries have lower GDPs, climate economy models understate the destruction climate change will visit on them, quite literally valuing the lives of Basrans, for instance, less than those of New Yorkers. This is true in a strictly economic sense if GDP were the only outcome worth considering. It's also a moral atrocity to consider the tens of millions of lives on the hook from climate impacts solely in terms of their productive output. Yet, that's what models are set up to do.

Owed to the influence of IAMs, IPCC reports—and with it most climate policymaking—tend to assume things exist that simply don't. Some global price on carbon, technically a proxy for other policies, is ubiquitous across scenarios compiled by the IPCC. Nothing of the sort seems likely. Neither does the capacity for the large-scale deployment of so-called negative emissions technologies, factored though it is into every scenario compiled by the IPCC's special report on limiting warming to 1.5 degrees Celsius. In such modeling, both reality and capping warming at livable levels take a backseat to cost effectiveness.

Considered in context, IAMs—with all their limitations—are a crucial guide for understanding how the economy and the climate might interact, highlighting important relationships among data gathered across disciplines. "I'm not saying we shouldn't have integrated assessment models," Anderson said. "I'm just saying they shouldn't dominate how we understand the future." As of now, they help to constrain what policymakers consider possible. Carbon pricing, for instance, is the only mechanism for decarbonization that an IAM can consider. Models seldom integrate the opportunities of climate action either, like the potentially hundreds of trillions of dollars to be saved by keeping the New York Stock Exchange from being swallowed up by the Atlantic Ocean. Decarbonization is posed as an immediate *cost* in the short term, not a long-term *benefit* to society offering massive returns to parts of the private sector. The narrow pricing focus also forecloses on the models' ability to consider massive public investments in things like transit, which would dramatically lower emissions, or a slowdown in GDP growth by prioritizing other metrics of prosperity. Or the government bringing the fossil fuel industry under public ownership, enforcing a managed decline of the industry that would allow us to meet energy needs in the short term and also meet climate goals.

Baked into most modeling, as well, is the assumption that the global GDP will continue to grow by at least 2 percent indefinitely, a scenario that would double the size of the world's economy—and explode its energy and resource demand—every twenty years.[34] "They're saying we could do mitigation today—which would be quite costly—or we could rely on this technology to suck carbon dioxide from the atmosphere. Even if it's very expensive you've discounted those costs," Anderson noted, referencing the 5 percent discount rate employed across IAMs. "Those models all use this logic, and they've all got this neoclassical growth model at their core. So they can't really question that particular model. The model has to deliver growth. It can't do something that would unpick the paradigm of growth."

The overriding assumption behind treating market-based solutions as a panacea is borderline religious: that given the appropriate conditions for firms and individuals to act in their own self-interest, the markets they populate will achieve ends that leave humanity as a whole better off. There are plenty of monstrous behaviors and business models that have been perfectly rational from a capitalistic standpoint. Today, it's in the self-interest of employers to ensure their workers can't make a living wage, which is why companies like McDonald's have quite rationally fought back against campaigns for a fifteen dollar minimum wage. And as the Wages for Housework campaign and others have argued, it has historically been entirely rational for bosses to keep the work of childbirth and rearing, cooking, and other traditionally feminized labor out of the sphere of the market entirely, to produce more workers and keep them alive off company time. The kinds of prices that would be needed to fully and properly account for the externality of greenhouse gas emissions and cap warming at anything short of catastrophic levels would amount to a ban not a nudge to polluters to gradually conform to over time. Fossil fuel companies' business model is to dig up and burn as many fossil fuels as possible as quickly as possible, and there's precious little evidence to suggest that will change anytime soon. This puts their self-interest and continued existence plainly at odds with that of the rest of humanity.

It's only been through massive pressure from below—slave uprisings, wars, the threat of open revolt, strikes, broad-based social movements—that capitalists have been compelled to act *against* their own self-interest. And it has been government intervention, not some great change of heart, that has spurred them to part with business as usual.

Fossil fuel companies themselves admit that putting a mild price on carbon, at least any one low enough to not spark massive popular backlash, won't slow them down—it's why they support them. While ExxonMobil has publicly endorsed calls for a $40 per ton carbon price, scientists at its own Canadian subsidiary found that a $75 per ton price would have been necessary to stabilize that country's emissions all the way back in 1991.[35] Any carbon price that could cut into carbon profits would be so regressive that it would have little chance of passing. Capitalist markets are both deeply dependent on fossil fuels and have been built around them, from the coal-powered Satanic Mills of the Industrial Revolution to markets' acute responsiveness to even modest changes in the price of oil today. That's not to say that dealing with climate change means we have to start from scratch, crashing out of either fossil fuel usage or capitalism entirely before making any progress; it also doesn't mean that carbon pricing, where it can pass, isn't a valuable climate policy for changing some behaviors in other sectors and at the consumer level. But it does mean seeing market mechanisms as tools in a toolbox rather than a silver bullet and reasserting a critical role for the state.

Seeing climate change solely as a market failure to be solved with market tweaks assumes the goal that markets and the actors within them are ultimately chasing—the boundless accumulation of private wealth—will align with society's best interest if only provided the right set of rules. At the heart of economists' fixation on growth has been the argument that rising GDP, fueled by rising corporate profits, lifts all boats: that while there may be inequalities within the system—some will accumulate more than others—making the world richer is a sum positive, lifting millions out of poverty and if not into the middle class then at least to the level of being able to buy a TV or refrigerator. If all you have is a hammer, everything looks like a nail.

In 2007, THE Supreme Court's decision in *Massachusetts vs. EPA* mandated that the EPA begin to regulate carbon dioxide as a pollutant. The Bush administration mostly ignored the ruling, but shortly after Barack Obama took office, OMB director Cass Sunstein and Council of Economic Advisers chief economist Michael Greenstone convened the Interagency Working Group on the Social Cost of Carbon, intended to determine a carbon price that could be used by government officials to calculate cost-benefit analyses for regulating carbon. During Obama's presidency, the working group developed the methodology for arriving at such a figure, using DICE and two other simple IAMs to calculate the cost imposed by each additional ton of carbon put into the atmosphere. It was eventually used to design seventy-nine different regulations and the administration's since-repealed Clean Power Plan, in which carbon pricing was offered as a way for states to comply. Before Trump disbanded the working group, it had settled on a cost of $36 per ton.[36] Since leaving the White House, Greenstone has been attempting to carry on his work on the social cost of carbon through the Climate Impact Lab. After a stint at MIT, he is (appropriately enough, as of writing) the Milton Friedman Distinguished Service Professor of Economics at the University of Chicago and directs both the Becker Friedman Institute there as well as the university's Energy Policy Institute.

The overarching goal of carbon pricing for him, Greenstone clarified, isn't to bring down emissions at the scale and speed science says is necessary. It's to have the price of carbon dioxide reflect its true value, as embodied—for him and others—in the social cost of carbon. "There's a disconnect between the way economists and scientists think about it," he told me. "The way the scientist thinks about it is, 'There's a level of emissions reductions we need to achieve. Let's work backwards from there to figure out what the best price might be.' Economists would think about it as, 'Let's understand the level of damage that climate change would cause. That should be reflected in the cost, without prejudging what the level of emissions reductions should be.'"[37]

It's hard, given statements like this, to see carbon pricing as the kind of panacea many economists imagine it to be. Treating it as a primary emissions reduction strategy hinges on being able to cost out the full scope of the damage climate change and the other negative effects of carbon are likely to cause, which is difficult at best. Can anyone, much less someone using a DICE model, accurately estimate the cost of a low-lying island state being permanently inundated by the sea? And what about all the impacts of rising temperatures we don't yet know exist—the "unknown unknowns," as David Wallace-Wells has called them.[38] Our predictions for what's likely to happen as a result of climate change are limited in the sense that we only have data available from the world as it is and has been—not the one that's coming.

That climate impacts are so hard to accurately quantify is part of why there's so much debate about what the ideal price on carbon should be to avoid climate chaos. In a 2017 report, Nicholas Stern and Nobel laureate Joseph Stiglitz argued that limiting warming to 2 degrees will require a price on carbon between $40 and $80 per ton, which would then rise to $100 a ton by 2030—lower than the IPCC's suggestion in SR 1.5.[39] At the left-leaning People's Policy Project, economists Mark Paul and Anders Fremstad have suggested a price of $230 per ton to keep temperatures from rising above 2.5 degrees.[40] Using an approach closer to the one used by financial economists, climate economist Gernot Wagner, Columbia Business School professor Kent Daniel, and former Goldman Sachs risk-management expert Robert Litterman worked out their own model in an article for the *Proceedings of the National Academy of Sciences* journal. While they didn't settle on an exact, ideal price per ton of carbon dioxide, Wagner told me they couldn't "in good conscience get the price lower than $125." The most important thing, he emphasized, is that there is no magic number. "It could be $200 per ton or $400 per ton," he said. "There are so many uncertainties that presenting one number is just insane. It's uncomfortable for an economist to say, but the grand conclusion is a bit of humility. We can't tell you the grand solution. Everything we know about how to price a ton of carbon dioxide tells us that it seems to be much, much worse than the standard climate economy models tell us."[41]

Actual prices skew quite low, averaging at just $8 per ton. Of the forty-two pricing systems in place worldwide, the OECD has found that the vast majority of countries and regions haven't set prices high enough to meet the goals laid out in the Paris Agreement; Sweden has the world's highest price, at $126 per ton.[42] As OECD secretary general

Angel Gurría put it, "The gulf between today's carbon prices and the actual cost of emissions to our planet is unacceptable." Carbon prices stateside remain few and far between, and the two systems currently in operation—California's cap-and-trade system ($12–15/ton) and the Regional Greenhouse Gas Initiative in the Northeast ($2–3/ton)—have been much better at raising revenues than curbing carbon.[43]

All that means is that carbon prices' track record for actually bringing down emissions has been shoddy at best and difficult to track. Since pricing is often enacted alongside other climate rules—including things like auto efficiency and renewable portfolio standards, requiring utilities to source a certain percentage of their emissions from wind or solar—it can be hard to parse out which reductions are coming from what policy. "It is safe to say that policies other than carbon pricing have driven the majority of emissions reductions to date," Jesse Jenkins, of Princeton's Andlinger Center for Energy and the Environment, told the *New York Times*.[44] However you define it, carbon prices have thus far failed to internalize the cost of carbon into the market. For now, those that do exist mainly in models.

To his credit, Nordhaus has changed his tune on several points over the course of his career. He no longer spouts lines about how people prefer warmer weather. As a result of several productive debates with Martin Weitzman, he's now interested in accounting for the risk of catastrophic and difficult-to-predict climate impacts—something DICE and other IAMs have repeatedly failed to do. He's also been quick to fire back, too, at the old-school deniers who have invoked his research over the years as an excuse for inaction. In his latest book, *The Climate Casino*, Nordhaus writes, "Markets alone will not solve this problem. There is no genuine 'free-market solution' to global warming. We need new national and international institutions to coordinate and guide decisions about global warming policies. These mechanisms can use the market, but they must be legislated and enforced by governments."[45] Still, he goes on to argue that approaches to curbing emissions beyond carbon pricing (fuel efficiency standards, building codes, subsidies, etc.) "are all more expensive and less effective than the ideal policies," calling some "horribly ineffective."

The day he received the Nobel Prize, Nordhaus walked into the Yale macroeconomics class he was scheduled to teach that day to booming applause, the words "Congratulations Prof. Nordhaus" scrawled on the blackboard behind him and eager undergrads at the ready with a bouquet of flowers. After a little fanfare, he joked: "Special rule for today: You can have your cell phones. As Yale students and faculty you learn how to deal with distractions. Don't let anyone distract you from the work at hand, which is economic growth."[46]

WHILE HE'S A self-described "big fan" of Milton Friedman, William Nordhaus isn't a climate denier in the mold of the Cato Institute or CEI. His flawed ideas were published in the right place at the right time. The neoliberals changed the foundations of what constitutes economic sense, to the point where they eventually stopped calling their ideas neoliberal altogether. Eventually, that just become the way things were. It was in this context that Nordhaus's theories gained prominence.

After years in the wilderness—or Switzerland and Chicago, at least—neoliberals had found an opportunity in the cascading crises of the 1970s. The old Keynesian tricks had stopped working. Whether for stagflation or the oil crisis, the New Deal order didn't seem to have the answers to what ailed the economy. Thanks to years of careful planning and with friends in high places, Friedman, Hayek, and company stepped in with solutions.

And now, we find ourselves in a very similar moment. The common sense that has reined in politics and economics for the better part of forty years offers no road map to this crisis. Its backers can try to hold on to their relevance and will—flanked by powerful industries. There are plenty of paths for neoliberalism as we know it to survive the climate crisis and profit handsomely off it.

In this crisis-filled context, mainstream economics offers a set of tools that has spent too long masquerading as a solution. The neoliberal revolution endowed this beast we now know as the market with apparently superhuman powers to coordinate society. In the process, its stewards anointed a tiny set of technocrats as the only ones capable of understanding it, trading in the idea of political economy—that the market can't be understood independent of the social and political relationships that shape it—for the allegedly harder science of economics. In practice, the field has become more of a religion attempting to silence heretics, now cloistered in a handful of departments that dare to challenge the reigning orthodoxy. Unlike after World War II, the challenge for those still clinging to neoliberal ideas isn't to displace the reigning order; it's to keep it going. They're the ones playing defense.

As an era-defining project, though, neoliberalism has never been solely or even primarily about economic theories. The neoliberal thought collective's ability to shape conventional wisdom across the political spectrum has always been owed more to the enduring strength of their political influence, alliance with corporations, and dogged institution building than that of their ideas. The doomed fight to implement a nationwide, polluter-friendly carbon-pricing system a decade ago showed just how successful that long march had been. It should serve as a warning, too, for those still eager to bend over backward to accommodate fossil fuel companies claiming to care about the climate for the sake of passing something, anything, called climate policy.

CHAPTER 3

FIRST AS TRAGEDY

UN CLIMATE TALKS tend to be a bit of a bubble. Dehydrated, over-caffeinated attendees—governmental negotiators, visiting dignitaries, and civil society representatives—dart between the wings of sprawling conference centers, ducking in and out of meetings and subsisting on stale pastries and scant sunlight. The physical experience of being at the Conference of the Parties of the UN Framework Convention on Climate Change (UNFCCC), referred to by attendees as a COP, isn't so different from being inside an airport: the food's not great, there are never enough places to sit down, and most everyone is stressed out. Invariably, there are delays.

COP1 took place in Berlin in the spring of 1995, three years after the adoption of the UNFCCC at the Rio Earth Summit in Brazil. Those first meetings worked toward what would become the Kyoto Protocol, the outline of a plan to finally do something about the climate crisis. Arrived at in Japan in 1997, the Kyoto Protocol was signed by the US under President Bill Clinton. Implementing it, however, would require legislative approval. Just before the next COP, Senators Robert Byrd and Chuck Hagel circulated a resolution through the Senate stating their opposition to any such thing. It was nonbinding but passed 95–0 and sent a clear signal that international cooperation on climate wasn't forthcoming from Capitol Hill. Momentum didn't pick up for the rest of Clinton's term and stalled out entirely under George W. Bush. As then EPA administrator Christie Todd Whitman bluntly told a reporter shortly after he took office, "We have no interest in implementing that treaty."[1]

A deal was eventually reached among the remaining 192 signatories once Bush Jr. had made clear he wouldn't join in 2001. But an agreement without the participation of the world's reigning hegemon and one of its top polluters wasn't worth much. And with its focus on long-developed nations, the policy regime the Kyoto Protocol set up also didn't include a natural mechanism for curbing emissions from countries like China and India, which in the decade-plus after its signing had emerged as industrial powerhouses. So with the Kyoto Protocol's "top-down" framework as a guide—wherein the countries who are more responsible for climate change have more work to do—negotiators for the next few years hotly debated the draft of a new text that would fill the gaps left by the US. It looked like they might be nearing consensus eight years later. Then a small handful of countries—nearly all from the Global North, including the US—huddled and put forward their own three-page document to the world to be voted on without further negotiation. Those talks collapsed under objections from the parties about the content of that document and the process (or lack thereof) that created it.[2] From there, it would take another six years to arrive at the world's first international pact on climate change, following much of the scaffolding those three pages laid out in Copenhagen: 2015's Paris Climate Agreement.

If we count the Rio Earth Summit as its birthday, the UNFCCC process is as old as I am. When I was born—in February 1992—sensors at the NOAA observatory on Mauna Loa recorded that 357 parts per million of carbon dioxide were in the atmosphere. After I attended COP25 in Madrid in 2019, it had reached 410 ppm and is climbing still at the time of publication. That makes catastrophic impacts more likely. The last time concentrations of carbon in the atmosphere were that high, thousands of years ago, trees grew in Antarctica. The ten hottest years since 1880 have passed since 1992; the earliest and coolest on the list was in 1998. Why have all these years of talks and meetings yielded an agreement without any sort of binding enforcement mechanism and that—if all of the pledges therein were perfectly honored—would still likely warm the world by more than 3 degrees? What has gummed this international process up? And why has the United States failed to pass national climate policy within its own borders and gone out of its way to block ambition internationally?

There are reasons that have to do with power and politics, and then there are the ones ascribed to human psychology. Despite the UNFCCC and other institutions' efforts to "confront the problem of global warming," writer Roy Scranton despairs that "we seem no more capable than were the people of Uruk when it comes to rescuing ourselves from imminent catastrophe," referencing the ancient and doomed civilization that inhabited the land now known as Iraq. And Scranton points out that "we," unlike them, have the benefit of "tremendous resources, the knowledge of thousands of highly trained scientists and engineers, and the support of hundreds of thousands of dedicated activists and concerned citizens." So what's the matter with us?[3]

Over the years, plenty of theories to explain the world and its problems have leaned on so-called human nature. Historian Ellen Meiksins Wood begins her seminal book *The Origins of Capitalism* by noting that "the 'collapse of Communism' in the late 1980s and 1990s seemed to confirm what many people have long believed: that capitalism is the natural condition of humanity, that it conforms to the laws of nature and basic human inclinations, and that any deviation from those natural laws and inclinations can only come to grief."[4] One take on our predicament is that the flipside of this supposedly biological longing for capitalist markets is our capacity for reckless consumption and deadly competition. This view that our current environmental disaster is rooted in individual choices is exemplified by a 1970s *Pogo* cartoon, drawn at a high point of public concern about ecological and nuclear destruction: staring out at a field blighted with litter, Pogo mourns, "We have met the enemy and he is us."

Every time you turn on a light, fill your gas tank, or fly cross-country on vacation or business, you play some small role in lofting more greenhouse gases into the atmosphere. But unless you own a private jet or run a fossil fuel corporation, your family vacation didn't cause the climate crisis—any more than the struggling couple buying their first house in 2006 caused the financial crisis because their bank sold them a subprime mortgage. This corporate-fueled fixation on individual carbon footprints reinforces the idea that some inborn defect in each of us is to blame for world governments' inability to handle this mess.

As concentrated as emissions are at the top, decision-making power over what to do about the problem is still more so. The world's most prolific polluters have spared no expense over the last several decades to obstruct climate action and stymie efforts at every level of government. As discussed in Chapter 1, prodigious amounts of fossil-fuel-funded climate denial—put out by companies who knew the grave threat rising temperatures posed—helped pull the conversation about climate change in the US, especially, into a circuitous discussion about whether or not the earth was warming instead of what should be done about it. Denialism, though, has never been the industry and its abettors' only tactic; the triumph of neoliberalism laid the groundwork for decades of delay.

There's a comic book version of how climate policy failed in the US in which diabolical, fossil-fueled billionaire villains call on their trusty henchmen in the GOP to swoop in and snatch away the country's best chance for climate action. There's plenty of truth to that tale, but identifying the hero isn't nearly so cut and dry. A decade ago, Democrats controlled the House, Senate, and White House and—for a time—seemed poised to pass legislation that would curb emissions and build a clean energy economy. Corporate meddling wasn't the only thing that torpedoed climate legislation in 2010. Neither was it only big donors who cost Democrats the legislature and presidency over the next decade. The saga of cap-and-trade—how the idea gained prominence among DC insiders, and the political calculus they employed to try passing it—should be a cautionary tale for those interested in seeing any sort of comprehensive climate legislation emerge from the US Congress in their lifetimes.

Arguably as important to the fossil fuel industry's victory in defeating climate policy during the Obama administration was the fact that companies *didn't* uniformly oppose its efforts. Having bought into right-wing nostrums about the dangers of regulations and the superhuman powers of the market, Democrats and Big Green eagerly treated the fossil fuel industry as a good faith ally in the climate fight, regardless of the industry's material interests and activities, which would keep driving greenhouse gas emissions up and ward off the threat any reasonable climate policy would pose to it. Despite the painful recession, the Democrats' drive to win over corporate support trumped both the urgency of rapid mitigation and the need to make the benefits of climate policy clear to ordinary people. By the time the Waxman-Markey cap-and-trade bill was introduced in 2009, polluters had already won their biggest fights. Democrats, in turn, had abandoned their best tools for fighting back—and the idea that fighting was even a good idea.

IT WAS WITH a heavy heart that BP and ConocoPhillips—two of the world's largest oil companies—announced their separation from the US Climate Action Partnership (USCAP), the coalition of greens and industry groups pushing a carbon pricing measure called cap-and-trade through Congress. By the winter of 2010, the measure USCAP was supporting had passed through the House, but the effort ran aground in the Democratic-controlled Senate, as supportive lawmakers attempted to salvage a deal by any means necessary, piling on more giveaways to polluters. It was all for naught. Though the two oil majors had tried to reconcile their differences with the coalition, they ultimately found it better to part ways on good terms.[5]

A marriage of Big Oil and Big Green may seem like an odd one today, when the fossil fuel industry seems to sit clearly on the side of the GOP. But bringing these strange bedfellows together was the founding ethos of the push to pass climate legislation at the start of Obama's first term. It was the closest Congress had ever gotten to passing a comprehensive bill to limit greenhouse gas emissions—which is to say, not very close at all. As much of the story of cap-and-trade's implosion is one of industry meddling and the radicalization of the Republican Party, it is every bit as much about the transformation of American liberalism.

A type of carbon-pricing system, cap-and-trade sets a "cap" (limit) on the total amount of carbon that can be emitted by the entities under its jurisdiction, typically industrial polluters like factories and power plants. Those polluters are allocated a set of credits corresponding to their share of those emissions over some set amount of time, which can be traded on a carbon market between companies depending on which of them need credits (i.e., to pollute more) and which have credits to sell, having polluted less. The idea is that the cap will decline over time, bringing down total emissions while giving firms time to adjust and flexibility in how they meet the cap.

The approach USCAP and allied lawmakers adopted in trying to establish such a system in the US was the culmination of a gradual if dramatic transformation within the Democratic Party on both style and substance, embodied as much in cap-and-trade debates as in the Obama presidency it inaugurated. The party's great, earnest hope was to see environmentalists and Wall Street bankers and union members and fossil fuel executives as interchangeable stakeholders in the climate debate, but this clouded a reality the right never lost sight of: that it is often the people with the most power who win, and those people generally act to further their material interests. In the case of the climate crisis, the fossil fuel industry has ample power—and the most to lose. But cap-and-trade offered a perfect opportunity for industry leaders to take full advantage of a situation—a Democratic Congress, a White House with a mandate, and rising public concern about the climate crisis—that might otherwise have been a death sentence for them.

Two weeks after ConocoPhillips and BP left, another high-up participant in USCAP jumped ship: Deryck Spooner, campaign manager at the Nature Conservancy, a USCAP member and one of the largest environmental organizations in the world. Having worked for the AFL-CIO and NARAL previously, two groups with ample sway in the Democratic Party, Spooner told Greenwire he was leaving to take a position that would allow him "the opportunity to further [the] conversation" about climate change: with the API. At the time, API was engaged in an all-out war against the climate bill USCAP was pushing—a war Spooner would now help wage. "The bottom line," he said in another interview, "is it's all about advocacy, that's what I'm passionate about. Mobilizing and organizing people to influence the public process and public policy is what I truly love to do."[6]

A few years later, in 2015, Spooner gave a lengthy address in Calgary to the Canadian Association of Petroleum Producers, API's sister organization in that country. The purpose of the talk was to help his counterparts up north replicate his group's success. Starting with an ode to the "passion" that led to Obama becoming president, Spooner went on to describe API's multitiered, state-by-state grassroots mobilization strategy, at one point naming in his PowerPoint presentation the 275 House members and 68 senators the group enjoyed influence with, including prominent Democrats like Senate minority leader Chuck Schumer and environmental champion Jeff Merkley.[7]

On everything from the repeal of the crude oil export ban to defeating climate legislation, Spooner declared, "We have won."

They didn't do it alone.

NOW IN HIS midsixties, Fred Krupp took over as CEO of the Environmental Defense Fund (EDF) in 1984 at the ripe age of thirty. Soon after he started, he looked to herd the dwindling organization in a different direction, away from the aggressive legal fights that had defined it to date. He was eager for EDF to broaden the tent of those interested in preserving nature beyond the usual scientists, treehuggers, and activist lawyers. With the help of environmental economist Dan Dudek, one of Krupp's first hires, he landed on market mechanisms as a way to fight pollution. As Eric Pooley describes in his colorful account of the cap-and-trade fight and its prehistory, Dudek envisioned "a world in which emissions reductions were traded like securities in a green market, and where if a company did something of value for humanity, such as cut the pollution it spewed into the air, it would profit."[8] More lofty than a straightforward Pigouvian tax, emissions trading systems could create new commodity markets whole cloth. Unlike the bureaucratic regime created by the Clean Air Act, Pooley adds, "a market mechanism wouldn't require EPA to pick winners and losers; the market would decide which technologies worked best."

Sensibly enough, Krupp reckoned that the "sue the bastards" approach that had to that point defined the EDF—a group that made its name on feisty anticorporate legal fights to protect wilderness—was also running up against its own limits, as reflected in the dire state of the group's finances. With Reagan in office and a growing number of conservatives in the courts, Washington had soured on the command-and-control approach as the gospels of cost-benefit analysis and deregulation proliferated.

For similar reasons, Democrats' own shift rightward was well underway. The so-called Atari Democrats—including Al Gore—had begun trying to appeal to Wall Street and suburban tech workers to bolster the party's electoral chances against Republicans, shifting their locus of concern away from traditional bases like organized labor and toward middle- and upper-class voters, whose hunger for low taxes and limited government created grounds for broad compromise with both the GOP and corporate America, the financial sector in particular.[9] As the party catered to well-off professionals, it started to sound more and more like them.

To keep up with the technocratic times, Dudek researched what an emissions trading system could look like in the US, and Krupp built out a well-heeled board that included both Republicans and Democrats. The new approach, Pooley writes, "was hardheaded, results-oriented, and politically incorrect—so it appealed to hardheaded, results-oriented, politically incorrect people who believed in markets because they had made their fortunes in the markets: Wall Street people, who would become the most important trustees and benefactors for EDF over the years."[10] Not everyone at EDF was happy about the strategy shift, but in 1986 Krupp put an end to the debate with an op-ed in—appropriately enough—the *Wall Street Journal*.

Krupp laid out what he called the third stage of environmentalism. Building on Teddy Roosevelt and John Muir–style conservationism (the first stage) and the more activist, *Silent Spring*–inspired second stage that ushered in the Clean Air and Water Acts, the third stage would chart a new path. "To move beyond reactive opposition demands a high level of economic and scientific expertise," Krupp wrote. "Growth, jobs, taxpayer and stockholder interests, agricultural productivity, adequate water and power for industry and consumers—all these are part of the third-stage agenda."[11] As key to the substance of the policies of Krupp's third stage environmentalism were the politics that accompanied them—persuasion, not confrontation, and a faith in skilled experts to get things done and communicate the truth to the people who mattered. Critics, he wrote, "assume that the issue is 'either-or': Either the industrial economy wins or the environment wins, with one side's gain being the other's loss. The new environmentalism does not accept 'either-or' as inevitable." The GOP had long been the party of big business, the theory went, so by giving them a seat at the table, greens could neutralize Republican opposition and work toward bipartisan support.

It's hard to overstate just how of the moment Krupp's thinking was. Before neoliberalism became a jab at centrists after the 2016 election—and long after the salad days of the Mont Pèlerin Society—prominent Democrats identified proactively with the word, a reaction by a new generation of politicians to the corruption of the Nixon administration exposed in the Watergate scandal and the perceived failure of New Deal liberalism to deal with the various crises of the 1970s. Its central tenets, fittingly, were a rejection of inefficient, corruptible bureaucracy and a tremendous faith in the power of economic growth and the private sector to deliver on the progressive aims the party had chased for decades.

In a 1982 piece entitled "A Neo-Liberal's Manifesto," *Washington Monthly* founder Charles Peters wrote that "our primary concerns are community, democracy, and prosperity. Of them, economic growth is most important now, because it is essential to almost everything else we want to achieve. Our hero is the risk-taking entrepreneur who creates new jobs and better products. . . . We want to encourage the entrepreneur not with Reaganite policies that simply make the rich richer, but with laws designed to help attract investors and customers."[12] The enduring success of the Reagan Revolution through the 1980s offered further proof that a new style of big-D Democratic politics was needed if the party was to survive and not repeat George McGovern's blowout loss to Nixon in 1972, still weighing heavy on party consciousness. That sentiment eventually crystallized in the Democratic Leadership Council (DLC), founded in 1985, which would define the policies and politics of the Clinton era, seeking a third way between left and right, when Republican and Democratic priorities would drift together.

The Democrats who came to embrace market-based approaches on climate weren't simply trying to make themselves look more like Reagan to steal his voters or swallowing talking points wholesale from corporate interests—although both dynamics were certainly at play—they believed in them as the most effective means to push through progressive priorities in a political landscape that was shifting toward the right. They believed, as well, in the need for corporations to be allies in doing so. "The principles and policies Clinton and the DLC espoused were not solely a defensive reaction to the Republican Party or merely a strategic attempt to pull the Democratic Party to the center," historian Lily Geismer has written. "Rather, their vision represents parts of a coherent ideology that sought to both maintain and reformulate key aspects of liberalism itself."[13]

In 1988, then senators Tim Wirth and John Heinz—a Democrat and a Republican, respectively—commissioned a report called "Project 88: Harnessing Market Forces to Protect the Environment," looking at several areas from water quality to the greenhouse effect to acid rain.[14] Drafted by a long list of economists, policymakers, and representatives from Beltway green groups (including Krupp), the project was spearheaded by Harvard economist and EDF alum Robert Stavins. With critics in mind, Wirth and Heinz write in their introduction: "We are not proposing a free market in the environment—far from it. This report," they assure, "is not about putting a price on our environment, assigning dollar values to environmental amenities or auctioning public lands to the highest bidder. What we are proposing is that once tough environmental goals are set, we should design mechanisms for achieving those goals which take advantage of the forces of the marketplace in our economy."

That's not inaccurate. Several of the recommendations the authors landed on include regulations, protections for public lands, and federal investment. But Project 88 was in no small part the extension of a broader, bipartisan push to use federal policy to help markets bravely go where they had never gone before. As Wirth would later tell journalist Steven F. Bernstein, he and Heinz "thought that economics was pervading everything else during the Reagan era and a lot of other issues were being looked at through an economic lens and why should environmental issues be excluded from that? . . . Environmental issues could not exist in a vacuum."[15]

It was in that context that Krupp's third stage op-ed caught the eye of C. Boyden Gray, Bush Sr. personal lawyer-cum-White House adviser, who was enthused by the Project 88 recommendations and tapped EDF to develop a market fix for acid raid, an issue the new president was keen to revive.

Jimmy Carter's administration had taken some initial steps toward dealing with the acid rain problem after it was discovered in the 1970s, and Reagan's advisers at first seemed friendly to the idea of continuing that work. Naomi Oreskes and Eric Conway document in *Merchants of*

Doubt that the mood quickly chilled. The Reagan administration rejected the findings of its own EPA on acid rain and convened a separate panel under the auspices of the National Academy of Science (NAS) and the direction of Marshall Institute cofounder William Nierenberg. NAS scientists' ensuing fight with the White House was an early testing grounds for old-school climate denial. That battle—waged within the administration, against experts it had funded—successfully kept anything from being done about acid rain through the Reagan era.

In the wake of that failure, Project 88 proposed an EPA-administered Acid Rain Program with a system of acid rain reduction credits, wherein a total number of credits for acid-rain-causing pollutants would be allocated and then traded among companies based on their emissions. Ironically, a version of the market-based policy that EDF and the administration ended up adopting had first been floated under Reagan by none other than S. Fred Singer, a member of Nieremberg's panel who would go on to be one of the country's most prominent climate deniers.[16] Acid rain may or may not be a significant problem, he argued, in contrast to the rest of the NAS panel. In any case, Singer recommended "a middle course: Removing a meaningful percentage of pollutants by a least-cost approach and observing the results, before proceeding with the more costly program."[17] Singer was at the time employed by the Heritage Foundation, which—with ample industry funding—had been instrumental in fleshing out Reagan's deregulatory agenda. Through his time working on acid rain, he reiterated Heritage talking points to whoever would listen.

With some haggling, the EDF, years later, sold the first Bush White House on a similarly minded approach, a 1990 Clean Air Act amendment (CAAA) to limit the particulates that cause acid rain. It seemed to work, with emissions of the acid-rain-causing pollutant sulfur dioxide reduced by 54 percent between 1990 and 2007. Moreover, the economic benefits of the program far outweighed its costs. And despite utilities' fearmongering that moves to reduce sulfur dioxide would drive up electricity costs, inflation-adjusted power prices declined over the same period.[18]

There were reasons to be skeptical, though. Thanks partly to a drop in the price of low-sulfur coal, sulfur dioxide emissions had already been declining years before Phase 1 of the Acid Rain Program was first implemented in 1995. Emissions at the units covered during that first phase remained flat through 1999, while emissions actually rose from the smaller plants that wouldn't be included until Phase 2, starting in 2000.[19] The European Union, notably, took a more traditional regulatory approach to dealing with the acid rain problem. Between 1980 and 2004, the fifteen countries initially included in the European Union reduced sulfur dioxide emissions by 78 percent, compared to the US's 39 percent reduction over the same period.[20] By 2010, biologist Gene Likens—part of the team that discovered acid rain—warned that "this threat to the environment has not been solved, and arguably is now worse than previously thought when the 1990 CAAA were passed," noting the continued negative impact of acid rain on forest growth.[21]

Owing to the modest or at least perceived success of the Acid Rain Program, the approach employed by Bush Sr.—a market-based pollution control system cooked up by experts—would become a go-to strategy in mainstream and mostly Democratic-leaning green circles for years to come, as the GOP and carbon-intensive industries enthusiastically embraced denialism through the 1990s and early 2000s. Still, the idea of a truce with industry and across the aisle around a Republican idea remained a powerful one. Bill Clinton quipped on the campaign trail in 1992 that "Adam Smith's invisible hand can have a green thumb," praising market-based policies' promise to "cut compliance costs, shrink regulatory bureaucracies" and "enlist corporate support," while advocating for an emissions trading system to deal with greenhouse gases.[22]

"Many of our environmental efforts in the past were based on a 'command and control' approach to regulation that told firms how much pollution to produce and what kind of technology to use," he said in the same Earth Day speech. "While that approach produced important successes, it sometimes stifled innovation by locking firms into a specific kind of equipment and increased regulatory costs and burdens by taking such a detailed and inflexible approach." Heritage Foundation analyst John Shanahan approved of the message, addressing Clinton in a detailed (if unheeded) policy brief days before the Arkansan assumed office: "If you really intend to use market forces to reach reasonable environmental objectives, and make results the test for legislative or regulatory action, that truly would be a welcome and positive change."[23]

In a similar spirit, the New Democrats empowered experts to engineer the most efficient solutions to the problems of the day and saw the involvement of the general public and institutions like labor unions in policymaking as either unhelpful or irrelevant. The Mandate for Change that the DLC's Progressive Policy Institute handed to Clinton to shape his first one hundred days echoed the Mandate for Leadership that the Heritage Foundation delivered to Reagan at the start of his first term, which had once seemed so radical. It called on him to introduce "choice, competition and market incentives into the public sector" and "emphasize economic growth generated in free markets as the prerequisite for opportunity for all."

Indeed, he tried to do just that in 1993, proposing a modest fuel tax intended to both reduce pollution and the deficit in one fell swoop, having been written into a larger deficit reduction package. Because West Virginia senator Robert Byrd staunchly opposed a carbon tax, the administration instead opted to levy a fee on British thermal units, measuring an energy source's heat content.[24] Democrats readily handed out concessions to get the tax through, none of which stopped the National Association of Manufacturers from galvanizing what they described as the largest coalition of business interests to oppose a single piece of legislation. They aligned the GOP and several congressional Democrats in opposition and took out aggressive ad campaigns in energy-producing states to brand it as a job killing, regressive burden on the middle class.[25] It made it through the House, only to be killed in the Senate. In his obituary for the fuel tax, the *Washington Post*'s David S. Hilzenrath noted that Clinton "compromised not only on the substance of his proposal, but also on his stated aversion to the special-interest horse-trading emblematic of politics as usual."[26]

But if the BTU tax had been an early fumble in this brand of politics, the New Democrats would have other successes. Boosted by a strong economy, their faith in markets and the private sector would only grow stronger through the rest of the decade as Clinton moved to "end welfare as we know it," get tougher on crime, and champion the North American Free Trade Agreement (NAFTA), which left labor on the losing end.

Meanwhile, Clinton-era Republicans like Newt Gingrich and Tom Delay prided themselves on dragging the party and the American political landscape toward their side, obstructing Democrats as much as possible. Mostly, it worked. As Delay bragged, "We moved the whole of American governance to the right." Among Clinton's top advisers through this first term was Democratic operative Dick Morris, who'd managed two of Clinton's gubernatorial campaigns in Arkansas and worked as an adviser to Republican senator Trent Lott, who became majority whip once the GOP regained control of the Senate in 1994 and later the majority leader during the second Bush administration. At Lott's home in 1995, he and Morris only half-jokingly reached an understanding: "You take over the Senate, I'll take over the White House, and we'll pass everything!"[27] Lott would describe welfare reform—a centerpiece of Clinton's tenure—as the "holy grail" of his party's "legislative master plan."

When the time came in 2009 for Democrats to govern with all three branches of government, they did so largely with the Clinton-era dogma of bipartisanship firmly intact; two Bush terms hadn't convinced them otherwise. If anything, Bush's buffoonery and eventual unpopularity in his second term had cast a sheen over the Clinton administration, with liberals pining for the days when a Rhodes Scholar had overseen a prosperous peacetime economy. Obama, of course, was not Bill Clinton. But his top advisers—including transition team head and longtime Clinton ally John Podesta—were plucked largely from the Clinton camp, a peace offering to the establishment after he trounced Hillary Clinton in the 2008 primary. What Obama's top advisers preserved from the Clinton era, among other things, were deep worries about appearing fiscally irresponsible and a belief in the inherent virtues of bipartisan cooperation and governing by expertise.

Although Republicans certainly worried about their path back to power after Obama's election in 2008, they had little interest in making nice—least of all to curb global warming. Over the course of partisan polarization on climate and many other fronts—and thanks to a few wealthy patrons—the GOP was on the tip of solidifying into the political arm of the fossil fuel industry after years of grooming; eight years of two former oil executives occupying the executive branch hadn't hurt.

We've got the benefit of hindsight, though. In the lead-up to the 2008 US election, it seemed all but inevitable to Washington insiders that Congress would come together to do something about climate change. John McCain—running against Barack Obama and distancing himself from the Bush administration—had adopted it as a pet issue years earlier, working with the EDF and the Natural Resources Defense Council (NRDC) to craft legislation. Both candidates described it as a top priority and said the Bush administration hadn't done enough to address warming. McCain was one of several lawmakers on both sides of the aisle to propose climate measures through the early 2000s. None resulted in curbed emissions, but some—including cap-and-trade-type measures—found some Republican support.[28] Since 2007, a House Select Committee on global warming had been holding an onslaught of congressional hearings to drive attention toward the problem. Moods seemed to be shifting outside the Beltway too. Months after the release of former vice president Al Gore's 2006 documentary about global warming, *An Inconvenient Truth*, a Pew Research Center survey found that the percentage of Americans who attributed global warming to human activity had jumped from 41 to 50 percent.[29]

Sensing another opening for bipartisan cooperation, in 2006 EDF began meeting with the leadership of corporations like GE and Duke Energy—at the time, the country's third-largest burner of coal—and other environmental groups to cobble together a coalition to push cap-and-trade over the finish line.[30] They eventually launched the US Climate Action Partnership in 2007, featuring EDF and other big Beltway greens and a who's who of Fortune 500 companies, including a number of other fossil fuel firms and utilities. The cap-and-trade idea they coalesced around was partly modeled on the Acid Rain Program that was then and to this day remains the crowning achievement of Krupp's career. It now enjoyed the added credibility boost of the giant emissions trading system, which had been rolled out in the European Union a few years earlier, the market-based Regional Greenhouse Gas Initiative (RGGI) in the Northeast, and ongoing conversations about implementing similar programs among a number of states, including California.[31] The White House wanted cap-and-trade too, with Clinton advisers still sore over the BTU tax's failure. The wonks were converging around carbon pricing, so insiders in the Obama administration thought Congress could too. Obama climate czar and former EPA administrator Carol Browner relayed to transition team member Reed Hundt years later that it took about 10 years to get to the passage of the 1990 Clean Air Act. Following that pattern from around 2000 to 2008, environmental leaders like Fred Krupp and Republicans like Boyden Gray, George W. Bush, and Bill Reilly [George H. W Bush's EPA administrator] to some degree were coalescing around the idea of using market mechanisms to achieve environmental benefits. So by the time Obama comes to office, it is widely accepted that the cap-and-trade method achieved the environmental goal of reducing acid rain at significantly lower cost than anyone had ever anticipated.[32]

Pushing cap-and-trade domestically also aligned with an approach Big Greens and Democrats had already started to take internationally. Along with the World Resources Institute and NRDC, EDF eagerly worked with BP through the 1990s to sell emissions trading as a compromise measure within the UNFCCC. BP had first partnered with EDF to create an internal, company-wide emissions trading system to announce their environmental enlightenment. Along with Shell, BP severed ties with the denialist Global Climate Coalition (GCC) in 1996 for greener pastures as the loudest corporate voices for climate action in the UN.[33] The Clinton administration was receptive. "It is not the case that business and NGO advocacy can fully be credited for the change in U.S. foreign policy" on climate, political scientist Jonas Meckling writes. "Rather, most observers say, it was a reciprocal process, in which the business coalition lobbied the administration, and in turn the administration was actively seeking business support." Through the 1990s, then, European producers had learned the value of engaging with a "pro-regulatory" coalition, as Meckling calls it, that sought to shape climate policy rather than block it outright, as the GCC had done in killing US involvement in the Kyoto Protocol. American producers would join them soon enough.

Another motivator for fossil fuel companies to get behind cap-and-trade was a Supreme Court decision giving the EPA a mandate to regulate carbon dioxide, *Massachusetts v. EPA*. Republicans, weary of new regulations that might result from that decision, were now more willing to come to the table and eke out a compromise that could forestall more stringent rules. In his account of that period, political scientist Matto Mildenberger notes that "stripping the EPA of this new authority became an immediate legislative priority" for industry and the GOP, something Democrats knew well. "Regulatory relief would become a significant bargaining chip for proponents building a reform coalition."[34]

Obama's election shortly after made climate legislation look possible in a way it hadn't been before. The time still seemed right to start moving something forward, ideally to present to the world by the time COP15 convened in Denmark in late 2009. In that context, the combined staffs of Massachusetts' Ed Markey and newly installed House Energy and Commerce Committee chairman Henry Waxman got to work drafting up a bill for the House (Waxman-Markey), adapting a measure previously introduced by the latter, in consultation with USCAP and meant to bridge gaps between earlier proposals. "The think tanks in town and everyone in the talking head community," one of Markey's staffers later said, "no one was talking about a carbon tax. Everyone was talking about cap and trade as being the vehicle. At the time, there was sort of this consensus that it was the moderate, most economically efficient way of dealing with pollution."[35]

When Browner began working to get the White House's blessing on the bill, she found that senior administration officials were wary of taking it on—especially chief of staff Rahm Emanuel. In a meeting with USCAP in the late spring of 2009, he explained his main priority: "We need to put points on the board. We only want to do things that are going to be successful. If the climate bill bogs down, we move on." The substance of the bill didn't much matter, in other words, so long as it would allow the administration to claim a win. Emissions reductions would be a nice perk.

Waxman-Markey—officially, the American Clean Energy and Security Act—was more expansive than its critics on the left claimed. While the public face of the bill was its cap-and-trade program, it outlined $190 billion worth of investments in clean energy, more than double the American Recovery and Reinvestment Act's green investments in 2009. It would have established a nationwide renewable electricity standard, a sizeable fund for transitioning workers out of carbon-intensive sectors, and a "cash-for-clunkers" program, offering vouchers to drivers who trade in their gas-guzzling cars for more fuel-efficient ones.

But the giveaways were extensive. Thanks to a concerted push from the Edison Electric Institute—the trade association for investor-owned electric utilities—the bill granted free polluting permits to coal-fired power plants, as well as oil refiners and automakers. It also would have kneecapped the EPA's ability to regulate greenhouse gases, absolving carbon-intensive industries of any worries created by the endangerment finding. For these and other reasons, many justice-oriented climate groups, including Greenpeace and Friends of the Earth, came out against Waxman-Markey.

Critics had good reason to worry about the bill's ability to curb emissions. By the time negotiations on the Hill were happening, the European Union's emissions trading system—seen as proof positive for a US system—had already effectively collapsed, and prices wouldn't rebound until nearly a decade later. Because the price of credits was so low, many utilities found it cheaper to keep operating coal-fired power plants under the new system than switch to gas—itself a dubious goal.[36] Carbon-saving behavior changes that did happen during that time were almost all in response to direct regulations, not the emissions trading system market. And the RGGI program in the Northeast was (and remains to this day) mostly a way to raise revenue.

Beyond those tracking it closely, though, few people understood what cap-and-trade was at all. Its supporters' strategy didn't aim at informing them, expecting the public would trust policy wonks to figure out a solution. Messaging focused on the more oblique concept of climate action. While well-funded field organizers and advertising campaigns aimed to raise a generalized concern about global warming among voters in the hopes that would compel them to support or at least not reject cap-and-trade, the details were kept mostly to the behind-closed-doors meetings among Beltway insiders in USCAP, where environmentalists were continually losing ground to corporations. In hiding the ball, they also obscured all but the vaguest reasons for why people should be excited about their proposed policy. As sociologist Theda Skocpol notes, "It is not clear that the climate change ad writers ever tried to spell out concrete benefits that new legislation could bring to ordinary families."[37] In failing to make that case, they lost crucial ground to the GOP and the fossil fuel industry, who were happy to spell out in often misleading detail what the bill would mean beyond the Beltway: more costs. Amid a painful and deepening recession (unemployment was hovering just below 10 percent), with the memory of high gas prices still fresh, industry fearmongering about how cap-and-trade would kill jobs and raise fuel costs by thousands of dollars hit home.[38] To the extent the White House focused on the bill, it did message cap-and-trade as a green jobs and energy security program—albeit mostly as a means of avoiding a discussion about the details of climate change itself, which polled poorly.

That Waxman-Markey got through the House that summer was largely due to what Skocpol dubs Speaker Nancy Pelosi's "near-Leninist discipline" in whipping votes, as well as a battery of compromises. Hours before it passed (219–212), Wall Street—via New Democrats and USCAP members AIG and Lehman Brothers, in particular—successfully nixed safeguards on the $2 trillion carbon derivatives market the bill would have created, opening it up to the kind of rampant, profiteering speculation that had just triggered the financial crisis.[39] Optics be damned.

Champagne was popped, and the fight moved to more hostile ground in the Senate—but not before industry had its say.

NONE OF THE membership criteria for companies to join USCAP—including staff capacity and a $100,000 annual contribution—prevented them from also supporting organizations that actively lobbied against climate action. Political scientist Jacob Grumbach pointed this out in a 2015 article in the academic journal *Business and Politics*, where he detailed several members' simultaneous membership in USCAP and trade associations actively working against it.[40] These companies, Shell, BP, and ConocoPhillips among them, negotiated with the EDF and other greens to weaken the EPA and expand free credit allowances. Simultaneously, through their membership in API, they helped to finance API's astroturfed Energy Citizens campaign against the bill, which organized rallies against Waxman-Markey attended mainly by oil industry employees bussed in by their bosses.[41] Deryck Spooner, a former Nature Conservancy staffer, eventually headed up API's grassroots activist arm, of which Energy Citizens was a part. According to an email obtained by Greenpeace, API retained "a highly experienced events management company that has produced successful rallies for presidential campaigns, corporations and interest groups" to host rallies around the country in opposition to cap-and-trade, targeting senators back in their districts for the summer recess.[42] Those events were planned in coordination with the US Chamber of Commerce, of which USCAP members Chrysler, Deere, Dow Chemical, Duke Energy, GE, PepsiCo, PNM Resources, and Siemens were members.

In other words, the same corporations touted as industry leaders on climate were playing on both sides of the fight. "When presented with a limited menu of options—especially when all are somewhat distasteful—the optimal strategy for firms and business lobbies may be to water down undesired regulation that is likely to occur or propose more moderate alternatives," Grumbach writes. As some of the world's most profitable companies, he explains, the Fortune 500 firms involved in USCAP could easily hedge their bets to maintain something as close to the status quo as possible: they worked within USCAP to ensure any legislation that did pass was favorable to them, while also funding trade associations that would try to make sure it didn't pass at all.

In 2009 and 2010—the years cap-and-trade was being debated on Capitol Hill—sociologist Robert Brulle found that lobbying on climate change represented over 9 percent of all lobbying expenditures in both years, peaking at $362 million and dropping off precipitously in 2011.[43] The biggest spenders by far were investor-owned utilities and fossil fuel companies, followed by the transportation industry.

Some of the most forceful opposition to cap-and-trade in the summer of 2009 came from outside the publicly traded oil majors: namely, Charles and David Koch's privately held fossil fuel empire. Ironically, journalist Christopher Leonard notes in *Kochland*, fairly senior members of their inner circle had spent months studying cap-and-trade legislation and found there was money to be made in its emissions trading market. They were handily outgunned. In addition to seeing it as an unacceptable government interference in industry profits, the brothers were especially peeved that oil drillers would get more free allowances than their oil refinery arms.[44] "Koch's political machine was deployed, in 2009, in ways that it had never been deployed before," Leonard writes. "In the fight that Charles Koch was about to wage, there would be no compromise." His main targets weren't liberal Democrats or environmentalists but the moderate Republicans who had voted for Waxman-Markey.

As it funded bogus studies into the allegedly outrageous costs of climate legislation, the Kochs' main vehicle for channeling popular outrage against it would be Americans for Prosperity (AFP), the Koch empire's organizing arm, with state directors spread out around the country. Between 2007 and 2009, the Kochs nearly doubled AFP's budget. By 2010, it had ballooned to $17.5 million.[45] With its help, Tea Party activists the summer after cap-and-trade passed through the House mounted primary challenges to unseat RINOs (Republicans In Name Only)—starting with those who had supported the bill, or even been vaguely open to the idea of doing something about climate change. Irate protesters showed up at usually sleepy town halls around the country, shouting to anyone who would listen about the "crap-and-tax" bill and how climate change was a hoax.

To ignite opposition against Waxman-Markey, the Kochs found white supremacy to be easy kindling. Before they got more seriously involved, the fervor of the Tea Party's predominantly white, mostly middle-class rallies—concentrated in former Confederate states—was directed in large part at the country's first Black president, not carbon pricing. Koch funding helped mold that anger around their particular policy concerns and injected it with an electoral muscle it might have lacked otherwise. One of its earliest wins in that vein was Trey Gowdy's blowout primary victory against South Carolina congressman Bob Inglis, a Republican with a strongly conservative voting record who had taken money from the Kochs for years. Though he hadn't voted for cap-and-trade, he was also a self-proclaimed "heretic" on climate within the GOP who believed the science of global warming after visiting the Arctic and campaigned on calls to scale up renewable energy. He paid dearly. Others did as well. The same summer, Republican Scott Brown flipped Ted Kennedy's Senate seat. Racism wasn't the only cause that animated the Tea Party, but its fossil-fuel-funded rise went a long way toward bringing more openly white supremacist ideas into the mainstream. As discussed in Chapter 1, the convenient alliance between racism and fossil fuel interests wasn't a new one—least of all for Charles Koch.

The summer's events chilled ambition by lawmakers at every level of government to talk or legislate about climate or work with Democrats on just about anything. With Tea Party primary challenges gaining steam,

Republicans who had previously been friendly to climate action in some rhetorical sense started either staying quiet or peddling denialist talking points to woo wealthy donors and keep any right-wing ire at bay. The bill also faced increased resistance from Senate Democrats who represented states with big carbon-intensive industries. The Senate companion bill—the combined effort of Joe Lieberman, John Kerry, and Lindsey Graham—haphazardly added even more business-friendly compromises to bolster support: building in more allowances for polluters and expanding offshore drilling, which became a PR problem when a BP rig off the Gulf Coast killed eleven people and became the largest environmental disaster in US history.

Having doubled down on their inside game, cap-and-trade advocates lacked both leverage and outside forces to mobilize for an alternative measure. Lindsey Graham withdrew his support, and Senate Republicans united against it. Harry Reid announced he'd rather focus on immigration. The White House had already moved on, pivoting away from hope and change and toward austerity.[46] Negotiations over the companion bill died with a whimper that spring, having never come to a vote. Attempts at some non-cap-and-trade climate compromise legislation floundered. "The assumption and hope," Greg Dotson, then chair of the Environment and Public Works Committee, later told E&E News, "was that if you got businesses on, you got Republicans on. So we did a lot of work to address business concerns, and we got a lot of business support, and that did not translate to broad Republican support."[47]

Cap-and-trade was far from the only contributing factor, but the broader strategic approach it embodied cost Democrats dearly in the 2010 midterms, when they lost control of the House. Amid a sluggish recovery from the Great Recession, Republicans saw their biggest gains since 1938, picking up sixty seats in that chamber.[48] Forty-four candidates with Tea Party backing made it to Washington.[49] All told, at the federal level and in statehouses around the country, Democrats lost one thousand seats through the Obama administration, eventually losing the Senate and the White House as well.

Big Green's and Democrats' approach to passing climate legislation had hinged on two hopes: first, that industry would come to the table to negotiate in good faith if advocates could make a compelling business case for cap-and-trade; second, that their support was the key to attracting GOP votes. Neither turned out to be true. With the Tea Party gaining momentum and the backing of wealthy donors, Republicans didn't have much to gain from siding with Democrats.

Having skillfully played both sides, the carbon-intensive industries that were a part of USCAP had an easy out once it looked like nothing was going to pass. Nevermind the fact that—beyond its at the time relatively ambitious target to scale down emissions by 83 percent below 1995 levels by 2050—any bill that *would* have passed was so riddled with sweeteners for fossil fuel executives that it likely wouldn't have gotten anywhere close to the level of emissions reductions needed. Erecting a market-based, corporate-friendly Rube Goldberg machine to deal with climate change wasn't the win-win Beltway insiders thought it would be.

It would be a mistake, though, to see the saga of the cap-and-trade fight as an isolated phenomenon. After decades of red-baiting and industry attempts to dismantle both unions and the New Deal Order, tangible alternatives from the left have been pushed out of the mainstream. Fringier right-wing proposals, meanwhile, have gradually entered it. If the founding principle of the third way had been to extract New Deal ends through neoliberal means, its greatest successes were in achieving neoliberal ends through neoliberal means, from welfare reform in the nineties on through to the Affordable Care Act. If that strategy had paid off politically in the nineties, it wasn't now. Proponents of climate action on the Hill had been optimistic that—*even during a recession*—climate would somehow rise above other issues; that climate politics, confined as they seemed to be to the realm of DC policy wonks and expert scientists, simply operated on another plane of existence from messy issues like welfare or criminal justice. Democrats, it turned out, had to make the case for why people beyond the Beltway should want climate policy to pass. "Environmentalists," Skocpol wrote in her 2013 postmortem, "can no longer presume that most officials in Washington DC or in many non-coastal state capitols are looking for expert solutions to an agreed-upon problem."[50]

As doomed as it may seem looking back, there was an internal logic behind Krupp and other cap-and-trade advocates' thinking that made a certain amount of sense in context: there really were Republicans who were talking about climate issues in a way they hadn't been before, however disingenuous they turned out to be. Certain elements of the corporate world seemed to be changing their tunes, however two-faced. And some of those corporations really did have close ties to the GOP. There was no dramatically better proposal on the table. With a defensible enough track record and rumblings of bipartisan support around cap-and-trade, it looked like it might just work. It didn't, but it wasn't so easy to see that in 2009.

What's astounding is how many powerful people in DC are eager to do it all again. Savvy fossil fuel companies are just as eager to repeat history.

IN SLICKED-BACK HAIR, round glasses, and a European-cut suit, Nick Schulz—ExxonMobil's director of Stakeholder Relations—looks like he'd fit in better at a Bay Area start-up than among stodgy fossil fuel executives. In September 2019, he's somewhere in the middle, at a Carbon Tax Forum near the San Francisco waterfront organized by a group called Business Climate Leaders, a project of the Citizens Climate Lobby. Schulz, it's true, is no career oilman. But he's worked at lots of places that took their money. Now in a line-up with senators and green group executive directors talking about the merits of a carbon tax for dealing with the climate crisis, it hadn't been that long since ExxonMobil had paid Schulz while he said it wasn't much to worry about.

Before joining ExxonMobil in 2013, Schulz had done stints at the American Enterprise Institute and the US Chamber of Commerce. On the think tank circuit, Schulz's oeuvre was typical conservative grist: cutting regulations, breaking up unions, and extolling the economic virtues of the nuclear family. On occasion at AEI he flirted with soft-core denialism, calling out "alarmists" and complaining about greens who assert that their "intellectual opponents in the climate fight are industry stooges." During the cap-and-trade fight, though, he preferred a carbon tax as "simpler, more efficient, easier to implement, and fairer." Most importantly, it was an alternative to the kind of big government "planning" inscribed into Waxman-Markey, a word which—by his count—appeared in that bill sixty-nine times.[51]

From 2001 until 2008, Schulz served as the editor in chief of a website by the painfully early aughts name of Tech Central Station, later TCS Daily. The site described itself as "a cross between a journal of Internet opinion and a cyber think tank open to the public," with a commitment to free-market ideals.[52] Its publisher was a conservative lobbying outfit called DCI Group, a veteran of tobacco-industry-funded campaigns against smoke-free laws that had also represented Myanmar's military junta. Between 2005 and 2016, the firm represented ExxonMobil. Before then, the oil giant—along with General Motors, PhRMA, Freddie Mac, and a number of other companies—sponsored TCS, a fact the site boasted about proudly.[53] Exxon's 2003 giving report states that it gave $95,000 to the "Tech Central Science Foundation"—the group's nonprofit arm, directed by DCI CEO Doug Goodyear—for "Climate Change Support."[54]

ExxonMobil got its money's worth out of TCS. Alongside content from conservative heavyweights like Newt Gingrich, TCS during Schulz's tenure published a who's who of the climate denier set: Patrick Michaels, Stephen Milloy, Bjørn Lomborg, Willie Soon, Marlo Lewis Jr., and many more. TCS writers took particular aim at the Kyoto Protocol, praising the work of denier groups like the Competitive Enterprise Institute. TCS targeted everything from the links between climate change and superstorms to pre-Waxman-Markey bipartisan climate initiatives from John McCain and Joe Lieberman. Schulz himself readily peddled climate skeptic talking points, jumping on the "Climategate" bandwagon in 2003 against climate scientist Michael Mann with an op-ed in USA Today.[55] "For a decade now," he wrote in April 2006, "the alarmists have been jumping up and down and saying, 'Look at me! The Earth is burning up!' Research has continued into climate change, but precipitous steps at mitigation have wisely been avoided." In the same piece he praised pioneering denier Fred Seitz as "one of America's highly regarded scientists."[56]

Schulz purchased TCS from the DCI Group in 2006, ending its corporate ties but keeping the deniers intact; it waned shortly thereafter. That same year he helped to start AEI's online publication The American and would become its editor in chief in 2008. While he worked there, AEI was (and remains) flush with Exxon cash.[57] It's unclear whether any of these funds made their way to Schulz personally, but between 1998 and 2017, Exxon gave AEI nearly $4.5 million, more than half of that after 2008; he left to go work for the company in March 2013.[58] It was at AEI, however, where Schulz seemingly learned to stop worrying about the alarmists and love the carbon tax, a supposedly noble but sadly hopeless cause. As he lamented in late July 2007, in an op-ed for *The Hill*, "There is little enthusiasm for an explicit carbon tax, even though this is the simplest and most transparent way to begin reducing greenhouse emissions."[59]

Did Schulz have a change of heart sometime between the spring of 2006 and the summer of 2007? Did he decide the climate crisis was worth tackling after all? An article he wrote for the spring 2007 issue of a journal called the *New Atlantis* sheds some light on his thinking. Discussing Daniel C. Esty and Andrew S. Winston's corporate social responsibility bible *Green to Gold*, Schulz—still ambivalent about the nature of the problem itself—concedes that "the environment is winning in the court of public opinion." Whatever the facts of global warming might happen to be, companies would do well to cash into the fervor around it. Noting Toyota's success with the Prius, Schulz notes that whether the car is actually green or not—critics argue otherwise—"is largely irrelevant from the green-to-gold point of view. For Esty and Winston, the *perception* that the Prius is eco-friendly is what matters in the marketplace."[60]

If it's appearances that matter, Schulz—by then in his fifth year at ExxonMobil—put on a great show at the forum in San Francisco that September in 2018, having long since swallowed any misgivings about the green-to-gold strategy he might have once felt. His former publisher, the DCI Group, had in 2016 been subpoenaed by the US Virgin Islands and New York State as part of investigations into whether ExxonMobil had misled the public about climate change. Two years later, Schulz was making the rounds at venues like the Aspen Ideas Institute for Exxon, cracking jokes about how eager he was to break out his fall wardrobe and sounding like a monkish environmental economist.[61] "Efficiency in policy is actually extremely important. There's a lot of policy in place nationally," he said, trailing off before naming any specifics. "We should spend some more time thinking about if this is optimal policy. If you have policies in place that impose a very high cost for low benefit, it may feel like we're doing something about this challenge, but in reality we're not."

It's telling that a person ExxonMobil dispatches to talk about its commitment whose employers accepted their checks as he spread misinformation about it. Exxon has a savvy understanding for just how far perception can be from reality. In the upswing of public concern about climate change in 2007, the company pledged that it would stop funding climate skeptics.[62] That stopped flows to groups like the Heartland Institute and the George C. Marshall Institute, which had gotten people like Stephen Milloy down in the dumps. But they still funded bodies like AEI and the National Association of Manufacturers, both of which have vigorously fought emissions reductions measures and championed Trump's environmental rollbacks. Exxon continues to donate large sums to the Manhattan Institute, where Rebekah Mercer is a major funder and board member, and remains a member of the American Petroleum Institute, which has kept up its fight against environmental regulations and climate measures. The oil company only left the Koch-funded American Legislative Exchange Council in 2017, after it pushed model legislation in Oklahoma, Colorado, and Arizona that described global warming as a "theory."[63]

With more tact than the deniers, the main contention of groups Exxon still funds isn't that the climate isn't changing; most don't talk science. They take aim at the unfair burden proposed climate measures would place on carbon-intensive businesses, actively opposing Waxman-Markey, the Clean Power Plan, and the Green New Deal. Functionally, these two arguments—either that the climate crisis isn't happening or that any proposed policies to deal with it are unworkable—aren't so different. What major oil companies have figured out is that skipping the first part avoids the political hassle of being linked to unseemly conspiracy theorists but with a similar outcome: maintaining business as usual. In claiming some interest in solving the problem, they'll be welcomed with open arms by green groups and politicians who haven't scratched below the surface. In addition to energy centers at some of the country's top universities, Exxon also generously funds more centrist think tanks, including the DLC's Progressive Policy Institute, the Council on Foreign Relations, the Bipartisan Policy Institute, and the Brookings Institution, which got $50,000, $100,000, $250,000, $200,000, and $100,000 from the oil giant, respectively, just in 2019.[64] It's hard to measure just how industry funding shapes what these groups do—and hard to imagine it has no effect at all.

It's worth remembering that ExxonMobil, which was not a member of USCAP, stepped forward with its support for a carbon tax at precisely the moment that momentum for cap-and-trade was building. In a January 2009 speech in Washington, DC, Tillerson said that "a carbon tax strikes me as a more direct, a more transparent and a more effective approach" than the complex cap-and-trade bill that Congress was considering."[65] For his company and others, a carbon tax has always been a cudgel to wield against some more tangible policy, be that cap-and-trade or a Green New Deal.

Months before world leaders convened for the Paris Climate Talks, six of the world's largest oil companies—Total, Statoil, BP, Shell, Eni, and BG—issued a letter to the UNFCCC calling for a global price on carbon. As Clean Power Plan implementation was being considered under the Obama administration, Tillerson warned that the government "works best when it maintains a level playing field, opens the door to competition and refrains from picking winners and losers," touting the company's support for carbon capture and storage technology and carbon taxes.[66] Exxon is one of thirteen private and state-owned oil companies to have joined the Oil and Gas Climate Initiative, "a voluntary CEO-led initiative" founded in late 2014 to establish internal industry standards for bringing down emissions across their operations—and to attract good PR. CEO Climate Dialogue—launched in 2019, as the Green New Deal dominated the Democratic presidential primary—has similar goals and representation from a number of major oil companies, along with EDF.

In the spring of 2017, as investigations into Exxon by attorneys general in New York, Massachusetts, and the US Virgin Islands were ramping up, Exxon announced with a full-page ad in the *Wall Street Journal* that it had signed on with Shell, BP, and a number of other companies as a founding member of the Climate Leadership Council (CLC), the group pushing for a revenue-neutral carbon tax starting at $40 per ton and gradually rising to $65 per ton, well below the levels experts generally recommend, per Chapter 2.[67] Conveniently, that group's plan, touted as a "grand bargain" and silver bullet for reducing emissions, would phase out most of the EPA's authority to regulate stationary sources of carbon dioxide like power plants. A statute nixed in late 2019—just before Exxon squared off in court against the New York State Attorney General's office—would also have brought about "an end to federal and state tort liability for emitters" and prevented EPA regulations on cars and trucks.[68]

The day after the IPCC released its special report on 1.5 degrees in October 2018, Exxon announced that it would invest $1 million into pushing lawmakers to enact that plan via the CLC's lobbying arm, Americans for Carbon Dividends. The effort was headed up by Squire Patton Boggs lobbyists Trent Lott and John Breaux, former senators who have each received major campaign contributions from the oil and gas industry.

In San Francisco, Nick Schulz hailed the CLC plan's pragmatism. "If you talk to the CLC people, their operating theory is that any sort of real meaningful action on [climate] will require Republicans to be involved," he told me in a brief interview after the panel. "This is a policy solution that should work for them given what their political beliefs are and their ideological leanings. But the CLC is talking to Democrats and Republicans, and I think that's appropriate because you're going to need a bipartisan solution . . . Greg Mankiw, who is a Republican and worked in Republican administrations, supports it. Larry Summers, who's a Democrat and worked in a Democratic administration, supports it. They're very good economists, and that's also what they think the most efficient policy [is]. So that's where we come down on this." With friends like these!

FOSSIL FUEL INTERESTS' spending habits will tell you more about their priorities for the future than any press release. They're still spending prolific amounts of money to block even common-sense climate bills and regulations at every level of government. In the lead-up to the 2018 midterms in Washington state, API's regional equivalent—the Washington State Petroleum Association—collected more than $30 million from its oil and gas industry members to quash a modest price on carbon, starting at $15 per ton and maxing out at $55 in 2035. The revenue would have been spent on green infrastructure upgrades and investments in frontline communities.

BP alone contributed $13 million to stopping the ballot measure, despite its nominal support for a much higher price through its membership in the CLC. Asked about the disparity, a BP spokesperson forwarded along a letter from BP Cherry Point refinery manager Robert K. Allendorfer: the measure, he wrote, "would fail to preempt other state and local carbon regulations." As such, the measure would jeopardize "thousands of Washington jobs." In the following paragraph, Allendorfer noted that BP supports more than 9,600 jobs in the state, as if to issue a thinly veiled threat: it'd be a shame if something happened to them. ExxonMobil noted something similar. Asked about carbon pricing more generally, spokesperson Alan Jeffers wrote in an email that the "most significant common shortcoming" of such schemes "is their failure to preempt existing greenhouse gas regulations. The preemption issue is important because a properly-designed carbon tax that replaces the existing regime of emissions regulations would be a beneficial policy rationalization."[69] In other words, they'll oppose any carbon tax that doesn't also kneecap regulations.

Even more money was spent that same cycle in Colorado—then the country's third-largest oil and gas producer—where a ballot measure (Prop 112) from the grassroots group Colorado Rising would have created a 2,500-foot setback zone between drilling sites and homes, schools, and playgrounds. The industry outspent Colorado Rising forty to one, spending nearly $41 million to quash the measure and another $10 million backing an unsuccessful ballot initiative that would have allowed property owners, including oil and gas companies, to sue local governments and the state for infringing on their profits.

Patricia Nelson began campaigning for Prop 112 after a friend encouraged her to attend a meeting where she lives in Weld County, home to 23,000 wells, an F air quality rating from the American Lung Association, and infant mortality rates twice as high as those in surrounding counties. A new drill site had just been set up behind the playground of her son Diego's school, where 87 percent of attendees are students of color and 90 percent fall below the poverty line. "This isn't over for me, personally," she said when we spoke late that night, after news came in that Prop 112 had failed. "We have had a warning, that we either end our dependence on fossil fuels or things are going to get extremely rough for mankind. For me, it shows that it's just about greed and money for this industry." When a measure similar to Prop 112 made it out of a state legislative committee in California months later, the share prices of two Golden State–based drillers dropped by 10 and 13 percent overnight. Another such bill was killed by California Democrats in a state legislative committee hearing in 2020.[70]

The money activists like Nelson are up against is enormous. Robert Brulle found that between 2000 and 2016, $2 billion was spent lobbying on climate issues, which he reckons is a conservative estimate given that people who spend less than 20 percent of their time on lobbying activities aren't required to submit disclosure forms.[71] While such disclosures don't specify how lobbyists pushed lawmakers to vote, major business associations like the US Chamber of Commerce and the fossil fuel, utility, and transportation sectors have reliably outspent public interest groups, including labor unions, by an average of ten to one on lobbying. On public relations—which aren't captured in lobbying disclosures—the same group outspent greens nineteen to one. All that free-flowing fossil fuel cash warps what congressional staffers think their constituents want. As political scientists Alexander Hertel-Fernandez, Leah Stokes, and Matto Mildenberger found, the ubiquity of corporate influence in US politics means congressional staffers tend to underestimate support for a range of progressive priorities in the districts they serve, including regulating carbon dioxide.[72]

Beyond lobbying, oil companies' operational budgets are even more telling. From 2010 to 2018, none of the world's major oil companies had invested more than 4 percent of their capital expenditures into low-carbon technologies. A survey by the UK-based think tank Carbon Tracker, which focuses on the risk of stranded fossil fuel assets for shareholders, found that 92 percent of the over thirty oil and gas companies they surveyed in 2017 tied executive-level bonuses to growing fuel production, reserves, or both. Just nine companies—half of European producers and ExxonMobil—included pay incentives related to climate mitigation, but "where they are included," the report's authors note, "these metrics tend to affect a small minority of compensation, and most of these companies simultaneously encourage fossil fuel growth."[73]

Conservative estimates from the International Energy Agency (IEA) suggest that keeping warming below 1.8 degrees Celsius—aligned with the "well below 2 degrees" figure from the Paris Agreement—should see a 60 percent drop in companies' capital expenditure on new oil projects from 2018 through 2030. To keep warming below 1.6 degrees would mean an 83 percent drop along the same time line. In 2018 alone, even those companies that have at least paid more lip service to curbing their emissions—Shell, BP, and Equinor—had all sanctioned new drilling projects that plainly ran afoul of Paris goals. Altogether, fossil fuel companies that year approved $50 billion worth of investments in projects that undermine climate goals and would never be approved if they wanted to honor them.[74] While there would be no new oil sands development in a Paris-aligned world, ExxonMobil, through a Canadian subsidiary, sunk $2.6 billion in 2019 into the first new oil sands project in five years. As of the same year, expected development in US shale fields threatened, moreover, to be the single largest burst of new carbon dioxide emissions to enter the earth's atmosphere through 2050.

Mike Coffin, a senior analyst with Carbon Tracker, laid out three possibilities to me for what a Paris-aligned business plan for oil and gas companies might look like: don't invest in projects that aren't needed, return capital expenditure to investors directly rather than pouring it into new fossil fuel development, or shift a significant portion of new investment into low-carbon technologies. "For a 1.5 degree scenario, we can't factor in any new oil and gas," he said. Ninety percent of ExxonMobil's new projects in 2017, Coffin said of his report, were inconsistent with a Paris-aligned world, far worse than European producers like Shell, BP, and Equinor. "Getting smaller," he told me, "is not in the DNA of big oil and big corporate cultures."

For their part, top executives at these oil companies seem to agree. "Despite what a lot of activists say," Shell CEO Ben van Beurden told Reuters in a 2019 interview, "it is entirely legitimate to invest in oil and gas because the world demands it." He added that the company has "no choice" but to invest in projects whose life spans would reach decades into the future, well beyond when the IPCC recommends that 87 percent of oil usage be phased out.

In 2020, BP—under new leadership and as the sector as a whole faced production shut-ins from COVID-19 shutdowns (see Chapter 9)—

announced the most ambition plan of any oil major yet: to slash its oil and gas production by 40 percent over the next decade and increase spending on low-carbon energy to $5 billion per year by 2030. As part of its "ambition to become a net zero company by 2050," the company had announced earlier that year that it would leave WSPA (Western States Petroleum Assocation) and two other industry trade associations. These are dramatic moves, to be sure—with ample credit owed to activists ramping up pressure on the industry. Whether BP follows through and stops funding other trade associations like API, which block climate policies they don't like, remains to be seen. In any case, the rest of the industry doesn't seem to be following suit. However much fossil fuel companies talk about their token investments in low-carbon fuels and carbon capture and storage or support for a carbon tax, there is simply no reason to believe that the fossil fuel industry as a whole is willing to break with business as usual on the scale that science requires.

The actual substance of these policies is less important than the political calculus behind them. In the lead-up to the 2020 election, talk of hydrogen and carbon capture and storage seemed to be eclipsing that of carbon pricing among fossil fuel companies and the wonks and politicians they fund, emphasizing the need for "innovation" as opposed to, say, stopping drilling or implementing a Green New Deal. The bottom line for industry is that it would rather craft its own rules for navigating the twenty-first century than have any meaningful regulation imposed on its profits. The methods the industry uses to keep its business model intact have grown increasingly slick, having progressed from funding people to deny climate change exists to backing efforts meant, ostensibly, to deal with it. Given the prodigious amounts of money fossil fuel companies spend on advertising these green shifts, it would be easy for good-faith politicians and climate campaigners to miss the forest of a toxic business model—companies who've known about the problem for generations and continue fueling it through mountains of political spending—for the trees: a few shiny investments and CEOs claiming commitment to the climate cause. For the political establishment, entrusting climate policy to the rule of expert Beltway wonks has made those polluters' voices louder than majorities of US voters who in poll after poll now support rapid, big government climate action. How many more times can companies fool powerful people who should know better?

IN PITTSBURGH IN late October of 2019, Donald Trump gave a rousing keynote address to the Shale Insight conference, an annual confab for natural gas drillers and a who's who of fossil-fuel-friendly politicians, that is, the country's most prominent Republicans. "I was here three years ago," he began. "You're much happier now. And you're much wealthier."[75]

After a hearty shout-out to Trump donor and oil and gas magnate Harold Hamm ("he can take a straw, and he can put it into the ground, and oil comes out"), the president went on to praise several of his cabinet members and various fossil fuel trade associations and boast about the administration's success in peeling back regulations and pledge to withdraw from the Paris Climate Agreement. To thunderous applause, he also talked up building a border wall in Colorado. Besides "foreign polluters," Trump's main attacks were reserved for the "do-nothing Democrats," at that point engaged in a "witch hunt" of an impeachment proceeding against his administration. Talk of then House Speaker Nancy Pelosi riled up the crowd so much that "booo" made it in the official White House transcript after Trump mentioned her name.

"When I last spoke at this conference in 2016, American energy was under relentless assault from the previous administration," he boomed. "Federal regulations and bureaucrats were working around the clock to shut down vital infrastructure projects, bankrupt producers, and keep America's vast energies and treasures buried deep underground. They didn't want to let you go get them." Triumphantly, he had "ended the economic assault on our wonderful energy workers."

Another conversation was taking place some 2,600 miles away, but it may well have been in a very different universe featuring a very different Republican Party. The Hamilton Project—an offshoot of the venerable Brookings Institute, founded by Clinton-era treasury secretary Robert Rubin after a lucrative turn at Citibank—convened an event at Stanford University to talk about how economic policy could be used to fight the climate crisis. Like the Hamilton Project itself, the lineup featured a smattering of former Obama-era officials, economists, and Beltway policy wonks of varying shades—lots of people eager to get their old jobs back in the next Democratic White House.

Among them was Nat Keohane, an environmental economist who had come to work for the Obama administration in 2011 after being in the thick of the cap-and-trade fight. "This was right after the failure of cap-and-trade. We thought, let's scale down. Let's just do the energy and power sector. Let's take all the ideas from the senators on the Republican side and get something more reasonable passed," he said.

Needless to say, that didn't work. But like so many before it, that failure hadn't dampened Keohane's hopes for getting an elegant, market-based climate policy passed through Congress—or even prompted him to consider a new strategy. His new job was as senior vice president of the Environmental Defense Fund. He had just worked with Democrats on the Hill, he said, on a "100 Percent Clean Economy Act," that he hoped would set a standard for the next Democratic administration. By his own admission it was scant on details: "That's just a big goal. Then we'll have a couple of years to fill that out."

Keohane was excited about changes coming from the corporate world, too: "There are some businesses that are starting to get engaged and we need to nurture that and support that." He also had a few weeks earlier praised ExxonMobil and other oil companies' commitments to scaling back methane emissions. He did point out that organizations like the US Chamber of Commerce and ALEC (American Legislative Exchange Council), whose membership includes those same companies, had been detrimental to the climate fight. "We need to start calling that out a little bit," he acknowledged. "But fundamentally business is going to have a role to play, which can include a leadership role."

Hope in a Grand Bargain abounded among Keohane and the event attendees. Echoing a sentiment voiced by several other panelists, Keohane said, "We're starting to see some Republican voices come out and be cautiously in support," Republican congresspeople who are "seeing support" for climate policy "within their constituencies." As others would that day, he credited Republican representatives Carlos Curbelo—who lost his seat to a Democrat in 2018—and Francis Rooney—who had introduced a

carbon tax bill in 2019 before announcing he wouldn't seek reelection—
for being climate leaders in the GOP. Former senator Jeff Flake had also
cosponsored a carbon tax bill before leaving Congress.

Despite every piece of evidence to the contrary and a GOP that has
only become more partisan and entrenched in fossil fuels, Keohane suggested that Republicans might just come around to climate policy—so
long, it seems, as they're not planning to run for anything. And the fossil fuel industry could be ready and willing partners. To crib from John
Lennon, bipartisan climate policy is here if you want it. So why not keep
chasing the dream?

CHAPTER 4

PARALLEL WORLDS

If equity's in, we're out.

—Todd Stern, lead State Department negotiator to the UNFCCC (2011)

And equality. What of it?

—Daniel Patrick Moynihan, US ambassador to India (1975)

AFTER PASSING THROUGH a metal detector and checking my coat, a cheery young woman in a red, white, and blue polo shirt thrust toward me a book that billionaire and former New York City mayor Michael Bloomberg wrote with a former Sierra Club director called *Climate of Hope*. Foggy with jetlag and not exactly hopeful, I accepted. In more welcome news, a cart nearby was distributing free espresso.

An adjoining room featured wall-to-wall programming from a rotating cast of visiting dignitaries including US senators, electric utility executives, Obama-era climate negotiators, and hospital network CEOs, all manning talks with relentlessly positive titles ("US Climate Action: Businesses Leading the Way," "We're All In This Together").[1] Ordinarily, these kinds of displays would happen inside whatever conference center is hosting that year's UN climate talks, splayed out among other government, NGO, and trade association pavilions that put on daily panels and happy hours under blond wood scaffolding. But a few months earlier, Donald Trump had announced his administration would pull the US out of the Paris Agreement as soon as possible. Consequently, the US did not have a pavilion at COP23 in Bonn, Germany, in November 2017. In its stead was the U.S. Climate Action Center (USCAC), housed in an igloo-shaped tent with compost toilets. This privately funded alternative made a bold declaration on behalf of a renegade American delegation: We Are Still In. But who was we?

Soured on the experiences of cap-and-trade and Trump, US political types at the UN conferences now cleaved to the folk wisdom that there was a time, not so long past, when the climate debate wasn't so awfully *polarized*. They pointed to George H. W. Bush's promises to tackle the greenhouse effect in the 1980s or to Richard Nixon signing the Clean Air and Clean Water Act. Even Ronald Reagan worked with Margaret Thatcher to phase out CFCs, a noxious greenhouse gas, through the Montreal Protocol. Nowhere has this pining for better days been more present than in Trump-era discussions about US involvement in global climate negotiations. Fittingly, just about the only climate stance that nearly every congressional Democrat can agree on is that the US should recommit to the Paris Agreement; by the time this book comes out, it might already have done so. That the accord was brokered at all has been seen—at least in retrospect—as proof positive that the US has been a world leader on climate. Just look at how skillfully a team helmed by then secretary of state John Kerry fought for a good deal, among the greatest diplomatic achievements in world history. We were statesmen once, before that orange menace took office and ripped America's reputation to shreds. It's time we were again.

"The truth is," an exasperated Kerry said days before Trump's inauguration, "that climate change shouldn't be a partisan issue. It's an issue that all of us should care about, regardless of political affiliation." It's true, of course, that America's party politics—subject as it is to the whims of wealthy donors and institutional morass—are a poor lens through which to understand the crisis we're facing. But there's no route to dealing with it that can escape the questions at the root of politics: Who has the power to make and enforce decisions? Who gets to be in the room? And who bears the consequences? However much the most elite participants in Congress and at climate talks keep trying to make the climate problem transcend politics, politics keeps finding its way in.

Take the last few years of UNFCCC talks. In 2018 they came to Poland, where the ruling Law and Justice Party had recently consolidated power after promulgating a conspiracy theory that foreign powers were yet again plotting to destroy the Eastern European nation. Among its supporters were neo-Nazis who for years in the lead-up to COP24 had taken to the streets by the tens of thousands on Polish Independence Day to call for a "pure Poland, white Poland." Months earlier the country had passed a law allowing national judges to be sacked over political disagreements with the administration, as well as a sweeping surveillance measure to collect data on climate conference attendees, some of whom were stopped at the border, detained, and turned back.[2]

The more pressing news at COP24 was happening back in France, though. After sustained cuts to public services, president and aspiring god-king Emmanuel Macron had proposed a new fuel tax.[3] Floated as a way to lower carbon emissions, it would also conveniently fill a budget gap left by a tax cut for the wealthy that he had pushed through months earlier. Backlash was almost immediate and spread via Facebook groups to cities and villages around the country. Demonstrators donned *gilets jaunes* (yellow vests), which would become the movement's calling card. The *Wall Street Journal* cast its lot with the rabble in a editorial called "Global Carbon Tax Revolt"; the editorial board lauded the French tax revolt as the death knell of an "ecological transition"—scare quotes and all.[4] The yellow vests themselves weren't so resolute. Hardly monolithic, most protesters—many of whom joined climate demonstrations themselves—weren't rejecting emissions reductions; they were railing against growing inequality spurred on by rent increases and cuts to public transit. As their rallying cry put it, "They talk about the end of the world and we are talking about the end of the month."

COP25—in 2019—had originally been scheduled to take place in Brazil. Shortly after he was elected, far-right climate denier Jair Bolsonaro—who came to power in the aftermath of a coup—abruptly withdrew his country's offer to host. In short order he handed the Amazon rainforest over to corporations eager to burn it down in search of a quick buck, and then blamed actor and environmentalist Leonardo DiCaprio for setting the blazes as part of a "campaign against Brazil."[5]

As those fires roared, preparations were well underway to hold the talks in a replacement location, Santiago de Chile. That country had been a testing ground for the neoliberal policies of men like Friedrich Hayek and Milton Friedman. In the hopes of ousting Salvador Allende's demo-cratically elected socialist government, the CIA in the early 1970s began paying the University of Chicago to recruit promising Chilean students to learn economics there. The US also abetted a violent coup helmed by Gen-eral Augusto Pinochet, who called on the "Chicago Boys" to overhaul the country's economy as thousands were rounded up, tortured, and killed. The dictatorship had come to an end some three decades before COP25 was set to kick off, but its ideas still ruled life there. It was none other than economist James M. Buchanan, then at Virginia Tech, who helped Pinochet's government craft its punishing 1980 "constitution of liberty," intended to protect the junta's policies from both popular opinion and any democratically elected governments that might succeed it. As Nancy MacLean writes of his influence on the Chilean constitution, "the wicked genius of Buchanan's approach to binding popular self-government was that he did it with detailed rules that made most people's eyes glaze over. In the boring fine print, he understood, transformations can be achieved by increments that few will notice." Visiting with top leaders, he urged them to write in "severe restrictions on the power of government," pay-as-you-go rules to enshrine balanced budgets and provisions that legisla-tive supermajorities be required to make any substantive changes. They enthusiastically obliged, also curbing the power of unions, expanding Pinochet and the military's powers, and enabling severe punishments for Marxists and anyone deemed "antifamily."[6]

All this had helped Pinochet's legacy endure for decades after his death. By 2019, people who had been placed onto a shiny new pension system at the start of the Pinochet regime were now retiring and unable to make ends meet, thanks to the privatized scheme that kept their monthly income below the national poverty line. Santiago's metro system—the continent's largest—was itself a kind of showcase of Chile's neoliberal revolution: where planners in the Allende government had looked to extend transit to the city's working-class neighborhoods, the Pinochet regime prioritized middle-class riders who could afford higher prices, leaving most residents to rely on the poorly performing private bus system. When billionaire president Sebastián Piñera announced a modest fare hike in the fall of 2019, the country exploded, first in the form of high school and college students jumping turnstiles before blooming into a nationwide general strike. Echoing the *gilets jaunes*, protesters placed their fare hike revolt in a broader context: "It's not about 30 pesos. It's about 30 years." With tanks on the streets to quell demonstrators, Piñera canceled the talks—a first in the history of the UNFCCC. By November, protesters had forced the country's political parties to agree to a national vote on replacing the constitution. The next October, 78 percent of Chileans voted to do just that.[7]

Within a few days of Piñera calling off climate talks, Spain stepped in to host, weeks before its own elections. This also meant that talks would be held in Europe for four straight years, skipping the 2020 Glasgow talks canceled by COVID-19; the last-minute venue change made it more difficult for Global South civil society participants to attend, given the sheer cost of travel and lodging and the lengthy process required in some countries for getting travel visas approved.

As COP25 began, Spain's center-left Socialist Workers Party had agreed to form a governing coalition with Podemos, the left-leaning populist party created as the electoral outgrowth of the indignados movement of 2015—Spain's answer to Occupy Wall Street. Both had backed calls for a Green New Deal as climate concerns there and across Europe reached a fever pitch, with many voters around the continent citing rising temperatures as their number-one concern. At the same time, Spain—for the first time since the end of Francisco Franco's dictatorship in 1975—was witnessing the emergence of an openly xenophobic and ultranationalist electoral force called Vox. Just before the end of those talks, Boris Johnson won a blowout victory against socialist Jeremy Corbyn—another Green New Deal advocate—and Scotland was set to host COP26 as the UK looked poised to crash out of the European Union.

From inside the halls of these confabs, you'd have a hard time knowing that any of this was happening. As protests shut down the intersections of tiny French villages, sessions about fine-tuning carbon markets and ideal pricing mechanisms continued apace in Katowice, Poland. Negotiators at COP24 were tasked with deciding the rulebook by which the Paris Agreement would be implemented, and the main section of that document detailing how emissions would be reduced internationally— Article 6—outlined market-based mechanisms, carbon trading in particular, as the prime mover. Asked about what was happening in France, representatives from Western NGOs gave bewildered quotes to the press. "If France is putting a brake on the carbon tax," World Wildlife Fund France's Pierre Cannet told Politico Europe, "it puts a brake on energy transition."[8]

Bloomberg's unofficial US delegation to Trump-era COPs would also have preferred to keep politics out—anything, that is, that demanded more of them than looking better than the president. But they, too, had to face activist groups and movements calling for more. At the USCAC in Bonn, then California governor Jerry Brown—Bloomberg's partner on We Are Still In—lashed out when interrupted by protesters calling out his support for fracking and the state's oil and gas industry. Seated onstage beside Walmart's senior vice president of sustainability, Brown was interrupted from the crowd with shouts of "Still in for what?"

A group of Californians, indigenous organizers, and other less-coiffed delegates to COP23 from the United States stood up as Brown started speaking, giving short testimonies about his extensive ties to the fossil fuel industry and how it had impacted them. After they chanted, "Keep it in the ground," in opposition to fossil fuel extraction, Brown barked: "Let's put you in the ground so we can get on with the show here." They were escorted out by security.

In the mud and rain outside the USCAC, Dallas Goldtooth, of the Indigenous Environmental Network, relayed his frustration about the disconnect between reality and Brown's show: "There seems to be a lack of interest to engage with the qualifications of capitalism to fix the problem that capitalism has created. What we do here, like this action you just saw, is a means for us to have public engagement on the issue and not accept things as the status quo just because you have a so-called climate hero like Jerry Brown standing up and saying it's good."

A more dramatic scene played out at COP25. A group of some five hundred attendees watched from the courtyard of a Madrid conference center as a metal wall rose up seemingly out of nowhere, locking them quite literally out in the cold of the UNFCCC process without so much as a coat. Moments earlier, some had their entry badges snatched off them by UN guards in skirmishes outside the main plenary hall before they were cordoned off. Security prevented them from speaking even to the press. Now blocked from the venue, these protesters—who had been calling for equitable climate financing from wealthy Global North countries—marched out the back entrance, where they were greeted by Spanish police. Civil society observers had been barred access to the conference center. The UN secretariat's office finally restored their credentials the next day after lengthy negotiations. All the while, observers from some of the world's biggest polluters like Shell and BP, along with fossil fuel financiers, spoke freely with negotiating teams inside. Ta'kaiya Blaney, a then thirteen-year-old indigenous campaigner from Canada, was among those locked out. "Security and police were protecting governments and polluters. In their eyes we are just something that needs to be removed so they can take more," she told Agence France-Presse.[9]

"They want to remain in a bubble," ActionAid International's Harjeet Singh, a New Delhi–based veteran of UN climate talks who joined the demonstration, told me. "They don't want to hear the reality we were trying to tell them: you are the ones responsible for the failure, and you are not responding to people's needs. You are only listening to polluting industries. People are dying."

As Singh noted, it's hard to shake the feeling that spaces like the COP are designed to keep certain people—or at least certain ideas—out. This isn't an accident. The UNFCCC isn't uniquely neoliberal as far as multilateral institutions go, indeed far less so than the World Trade Organization (WTO) or IMF. But it's bound up in an international order designed, as historian Quinn Slobodian has written, to "inoculate capitalism against the threat of democracy." In the process, that order has also inoculated itself from the reality of the crisis itself and what really tackling it will take.[10]

NEOLIBERALS DIDN'T JUST influence government within nations, shaping economics and constraining the realm of the politically possible; they also shaped the global order to suit the global 1 percent and protect capital from threats like democracy. Decades of these efforts have resulted in rules protecting corporate investments across borders that are leagues more powerful than the nonbonding documents meant to protect the planet. Far from being antigovernment, early neoliberals looked to construct an expansive global governance structure meant to encase markets within rules and institutions dutifully overseen by experts. Ideally, there would be what Hayek and colleagues referred to as a "double government" spanning the world: one set of rules for markets and another for politics—but the two should never meet. "Over, under and beside the state-political borders of what appeared to be a purely political international law between states spread a free, i.e. non-state sphere of economy permeating everything: a global economy," philosopher and Nazi jurist Carl Schmitt described.[11]

They would be "large but loose federations within which the constituent nations would retain control over cultural policy but be bound to maintain free trade and free capital movement between nations," Slobodian summarizes, as a means to "satisfy mass demands for self-representation while preserving the international division of labor and the free search for profitable markets." Any matters of real material consequence, that is, were to be protected from the unruly *demos*, and Hayek himself frequently recommended *limiting* democracy and sovereignty to keep it from becoming beholden to so-called special interests like trade unions—or newly sovereign governments looking to make claims on wealth, resources, and decision-making power.[12]

It's hard to understand this vision and the neoliberal project more broadly without acknowledging just how scandalized many of its chief architects were by the collapse of European empires and the threat that represented to their access to the world's most precious resources. The fall of the Hapsburg Empire headquartered in Austria—the homeland of Hayek and Von Mises—had given way to socialism in the form of Red Vienna, before fascism gobbled up much of the continent. Decolonization struggles after World War II brought into the world order dozens of newly independent nations seeking an equal footing on the world stage. Neoliberals of a certain age had seen it all happen. Not unlike James Buchanan, looking on with horror at desegregated schools and demands for equality within the US, they intended to fight back.

Newly sovereign states—asserting the right to natural resources held within their borders—threatened the imperial status quo that governed the economy and what fueled it. To avoid the chaos, speculation, and conflicts that had come from basing currency values on gold, thinkers like John Maynard Keynes sought to design the postwar order to orbit around the flow of real commodities, including oil. Western leaders proposed the formation of an International Petroleum Council to hold equal weight to the other Bretton Woods Institutions, the IMF and World Bank, and meant to both keep multinational oil companies in line and ensure Anglo-American control over reserves to be traded mainly in dollars. Keynes, for his part, thought it important to keep talks at this level among First World technocrats, complaining that participants from Colombia, Liberia, and the Philippines, for instance, "clearly have nothing to contribute and will merely encumber the ground."[13]

Functionally, Anglo-American control of oil was accomplished via a cartel created at the behest of the US State Department among the "seven sisters" oil companies: five were American, one was British, and another, Royal Dutch Shell, was British and Dutch. Despite the relative abundance of oil in the Middle East, the sisters collaborated with the poorly named Texas Railroad Commission (see Chapter 9) to concentrate drilling among US firms and strategically limit production worked closely with local governments to defeat left-wing politics, empowering conservative and sometimes autocratic leaders in resource-rich countries who were less likely to challenge foreign capital. The old empires were gone, but power was still concentrated in similar hands. "Sovereign power belonged not only to a handful of European states, but also to the colonising corporations," Timothy Mitchell notes in *Carbon Democracy*. And they developed a system of domination, Mitchell explained, "based on the exclusive control of oil production and limits to the quantity of oil produced—only an antimarket arrangement of this sort could guarantee their profits."[14]

By 1945, American companies produced two-thirds of the world's oil, and a full 85 percent of global reserves were controlled by the seven sisters companies by the end of the 1960s. But claims to that oil put forth by long sovereign governments in Latin America and the Caribbean, as the State Department's Herbert Feis put it in 1946, "made the American industry doubtful about the security and profitability of their ventures in Latin America." The industry, he added, "lacked equanimity in the face of governments which could, if they insisted, have the last say on the rules." Moreover, because US reserves had been so thoroughly tapped through the '30s and '40s under the seven sisters' cartel, the country's easily accessible sources grew more scarce starting in the midfifties, and production began shifting abroad. In the name of Cold War strategic interest, US petro politics shifted focus onto more nascent and, the thinking went, easier-to-control states in the Middle East.[15] "The oil companies could portray their role there as the 'development' of remote and backward peoples," Mitchell writes, "and impose less equitable arrangements."[16] By controlling what fueled it, often by force, Western powers shepherded the course of the global economy. In doing so, they hinged expanded prosperity within their borders on its suppression abroad. "Postwar democracy in the West appeared to depend on creating a stable machinery of international finance, an order assembled with the help of oil wells, pipelines, tanker operations and the increasingly difficult control of oil workers."[17] By around 1950, fossil fuels had become the world's largest man-made source of carbon dioxide. That share has continued to grow ever since.

During the 1960s and 1970s, however, more newly independent states were questioning US and UK companies' God-given right to spigots abroad, as part of a broader movement looking to meld national and economic sovereignty. The most successful effort to stymie Anglo-American control of the world's oil was the Organization of Petroleum Exporting Countries (OPEC), formed in 1960 as an alliance of Venezuela, Saudi Arabia, Iraq, Kuwait, and Iran. Through higher tax rates and price controls, it worked to increase the local benefits of oil extraction, which to that point had flowed largely to the West, modeling itself off the Texas Railroad Commission that helped maintain the seven sisters' cartel.[18] By the early 1970s, many OPEC member states were able to take over ownership of the oil fields from Western companies and become producers themselves, transitioning formerly UK- and US-based firms into public corporations like Saudi Aramco. Western companies were relegated largely to refining and marketing Middle Eastern oil, which carried a different set of economic incentives than production. They were importing more fuel. By 1973, just 16.5 percent of the world's oil came from the US.

The oil crisis of the 1970s emerged out of a power struggle over the lifeblood of the global economy. In response to the US taking Israel's side in the Arab-Israeli War, Middle Eastern producers announced a modest 5 percent supply cut. Further 5 percent reductions, the bloc added, would continue each month until the US backed a comprehensive peace agreement. Incidentally, a day before, several oil-producing countries in the Gulf had announced a 70 percent hike in the posted price of oil when privately held oil companies refused to cooperate with demands to transfer ownership. The US refused to budge on a peace settlement, clandestinely funneling arms to Israel. And private companies chose to advertise the accumulation of these events as a price hike at the pump being cruelly doled out by OPEC. Lawmakers who opposed peace talks designed emergency measures that only added to public panic over a global oil shortage, leading consumers to buy up more fuel than they would have otherwise. Drivers lined up on odd- and even-numbered days to buy gas that corresponded to the last number of their license plates, and in six months prices had ballooned by 400 percent.[19]

From protests against the Vietnam War to revolutionary Black freedom struggles, increasingly internationalist social movements and labor unrest only added to corporate America's sense of unease during a tumultuous decade. One in six union members went on strike in 1970, and the oil and gas business certainly wasn't immune to the era's uptick in shop floor organizing. January 1973 saw four thousand Shell workers at five refineries and three chemical plants embark on a five-month strike across the US, in which New York City taxi drivers refused to buy from Shell gas stations and longshoremen refused to unload Shell cargo. Organized by the Oil, Chemical and Atomic Workers Union, it was the first such action to demand health and safety language be written into a labor contract. Unions stateside, that is, were making their own claims on how oil wealth was shared and linking workplace disputes to their bosses' geopolitical stressors. "Shell's vast wealth," one Bay Area striker wrote in a pamphlet recounting the strike, "comes from its long history of exploiting workers not only in this country, but even more so the working people and re-sources of the Middle East, Asia, Africa and Latin America."[20] He cited OPEC as a key challenge to Shell's profits and calls out the industry's cyni-cal marketing around the "energy crisis": "For the oil monopolies, the 'gas shortage' serves to silence critics, drive out smaller competitors and raise gasoline prices—all to relieve the squeeze on their profits."

Beset by threats at home and abroad, capital feared the entire global order might be remade. As historian Adom Getachew describes in *World-making After Empire*, postcolonial leaders throughout the Third World recognized that, after centuries of underdevelopment—of colonial powers extracting labor and resources to continue expanding their economies—narrow political sovereignty could only get them so far. The economy was by that point truly global, and countries having their own flags couldn't offer an escape from economic relationships crafted over hundreds of years to funnel wealth from the Global South to the Global North. Many of these leaders had studied in the West and recognized both the promise of the postwar settlement and the extent to which it excluded people who looked like them. As Nigerian nationalist and president Nnamdi Azikiwe once put it, "There is no New Deal for the black man."[21]

Since the end of World War II, leaders of newly sovereign nations and with a wide range of politics had been meeting to stake out a Third World against the First (the US) and Second (the USSR), eventually form-ing an alliance of nonaligned states that, as Vijay Prashad writes, "came together in a political movement against imperialism's legacy and its con-tinuance."[22] Prime ministers Michael Manley in Jamaica, Kwame Nkru-mah in Ghana, and Julius Nyerere in Tanzania (among many others) set out after independence to build a "welfare world" that would extend the gains enjoyed mostly by whites in the Global North to the South. Their goal was to encourage broad-based economic development with strong labor protections and democratic decision-making and to place limits on multinational corporations operating within their borders. They deemed control over commodities that were integral to the global economy, in-cluding natural resources, central to their project. Genuine democracy necessarily demanded not just a redistribution of wealth but a redefini-tion of what and who generated it.

Third World leaders saw tremendous potential in the newly formed United Nations as a vehicle for accomplishing these aims and for crafting a genuinely postcolonial and democratic world order. From the UN's founding in 1945 through 1962, the number of member states exploded from 26 to 110. Every country was represented in the UN General Assembly (UNGA), technically putting former colonies on the same footing as their colonizers. Under such an arrangement, the bloc of developing nations formed in 1968—the G77—would enjoy a majority. This democratic arrangement, many hoped, would go hand in hand with the right of countries to control the resources within their borders. As Getachew writes, "This vision of an international order, premised on the independence and equality of states, which are to be free from domination, was not born in the Westphalian Treaty or the UN Charter. Instead, it should be understood as an anti-imperial project that went beyond the inclusion of new states to demand an expansive vision of an egalitarian world order."[23]

Bolstering that ownership would be a new set of OPEC-style alliances, in which resource-rich nations could collaborate among one another to keep Western companies from wholly controlling the markets they participated in and build strong domestic economies, constructing a New International Economic Order (NIEO). Led by an alliance of postcolonial states, these principles were inscribed in a resolution by the UNGA in 1974. Drawing heavily on the work of Argentine economist Raúl Prebisch's work with the United National Commission on Trade and Development (UNCTAD), the NIEO platform included calls for debt forgiveness, unconditional technology transfers, preferential trading arrangements, the regulation of transnational corporations, and a "permanent sovereignty of every State over its natural resources and all economic properties . . . including the right to nationalization or transfer of ownership to its nationals."[24]

The neoliberals were scandalized. This was all taking place against the backdrop of an energy crisis and the collapse of the Bretton Woods exchange rate system, when tectonic shifts seemed all too possible. They aimed to protect markets from democratic uprisings in their own backyards, be it socialism or fascism. Postcolonial leaders' expansive definitions of sovereignty and democracy—necessarily extending to the global economy—represented a threat on a new scale, all the more so because a number of those leaders were themselves socialists. Where NIEO proponents saw an opportunity in the democratic potential of the UN, the neoliberals saw a threat.

Karl Brunner—a Swiss member of the Mont Pèlerin Society best known for coining the term *monetarism*—was particularly aggrieved by one idea put forward by the NEIO: that rich governments should pay some kind of restitution to the G-77 for colonialism. Brunner believed that colonialism had in fact been a sum positive for the South. Writing on a stay at the Hoover Institution in the late seventies, he argued, "Justification of the NIEO in terms of an established 'right' based on 'past exploitation' should eventually be recognized as a theme without support in reality. It remains, however, a powerful ideological weapon to lure the support of a gullible western intelligentsia for the persistent raids on the wealth of western nations."[25]

Establishment US politicians were just as riled.[26] "Obviously we can't accept the new economic order," Secretary of State Henry Kissinger told fellow members of Gerald Ford's cabinet, "but I would like to pull its teeth and divide these countries up, not solidify them" in unified opposition to the US by being openly hostile. Kissinger's conciliatory divide-and-conquer approach would become that of the US. That didn't stop US ambassador Daniel Patrick Moynihan from being more publicly confrontational from his post as the US ambassador to India. "At this moment we have, arguably, complete and perfect freedom to commence industrial use of the high seas. This freedom is being challenged, however, and almost certainly some form of international regime is about to be established," he wrote in *Commentary* in 1975. "It can be a regime that permits American technology to go forward on some kind of license-and-royalties basis. Or it can assert exclusive 'internationalized' rights to exploitation in an international public corporation. The stakes are considerable. They are enormous."[27]

The multicommodity cartel network the NIEO imagined and the West feared would never materialize. In the lead-up to the 1973 crisis, the seven sisters controlled some 85 percent of reserves; now, public corporations such as OPEC members, Petrobas (Brazil), and Gazprom (Russia) control some 65 percent of the world's oil reserves.[28] Where those transfers of ownership did happen—in the case of Aramco, most infamously— US companies and their allies were happy to work with and empower stable, friendly partners in the region, particularly if they suppressed the left-wing movements they despised. With their access to drilling in the Middle East and Latin America strained, the seven sisters set their sights on Africa using similar tactics: helping to prop up friendly governments, feed corruption, and fight nationalizations. Coffee and copper, though, were not oil and claims to them wouldn't yield nearly the level of bargaining power that OPEC enjoyed.

Other factors ensured that NIEO lost its fight. The oil crisis itself also sowed divisions within the G77; oil producers won out from price hikes, while importers suffered. Wealthy economies like the US were eager to defeat NIEO proposals and exploited that and other divisions, helping ensure the UN lacked any meaningful enforcement power over economic questions.[29] Decision-making power over how valuable commodities moved about the world and the rights of private investors were wherever possible ceded to corporations and the expert technocrats who wrote the rules for them.

In the US, meanwhile, the Richard Nixon administration's import quotas and tax breaks for oil drilling spurred domestic fossil fuel production in the name of "energy independence." In an influential lecture to the American Economics Association in 1974, MIT economist Robert Solow proposed taking the governance of natural resources out of the realm of democratic decision-making, arguing that it "is far from clear that the political process can be relied on to be more future-oriented than your average corporation."[30] US lawmakers in turn created what eventually became the US Department of Energy and the International Energy Agency to stockpile information and better coordinate efforts among Western countries and energy companies, as an explicit counter to both OPEC and the NIEO.

The Iranian Revolution—against a government imposed by the US to secure BP's access to the country's oil—helped spark another energy crisis in 1979. In an early turn toward a neoliberal policy suite, President Jimmy Carter had lifted price controls on oil and other petroleum products, which allowed prices to spike again. But the message remained largely the same as it had years earlier: blame OPEC, only now with an austerian green flair. Like Nixon, he helped lay the groundwork for decades of presidents striving for energy independence against enemies abroad, across party lines. Carter called his iconic installation of thirty-two solar panels on the White House an attempt to harness "the power of the sun to enrich our lives as we move away from our crippling dependence on foreign oil."[31] In a speech the next month, Carter relayed a similar sentiment on foreign oil from a visitor at Camp David: "Our neck is stretched over the fence and OPEC has a knife."

"Energy will be the immediate test of our ability to unite this Nation," he told the country, "and it can also be the standard around which we rally. On the battlefield of energy we can win for our Nation a new confidence, and we can seize control again of our common destiny." He proposed stringent import quotas, mandates for utilities to derive 50 percent of their energy from domestic sources ("principally coal"), and—through a World War II–style "energy mobilization board"—a dramatic boost in the production of domestic fossil fuels and solar, including a promise to get 20 percent of the country's energy from the sun by 2020: "We will protect our environment. But when this Nation critically needs a refinery or a pipeline, we will build it." Toward the end of the address Carter urged patriotic acts of energy conservation, encouraging Americans "to take no unnecessary trips, to use carpools or public transportation whenever you can, to park your car one extra day per week, to obey the speed limit, and to set your thermostats to save fuel. . . . There is simply no way to avoid sacrifice."[32]

The "malaise speech" (as it became known) would be misremembered as a flop that showed how out of touch Carter's supposed liberal environmentalism sympathies were with Americans struggling to fill up their tanks and heat their homes. Really, it showcased Carter's most steadfast political commitment: not to environmental or energy conservation but to economic austerity. It was during his presidency, not Reagan's, when world politics would start to lurch in the neoliberals' favor.[33]

By sharply raising US interest rates to cut climbing inflation, Carter's Fed chair appointee, Paul Volcker, killed two birds with one stone in 1979. In triggering a painful global recession, the "Volcker Shock" dealt a "death blow" to the Third Worldist movement, saddling large parts of the Third World with cripplingly high payments on dollar-denominated debt as commodity prices tanked and tightening credit that had once flowed South. It helped break organized labor's power in the US and UK, especially, by cratering business investment and workers' bargaining power with it, gutting the institutions that had been so central to building those countries' welfare states and placing checks on corporate power.[34] Ninety percent of the job losses that resulted occurred in some of the most militant and heavily unionized sectors—mining, manufacturing, and construction.

For executives, the timing was perfect. As Mitchell points out in *Carbon Democracy*, the influx of profits to US-based fossil fuel producers in the wake of the oil crisis and high prices had been a boon to some of the biggest backers of neoliberal (and eventually climate-denying) think tanks. Koch Industries, for instance, benefited handsomely. And Gulf Oil heir Richard Mellon Scaife gave $340 million over four decades to the Heritage Foundation, the American Enterprise Institute, the Hoover Institution, and the Manhattan Institute. The ideas of Friedman and Hayek were further cemented on the world stage by Reagan in the US and Thatcher in Britain, these institutions' favorite politicians. Offering solutions to the crises of the 1970s, they ushered in a newly emboldened understanding of the role of global politics and international institutions in protecting markets.

Today, with more enforcement and decision-making power vested in trade agreements and related bodies like the WTO than the UN, the basic concept of the "double government" still holds. Indeed, modern trade agreements, including NAFTA and more under-the-radar bilateral deals, regularly include investor-state dispute provisions that allow corporations to sue sovereign governments should they impede on their profits with anything from buy-local provisions to environmental regulations.

Decades on, the asymmetrical North–South power dynamics gelled by the Volcker Shock remain intact. Neocolonial bodies like the IMF and World Bank—with leadership drawn largely from the North—impose punishing repayment terms on debt shouldered, for the most part, by the South. But who wins and who loses from the climate crisis is less a matter of fixed geography than an economic relationship in which there are plenty of losers within wealthy nations as well as a small group of fabulously wealthy winners among Southern elites.[35] The US, for instance, replicates its own forms of labor and resource extraction internally. If markets can insinuate themselves into every corner of the world, in other words, so can the destruction they bring with them. That helps explain why climate justice campaigners forced to live with leaky, explosion-prone fossil fuel infrastructure in their backyards have until recently been some of the only voices in the North calling for the level of ambition reliably demanded by those in the climate-vulnerable South, not unlike Bay Area Shell workers finding common cause with their counterparts internationally in 1973.

As oil companies discovered through their own cutting-edge research during the 1970s, creating a world of no constraints for corporations has been a disaster for the planet. As they transferred resource wealth from South to North, capitalism and imperialism have also transferred prodigious amounts of carbon from ground to sky. The lifeblood of the status quo, fossil fuels have distributed their winners and losers along similar lines. It's no accident that the poorest parts of the world—on the losing end of colonialism, slavery, and more—are also the places now hardest hit by climate impacts and struggling the most to adjust. Even the House of Saud's vast oil wealth can't protect it indefinitely from skyrocketing temperatures.

Perhaps the greatest tragedy of UN climate talks is that they only started happening after proposals for a more equitable and democratic world order—one that may well have tackled this crisis sooner—had been mostly squashed. In debates about how to finance climate mitigation, adaptation, and recovery from those impacts already unfolding, the questions of historical responsibility and debt that coursed through earlier eras of geopolitics are as live as they ever were during the heyday of the NIEO. Power imbalances, however, are more firmly entrenched, with the winners of history having choked out alternatives from its losers. When it comes to the climate crisis, campaigners at COP25 described the gap between the reality of the climate challenge and the remedies discussed or ignored there as a "parallel universe." They might as well have called it a double government, with ironclad rules protecting capital and nice statements for the planet and its people.

STANDING BEFORE A press conference in the first week of talks in Madrid in December 2019, House Speaker Nancy Pelosi—flanked by members of that body's Select Committee on the Climate Crisis—echoed the talking points of Michael Bloomberg, Jerry Brown, and others who had gathered in the USCAC and similar outfits two years earlier: "The United States is still in."

"Combating the climate crisis is the existential threat of our time," she urged, "and it was essential that our delegation stand with international partners, who are continuing to build upon and solidify their commitments to meet the Paris Agreement's goals."[36]

Pelosi's trip was short; there was real business to attend to back in Washington. That week she would bring articles of impeachment against the president. She also hustled to pass the US-Mexico Canada Agreement, or USMCA—otherwise known as NAFTA 2.0. Trump had first tried to pass through his pet trade deal a year prior and was shot down, with Democrats calling the proposal "flawed and dangerous."[37] Pelosi and other moderate Democrats now saw some kind of deal as necessary for proving to conservative voters that they could "do something" before heading into the 2020 cycle and that the party could "walk and chew gum at the same time."[38] Bending over backward to accommodate a White House it now alleged was illegitimate, House Democrats' strategy for passing the USMCA wasn't complicated or unfamiliar: bipartisanship by any means necessary. The result would have a bigger impact on the climate than any statement Pelosi and company made at COP25. Like just about every trade agreement on earth, it will have more enforcement power than the Paris Agreement.

Trade and climate talks have moved in parallel since the early nineties, as Naomi Klein points out in *This Changes Everything*. The former—through the World Trade Organization and regional agreements like NAFTA—has been armed with progressively more muscular enforcement powers, handing corporations the authority to halt and dictate state-level policy to their liking on everything from state subsidies to hiring standards. Climate agreements, meanwhile, have been watered down to become progressively more voluntary and more distant from increasingly dire scientific realities. And the same industry-funded politicians and think tanks that have pushed to beef up corporate protections also worked to weaken international climate agreements from the jump.

While the actual text of the USMCA agreement doesn't mention the climate crisis, it does plenty to fuel it. As under NAFTA, companies can still dodge climate, labor, and environmental regulations at the state and city level in the US in favor of more lenient laws elsewhere. It includes no binding limits on or enforcement mechanisms to deal with pollution and, like its predecessor, encourages fossil fuel exports. Even one of the USMCA's most promising elements—the elimination of NAFTA's investor-state dispute settlement provision—includes a massive loophole preserving oil and gas companies' ability to sue Mexico through private tribunals if government policies infringe on their profits.[39] Another novel measure lets companies challenge new regulations before they're finalized, mounting new hurdles to any future climate laws.[40] It was a deal Democrats, Pelosi said, should "take great pride . . . in advancing."[41] It was also an unmitigated win for Trump, who—along with the world's biggest polluters—got most everything he wanted out of the USMCA. "For the natural gas and oil industry, USMCA means more jobs, stronger energy security and continued economic growth," the American Petroleum Institute's president and CEO Mike Sommers boasted in praising the deal's passage.[42]

The apparent disconnect between Pelosi's enthusiasm for both the Paris Agreement and the USMCA may or may not be intentional. In any case, it's not unique. It's also climate denial. For most establishment Democratic politicians, climate change is simply an issue to have a good line on: We Are Still In. We Believe Science. We Will Rejoin the Paris Agreement. The climate, as a singular issue, is thought to float in the clouds above supposedly meatier and more pressing issues like jobs and trade and the economy, something to be rolled out when it's a political winner and stowed away when it's not. That's wrong on its face, failing to account for the urgency of the crisis and the carbon footprint embedded in all policy fields.

It's also a political loser for Democrats. As NAFTA freed companies to pursue cheaper labor and scarcer regulations elsewhere, America's manufacturing base was hollowed out. All the while, Republican and Democratic politicians—with hearty industry backing—competed to shave the public sphere, gutting not just the industrial workforce but leaving whole communities now cut off from vital federal programs. Wealth alleged to be trickling down was hoarded at the top, and wages stagnated with no tangible alternatives on offer. This global race to the bottom on trade has helped fan the flames of a global right-wing backlash that is making any kind of climate policy more difficult. Pre-Trump business as usual in today's international order simply won't get the job done.

"You can't really think of fairness and equity as just another objective that it would be nice to try and squeeze in if we can while we deal with this fantastically crazy emergency. It's actually something that we have to deal with if we want to have any hope of dealing with the climate emergency," said Sivan Kartha, senior scientist with the Stockholm Environment Institute's US office. "At the end of the day, it's a global problem that will require long-term cooperation across vastly different countries and people in terms of their contribution to the problem and their ability to deal with the problem. The only way you can sustain that kind of cooperation is if folks feel like it is fair."

With seating scarce, Kartha and I opted to sit on the floor of the hectic main hall of the COP25 venue. Initially scheduled to fly back to Boston that day, he had changed his flight to speak at a hastily organized press conference the next day, when the Swedish teenager quickly became the star of COP25. Thunberg's solitary school strike in Stockholm sparked a millions-strong international movement, a weekly strike known in English as Fridays for Future. Unassuming at the Katowice talks the year before, she spent much of her time in Madrid attempting to dodge eager hordes of cameramen and refocus attention away from her own celebrity to activists from climate-vulnerable countries and research like Kartha's, collating the historical responsibility of wealthy nations for the climate crisis and what an equitable response could look like. Their event was one more attempt to fulfill the mandate etched onto the boat Thunberg rode to the US for the UN Climate Action Summit in New York that September: "Unite Behind the Science."

That science—as Thunberg and Kartha each pointed out—is a long way off from the Paris Agreement. A 2019 report from the UN Environment Program, released a few days before, noted the gap between countries' existing commitments under the Paris Agreement (nationally determined contributions, or NDCs) and what it will take to meet its lofty goals.[43] Existing NDCs will shoot temperatures up by 3.3 degrees, leaving major coastal cities and some whole nations underwater and collapsing crop yields worldwide. To get back on track for 1.5 degrees, per demands from the Global South, global emissions need to decline 7.6 percent each year between 2020 and 2030. That's far greater than the largest single emissions drop in world history: the collapse of the Soviet Union. Every year. For a decade. "Common but differentiated"—per the UNFCCC—has generally been interpreted to mean this burden should be shared equitably.

A former lead coordinating author in the *IPCC Fifth Assessment Report*, Kartha researches how to make that happen. Between 1850 and 2002, the Global North emitted three times as many greenhouse gas emissions as the Global South, where some 85 percent of the world's population lives.[44] The US and countries like it, Kartha explains, now have the greatest capacity to do two things within the UNFCCC: rapidly decarbonize their economies with ambitious NDCs and provide financing that allows other less-developed countries to do the same as soon as possible.

Climate finance operates along a continuum: The less successful and well-funded mitigation efforts are, the more adaptation funds are needed. The more beleaguered the adaptation, the more funding is needed for loss and damage—effectively, compensation for when the worst happens. Now, without a major course correction on mitigation, the need for adaptation is estimated to reach up into the trillions as soon as 2030. Even so, there's no agreed-upon definition of what climate finance actually means. While the UN-created Green Climate and Adaptation Funds were set up to fund mitigation and adaptation efforts, respectively, there's still no means through which to distribute loss and damage financing, despite years of pressure. For their part, US negotiators have consistently tried to keep discussions of loss and damage off the table entirely, arguing that such funding should instead be housed under the umbrella of adaptation. By 2030, experts estimate that loss and damage costs alone in developing countries could easily reach $300 billion each year. The Civil Society Equity Review finds that the combined historical responsibility of the US and EU for greenhouse gas emissions should make them responsible for

54 percent of that.[45] Among climate justice advocates, this responsibility goes by the name of climate debt.

Equity in decarbonization, as Kartha notes, is as much a moral commitment as a pragmatic one. Centuries of the Global North extracting wealth from the South—whether through colonialism or one of its many antecedents—means there's vital trust to be rebuilt to make any kind of functional international agreement really work. On top of that are real material constraints: many places simply can't afford to build thriving, low-carbon societies, particularly as they try to deal with the climate-induced disasters that are disproportionately clobbering them. Wealthy countries built their massive economies off land, labor, and resources extracted from what is today the South. Now, they've got more than enough resources to make a transition possible for the places that furnished that wealth. After centuries of plunder, both bank accounts and trust need to be rebuilt if a global transition is going to happen as fast as we need it to.

"We view ambition as a package, not as a one-way street. Ambition needs to improve mitigation, adaptation, and means of implementation. If you ask me to climb a mountain and I don't have the muscles to do that, you are asking the impossible," Palestinian ambassador Ammar Hijazi, lead negotiator for the G77 plus China, said in Madrid. "An electric car is still an expensive commodity. I buy an electric car in Palestine; there are only four or five charging stations. Then I'm stuck with this car that doesn't take me anywhere."

Acknowledged or not, countries and corporations alike are all fighting for their share of our remaining carbon budget—the fixed amount of pollution that can be expended before the world crosses over into a certain threshold of warming. The biggest consumers of fossil fuels can leave space in the budget for other places to develop prosperous and ultimately low-carbon economies that are resilient against the kind of climate-fueled storms, floods, and droughts already headed their way. Or they can keep eating up that remaining carbon budget as they've been doing, without a second thought that others might want a slice of the pie.

The US has tended to prefer the latter option. It was Republican president George H. W. Bush who famously said that the "American way of life is not up for negotiation" at the Rio Earth Summit in 1992, where the UNFCCC was adopted. But it was the Obama's lead climate negotiator, Todd Stern, who echoed a similar sentiment a decade later in South Africa: "If equity's in, we're out."

The Third World Network (TWN) has been involved in the UN climate process since its beginning in the 1990s, consulting with southern countries to push for greater equity and ambition. TWN legal director and senior researcher Meena Raman, a veteran of these summits, possesses what can only be described as an encyclopedic knowledge of the UNFCCC process. She's quick to emphasize that bad behavior by the US at UN climate talks didn't start with Trump. Without a dramatic overhaul in American foreign policy, it won't end with him either. "The problem is the crime that Donald Trump is committing—to deny climate science now . . . you can't. Nobody should deny that. People are dying. That's really criminal," she said. "But the US has never been a leader. It's always taken everybody backward."

Documents leaked by whistleblower Edward Snowden showed that the CIA enlisted the National Security Agency to surveil private communications among other delegations at the Copenhagen climate talks in 2009.[46] Just before those talks, the US and other developed countries were widely suspected of influencing the Philippines' decision to pull the veteran negotiator Bernarditas de Castro Muller, one of the most notoriously fierce advocates for developing countries, from its team. The US negotiators, like others from developed countries, regularly dangle billions of dollars of aid to developing countries to extract support for their climate positions. In 2010, the US ended up cutting aid to Ecuador and Bolivia because they opposed the Copenhagen accord.[47]

As ever, the line between what constitutes an official US governmental priority versus that of its biggest companies is a thin one. Only state actors can officially negotiate over the text of climate agreements, including Paris, but corporations can be "observers" to that process, just as NGOs and trade unions can. The World Health Organization, by contrast, has a stringent conflict of interest of policy. Its Framework Convention on Tobacco Control bars the tobacco industry from entrance into negotiations, citing the need to protect those talks against the "vested interest of the tobacco industry."[48] The UNFCCC has never adopted a parallel conflict of interest policy. As calls for fairness and ambition from climate vulnerable countries have been shut down, the world's biggest polluters have had no trouble being heard. An analysis from the Climate Investigations Center found that between its 1999 founding and 2019, the International Emissions Trading Association—a trade lobby representing polluters at the climate talks—sent 1,817 delegates to UN climate talks. The International Petroleum Industry Environmental Conservation Association (IPIECA) sent 258 of its employees in the same time period. Industry groups regularly send delegations larger than those of many sovereign nations.

Their return on investment has been stellar. Tucked away at the IETA (International Emissions Trading Association) Business Hub in Katowice, Shell climate change adviser David Hone bragged about just how much a hand his company had played in making sure carbon markets were central to Paris through Article 6. "We have had a process running for four years for the need of carbon unit trading to be part of the Paris agreement. We can take some credit for the fact that Article 6 [of the Paris Agreement] is even there at all," Hone boasted. "We put together a straw proposal. Many of the elements of that straw proposal appear in the Paris Agreement. We put together another straw proposal for the rulebook, and we saw some of that appear in the text." Hone added that he had been "chatting with some of the delegations" and that "the [European Union's] position is not that different from how Shell sees this." In a perfect world, he told me after the session, those mechanisms would be the only government mitigation policies on the table, echoing Exxon's thoughts on carbon pricing. "The ideal for a cap-and-trade system is to have no overlapping policies . . . if you really wanted it to work as effectively as it possibly could. But I'm being a bit idealistic there, I suspect."

Shell—the world's seventh-largest emitter of greenhouse gases—isn't nearly as beloved outside the IETA pavilion. Philip Jakpor, head of media and campaigns for Environmental Rights Action in the Niger River delta, has seen the effects of Shell's oil and gas business firsthand. The company operates some two hundred gas flares in the region that burn for twenty-four hours a day, despite having been repeatedly declared illegal there.[49] Nearby communities, Jakpor said, deal with rashes, respiratory problems, and disruptions to farms and fishing as a result. They have been fighting for Shell to stop the practice for years. "Shell is gassing these communities out of existence," he told me. Rather than ending the practice and complying with national law, oil companies have sold carbon offset credits for new infrastructure to prevent flaring.[50] "The community is not saying make money from this. The community is saying stop the gas flaring," Jakpor added. "We have said time and again that the solution[s] are nonmarket mechanisms. We are against the commodification of the environment. If we allow this, even the air we breathe will be commodified. The way to go is to end fossil fuel extraction. And we don't want companies like Shell and their cronies crawling all over the place trying to influence the talks."[51]

For now at the UNFCCC, those who benefit from Shell's upstream profits have a greater say over how the world responds to the climate crisis than the people forced to deal with the consequences of its business model downstream. They've had a far bigger say over how the world's governments deal with that crisis, or don't. "A proposal for curbing emissions from the developed world so that the billion individuals who live without electricity can enjoy its benefits would probably pass in a landslide in a world referendum," Astra Taylor writes of climate change's democratic conundrum, "but it would likely fail if the vote were limited to people in the wealthiest countries."

For the most part, and by design, it has been.

FROM THE BTU tax to Copenhagen to cap-and-trade to the Paris Agreement, the script in the US is the same: one arm of industry spends millions battering the idea that anything should be done at all, speaking most effectively through the Republican Party. Another steps in with a seemingly reasonable alternative that is lauded by Democrats. Republicans and their allies in the fossil fuel industry hit that with everything they have, and the cycle begins again, with those at least nominally interested in passing something called climate policy bringing weaker proposals they think will be more amenable to an opposition whose main goal is not passing anything at all, triangulating themselves and the planet toward oblivion. It's a race to the bottom, leaving out not only what most of the public in the US actually wants but basic material realities about what is required to deal with the crisis, the worst impacts of which will be felt beyond US borders. For most politicians, it's been easier to neglect climate politics altogether.

To have any kind of fighting chance of meeting the Paris goals—living up to what most Democratic politicians say they want—the realm of what should be considered climate policy has to expand radically, starting with an honest accounting of where emissions are coming from. As of now, national emissions inventories considered by the UNFCCC and IPCC only track the emissions a country produces within its own borders, known as territorial emissions. This leaves out a whole range of activities: the carbon emitted elsewhere to produce products we import, the fossil fuels the US exports to be burned abroad, even the carbon costs of companies headquartered in the US with operations abroad—in other words, trade. When politicians here talk about their ambitions to reach net-zero emissions by 2050, for instance, that doesn't tend to include the millions of barrels of oil and gas the US exports to be burned abroad each day. Neither does it account for the emissions they hope to "offset" with land grabs to mass produce bioenergy halfway around the world, and speculative new technology. And as globalization has encouraged companies to move their factories elsewhere, their emissions have traveled too, out of sight and mind for politicians and corporations eager to put on a green face. Because of idiosyncrasies in how carbon is counted, the US has come out looking better than it should.

The seemingly miraculous trend of Global North countries "decoupling" GDP growth from emissions can more accurately be described as offshoring, shifting carbon-intensive manufacturing (for instance) abroad. Though US territorial emissions increased by 9 percent from 1990 to 2017, consumption-based emissions—those accounting for trade—increased by 17 percent over the same time period.[52] "Given that emission transfers via international trade are a significant and growing share of country, regional, and global emissions," climate modeler Glen Peters and several of his colleagues wrote back in 2011, "we suggest that policies that affect international trade should not be continually separated from climate policy."[53]

If technocratic fixes might have been able to ward off catastrophe a decade or two ago, they won't now. As each fresh disaster is making clear, the only safe path forward through the twenty-first century and beyond runs through a reimagining of what society values most and who it listens to. There was nothing inevitable about how the system we now live in was built, and plenty of calls for genuine democratic alternatives were stamped out along the way. History doesn't offer a blueprint, in that respect. But it can provide some solace that the powers that be have had to work diligently throughout the last several decades to maintain business as usual. That control is finally starting to crack.

There are good cracks and bad. Given the massive amount of change required, the prerequisite to getting anything done at scale will be to take back state power from the right and stop the Trumps and Bolsonaros of the world from gaining more ground. The 2020 election in the US thankfully saw Trump himself defeated. The minoritarian politics that brought him to power are alive and well—enshrined by the United States' own antidemocratic institutions, like the Supreme Court, Senate, and electoral college, and in its elites' enduring footprint on an international order built to protect markets from majorities. Doubling down to defend an establishment that has left most people worse off is a recipe to hand more of it over as the world burns, resurrecting the world that set the stage for the far-right ghouls who've taken power around the world these last few years and for the climate crisis itself. A winning climate politics has to be a politics of, by, and for the global 99 percent that fights for control at every level of government. To do that, it needs to offer a future worth believing in.

CHAPTER 5
NEW SCENARIOS

THE YEAR IS 2020, and a "new society of market states is in place."[1]

If the "social contract of the old nation state model was based on the welfare of its citizens, on taking care of them," today's now "seems to hold itself responsible for something else: ensuring minimal safety nets while maintaining a fair economic playing field for its citizens. The new model seems highly effective. Many problems that had seemed intractable in the 20th century are now being solved," including rising temperatures.

Institutions are strong. "Means-testing is widespread," and "well-functioning capital markets" provide "aging populations in the world's richest countries with relatively safe investment opportunities in developing countries." Improvements in health allow for a steady rise in the retirement age, and these factors combined have warded off a pension crisis as the elderly work for longer. A consensus sets in "that people are poor because they deserve to be poor." After leading on the implementation of the Kyoto Protocol, which is ratified across the board, multinational corporations become the key drivers in the creation of a WTO-style World Environmental Organization (WEO), arguing that the "complex and often inconsistent national rules governing the environment are not achieving environmental objectives and simply add unnecessary costs to consumers." Environmental objectives are met not through harsh regulations but through the creation of new markets and property rights, rules for which are enforced vigorously by the WEO and other international institutions. The price of carbon permits "is as widely quoted on nightly news reports as the price of oil and gold."

131

This is not the 2020 that came to be, or even, necessarily, the one Shell wanted. It was one that—in 1998—the company saw as eminently possible and prepared for accordingly. While officially neutral, per the bounds of this scenario-planning exercise, it's hard to imagine Shell top brass weren't at least a little excited about this "New Game" scenario of market rule. As discussed so far, it's the kind of stable, profitable world that decades of corporate lobbying and rule making had been striving toward, one where they call most of the shots.

Shell also instructed managers across departments and continents to ready themselves for another, perhaps less desirable future: the "People Power" scenario, where periodic crises and upheavals produce rapid, large-scale change—for better and for worse. A "Millennium Bang" kicks off with the biggest recession since the 1930s and puts the "inadequacy of institutions" on full display, giving way to a world in which people handle their problems at the local and regional levels through individual initiatives, NGOs, and corporations. "Above all, the new politics is one of creative experimentation, in which people have to manage a veritable 'portfolio of identities,' and where the national government becomes just one of a number of competing voices—and not a particularly powerful voice at that." After one particularly bad set of climate-fueled storms, young people wage "direct-action campaigns" against the fossil fuel industry.

As with other scenarios, the New Game and People Power were each crafted with the help of Betty Sue Flowers. Flowers—former director of the Lyndon B. Johnson Presidential Library, with a PhD in English literature—seems like an odd fit for one of the world's largest oil companies. Yet that's exactly where she landed in 1992 when she was recruited to join the company's London-based Group Planning Team; she still works with its modern equivalent, the Shell Scenarios team, now helping navigate the company through a world where the role of oil companies in the coming decades is increasingly up for debate. When she came on board, scenario planning had been happening at Shell for a quarter century, carried over from Cold War think tanks eager to outgun the Soviets into the future.

At the urging of Shell veteran Ted Newland, eccentric Shell Française economist Pierre Wack (pronounced "Veck") had joined him at a seminar for multinational corporations put on by the Hudson Institute in the

1960s. It was designed to help translate scenario techniques from futurist academics and defense planners to the corporate world. Through the sixties and seventies, dozens of the world's biggest companies attended and afterward dabbled in scenario planning. The technique would have its most storied life, however, at Shell.

Wack, Newland, and their colleagues became evangelists for scenario planning within the company. In the early days, Shell's brightest minds went "into the green" at chateaus in the South of France to write early scenarios, enjoying wine with long meals and walks between heady, marathon sessions mapping out the changing face of geopolitics and evolutions in the oil and gas business. Disagreements were encouraged, and fights were common. Wack, famously, split his time between the West and the East, where he had sought out spiritual guidance in scattered ashrams and monasteries from the age of twenty. His office smelled of incense. One member of the planning team recalled that his final interview for the job was with Wack, who conducted it in a "complicated yoga position."[2]

"He didn't want to feel trapped . . . ever. Not by the people in his life, not by companies, not by economic models . . . he wanted to be free. That was his greatest pursuit," Wack's wife, Eve, later recounted. Among his motivations for joining Shell was that the company would allow him to take extended summer breaks to go back and visit his favorite spiritual mentors, further honing what he called the "gentle art of reperceiving."[3] According to Shell lore, it was his combination of mysticism and hard-nosed technical expertise that enabled him to be their fearless leader, challenging those gathered (mostly economists) to think harder and smarter and shake loose their preconceived notions of how their sector and the world worked.

The goal was to upend the suits' "mental models," per Shell planning-speak. As Flowers told me, people respond "not to their reality but to their stories of reality."[4] Those who gathered in the green had their own mental models to interpret the world, and the scenario-planning process would need to break those in order to present something genuinely new to higher-level executives with control over the company's vast resources. Scenarios were seen as an alternative to traditional business forecasting, which projected existing trends out into the future. By contrast to narrower computer modeling used in such forecasts—what Wack called "the enemy of thought"—scenarios imagined disruptions and stark changes to business as usual. The Group Planning Team's big break internally came when a scenario in the early seventies predicted the oil crisis. The company was able to navigate that period more ably as a result and afterward started giving Wack's team more leeway.

Much of the spirit that fueled earlier teams was still around when Flowers got there. Compiling piles of notes and data, she and the Group Planning-cum-Scenarios team would, through months of discussions, consultations, and research, arrive at at least two paths. "It was so much fun. Shell at the time had a big cast of characters in London," she told me over the phone in a slight Texan accent. "If you were writing about the Middle East, you could drum up a conference and pay to bring people from all over the world in for a day. It was like being in graduate school but having the professor come to you."

The goal, she explains, was never to make value judgments about which future was preferable, simply to be prepared. "It's not that one comes true," she said. "It's that there are elements of both in the world. That scenario in 1998 was based on 2020. If it did its work right twenty years ago, Shell should be well positioned in this new world."[5]

Flowers usually produced several versions of each report, ranging from book-length accounts for internal use to ten-page pamphlets to hand out at UN meetings and other such high-level gatherings. For years, they were tightly guarded. "I had to lock my drawer of notes before I went to lunch in the Shell building. They were kept under lock and key but told to governments on a one-to-one basis," Flowers said. Shorter, public versions of scenarios starting in 1992 are available on the Shell website. But the 1998 TINA (There Is No Alternative) scenario appears to be the only full-length version accessible online.

Through the breathless work of business writers, Wack's own articles in the *Harvard Business Review*, and a slew of corporate planning bibles, he came to be celebrated among corporate types as a genre of person well known today: a thought leader, out to disrupt the stodgy status quo with bold new ideas. In his book on Wack and other iconoclast executives, *strategy+business* editor Art Kleiner used the term *heretic*. Before Steve Jobs or Elon Musk, Wack was a philosopher concerned with deeper questions of meaning and existence, not just the pedantic business of making money. He didn't just want to change Shell; he wanted to change the world.

You didn't need to be any great genius, though, to see that linear forecasting models would only work for so long for the oil industry of the late 1960s; business as usual wasn't going to cut it. Shell ran the first scenario in 1967, the same year oil-producing Arab nations imposed an embargo to discourage governments from supporting Israel in the Six-Day War. US dominance was by that point declining, and challenges to the colonial concessionary regime were starting to crystallize. Claims to oil profits and calls to limit the power of multinational corporations were being made across the Global South.

Shell's early scenario planners, moreover, would have felt decolonization viscerally. Like several other members of the team, Ted Newland had formative experiences working for Shell in Nnamdi Azikiwe's Nigeria in the 1960s, where the company had first obtained the right to explore for oil under British colonial rule in 1937. Before helping lead the country to independence, Azikiwe had fought Shell as an organizer against colonial rule, supporting communities who protested its unfettered access to oil-rich land under the Crown-imposed Minerals Ordinance.[6] During a 1964 strike by Nigerian oil workers, Kleiner recounts that Newland proposed building a moat around the company's compound to keep out the rabble. When his colleague suggested strikers might swim across, Newland quipped that "we can always put crocodiles in."[7] That was the same year that Nigeria attended its first OPEC meeting as an observer. Following a brutal contest over oil that effectively halted production during the country's 1967–1970 civil war, Nigeria would eventually join OPEC in 1972.

Newland's joke was a chilling preview of things to come. In 1995, Ken Saro-Wiwa—a prolific Nigerian writer and creator of a successful sitcom, *Basi and Company*—would be executed alongside eight other members of the Movement for the Survival of the Ogoni People (MOSOP). The group had for years led a high-profile, nonviolent campaign against Royal Dutch Shell's state-supported operations in Ogoniland, which, they argued, trampled both on the environment and the rights of the indigenous Ogoni people. The company has been ensnared in a slew of lawsuits since.

Plaintiffs—mostly relatives of those killed—have argued that the deaths were the result of close collaboration between the company and the country's military dictatorship, including allegations that Shell provided the Nigerian army with vehicles, ammunition, and logistical support for conducting raids on MOSOP organizers.[8] In 2009, the company agreed to pay a $15.5 million out-of-court settlement for its role in the killing of the Ogoni Nine and as of 2020 has continued to face legal action from several of their widows. The company denies having played any role.

It wasn't just pesky Third World governments and movements prompting Shell to consider new futures, however. *Limits to Growth,* an influential 1972 report commissioned by a group of academics, politicians, and business leaders called the Club of Rome, ignited a public conversation about unsustainable resource use that put the sector in the hot seat. A version of the type of planning Wack embraced, the report outlined "collapse" and "overshoot" scenarios that cast dreary forecasts for the oil business. Amid such pressures, economic historian Jenny Andersson told me that Shell "needed some more flexible way of dealing with the future that could show there is plurality in future scenarios and avoid falling into a determinism that is detrimental to their interest. They discovered they could use scenarios to actively narrate other versions of the future that were not catastrophic to them."[9]

They weren't alone in this shift. It was around the same time that Exxon, like Shell, began sponsoring cutting-edge research into climate science and explored the possibility of becoming a more broad-based energy company. The push emerged less out of a concern for the environment than out of a desire to profit off the alternative energy sources that higher fuel prices might prompt a shift toward. However grand Wack's vision may have been, Shell's scenarios, like the rest of its business, was never meant to do anything more than secure steady oil profits.

That has been an enduring theme. After a McKinsey-driven management restructuring in the midnineties to make Shell "more like an American company," per Flowers, scenarios were treated more forthrightly as PR. When David Hone—a member of the Shell Scenarios team—eagerly pushed Shell's 2018 Sky scenario into my hand in Katowice, he was helping Shell build its brand. As Flowers told me, scenarios aren't what they used to be, having narrowed to focus more on discrete energy issues. She still gets a lot out of working on them but acknowledges they've "morphed" into more computer-driven modeling than the narrative meditations on geopolitics she once drafted. As the scope of scenarios has narrowed, their role in the Shell brand has ballooned. Shell's website boasts that its scenarios work has done everything from confront the AIDS epidemic to envision a transition away from apartheid in South Africa, where they had continued doing business long after boycotts and sanctions started.[10]

With nature-themed names like Sky and Ocean, today's scenarios help to posit Shell not just as an inquisitive, good-faith actor on climate but as an integral part of the global energy transition. Its scenarios, the company's website boasts, have been instrumental in Germany's Energiewende and in helping China transition off coal. It was around the same time it started selling scenarios that Shell, like BP, began to shift gears from publicly denying climate change to painting itself green. A 1999 document unearthed by the Dutch publication *De Correspondent* presents a "Profits & Principles Advertising Campaign," intended to help present a kinder and gentler multinational oil firm. "We need to reassure people—publicly—of our commitment to the principles of sustainable development, balancing our own legitimate commercial interests with the wider need to protect and enhance the environment and contribute to social progress and stability," the booklet states, two years on from the Kyoto talks and four from the execution of the Ogoni Nine. From there it presents a series of magazine ad mock-ups, juxtaposing common perceptions of Shell with the more forward-thinking (alleged) reality.[11]

One ad shows a mass of clouds in an otherwise blue sky, the gray outline of Shell's logo superimposed on top. Text below lays out a challenge: "The issue of global warming has given rise to heated debate. Is the burning of fossil fuels and increased concentration of carbon dioxide in the air a serious threat or just a lot of hot air?" Another ad, entitled "Or Clear the Air," shows the path taken by a more enlightened multinational oil company. "Shell believes that action has to be taken now, both by companies and their customers" this ad clarifies. "So last year, we renewed our commitment not only to meet the agreed Kyoto targets to reduce greenhouse gas emissions, but to exceed them."

That ad never ran, but similar images have been a common feature of the company's PR strategy ever since. Shell plays up both its commitment to an energy transition and the prosperity that continued oil and gas exploration can unlock. Scenarios feature heavily, flanking the sponsored content it takes out in places like Politico and the *New York Times*.[12] One recent climate-concerned scenario promises "A Better Life with a Healthy Planet," premised on negative emissions and a global carbon price that disincentivizes downstream fossil fuel usage. In one 2017 scenario generated by the company's Global Supply Model, meanwhile, a rising price of oil can stimulate "improved/enhanced" fossil fuel production, which makes it more possible to access unconventional reserves out through 2100.[13]

After the Paris Agreement was signed, the company went a step further in its PR efforts. Amid its ad spending and scenarios work, they began hosting gatherings like Powering Progress Together, billed as "an action-oriented day of dialogue focused on accelerating the energy transition" for industry members, clean energy entrepreneurs, and environmentalists. Three such events were held in 2019, in Singapore, the UK, and San Francisco. Attendees that year included Transport for London's Michael Hurwitz and Nissan's EV director Gareth Dunsmore.

Climate scientist Peter Kalmus—who specializes in ecological forecasting and analyzing boundary layer clouds—was invited to speak at the one in California, on the advice of another climate scientist who wasn't able to attend. It didn't take long for him to find out the limits of Shell's commitments to a low-carbon world.

Having agreed to talk on the condition that they wouldn't censor him, Kalmus's presentation planned to lay out Shell's long history of fueling climate denial and pushing to delay action at the local, state, and international levels. The slideshow he made for the event pointed out that Shell "has made no significant shift away from developing fossil fuel" and has continued to lobby against rapid mitigation. Bringing the company in line with climate goals would require, in his words, planning "to really transition to carbon-free energy and become a market leader" and preparing "for substantial voluntary transition away from fossil fuels" as a means of regaining public trust. Kalmus also suggested that Shell refrain from making any policy recommendations on lowering emissions, instead focusing on changing its own practices: "Anything you recommend will set the activists against it." Hours after sending in his slides so they could be uploaded to the venue's AV system, he got a call while driving up the California coast with his family; he had been disinvited. In an email afterward, a Shell spokesman wrote to me that the "panel of which Mr. Kalmus was invited to participate differed from his presentation, which did not meet the objective of the session."

"I'm sure there are some very concerned people within the Shell organization who do want to make a difference," Kalmus told me later. "But as an organization, there are really strong forces—including legal forces—that force them to put their shareholders' interests first. Asking them to make the kind of change I was asking is a huge ask for them. But we're literally destroying the planet. That's where I was coming from: let's have that conversation."

Shell, it seems, doesn't want to. At London's Oil and Money Conference, CEO Ben Van Beurden was quick to disabuse any notions that the company lacked a commitment to its oil and gas business or was even making any significant strides away from it. His and the company's bullish messaging on climate and its token investments in renewables, he said, "might even make people think we have gone soft on the future of oil and gas. If they did think that, they would be wrong . . . Shell's spending on New Energies is, indeed, huge. But it is $1–2 billion out of total annual capital spending of around $25 billion."[14] To put a fine point on it, he reiterated: "Shell's core business is, and will be for the foreseeable future, very much in oil and gas." There hasn't been much evidence to suggest that he's lying. Yet what's needed, he insists, isn't to point fingers at energy companies like his. Instead, he argues for "a more mature debate where suppliers and users of energy join to figure out how to do things."[15]

Shell's evolving climate stance offers a window into how multinational oil companies work, and how it is that they're so effective: they're good materialists. Whether their executives ever believed in climate denial or now take the reality of the crisis seriously is irrelevant. While knowing full well the planet was warming, polluters funded climate denial so long as it was politically useful to keep profits flowing. When that stopped being the case, they moved on to selling themselves as climate champions. It's not that fossil fuel executives are individually loathsome, though that's probably true in some cases. It's their job.

As with any company, Shell has a constitutive block that keeps it from being an ally in the climate fight: an inability to envision a future without Shell. The company's overriding mission is to ensure an indefinite life for itself and its profits. The point at which those aims align with the goal to cap warming at 1.5 to 2 degrees Celsius may well never come, and if it does, it will almost certainly be too far off in the future to make either of those goals a reality. It's possible to imagine a Shell or ExxonMobil voluntarily becoming the broad-based energy companies of the low-carbon tomorrow, just as one can imagine flying cars in the skies above New York City and a future in which humans converse in full sentences with their house pets. It's worth asking whether these companies are really suited to do anything else, given the specificities of the oil and gas industry. And they may never want to: there isn't as much money to satisfy shareholders in an energy system utilizing freely available wind and sun instead of relying on hard-to-access hydrocarbons.

With precious little time left, though, whether a green Shell is theoretically possible matters less than their actual track record: there is no fossil fuel company on earth that—after having known for decades about the reality of the climate crisis and their central role in fueling it—has demonstrated a full-throated commitment to keeping temperatures below 2 degrees Celsius. Every shred of evidence suggests the industry is moving full speed ahead in the opposite direction, pushing more exploration and more production as temperatures rise, seas swell, and fires burn. Companies with the power to start wars, kill legislation, and end careers could be throwing their armies of lobbyists and multibillion-dollar advertising budgets toward a rapid low-carbon transition. They aren't.

In spite of all that, Shell's approach to scenario planning—in its most idealistic form, at least—may be one of the few parts of the company's business model not totally fit for the dustbin. "The point of scenarios is to make you flexibly oriented toward the future," Flowers said. "If you didn't write entirely new futures, you weren't doing your job."

By contrast, as neoliberalism took hold, writing new futures got harder and harder for all but the world's wealthiest. The idea that there was no alternative nuzzled in deep. Even after the financial crisis seemed to break open forty years of economic orthodoxy—fizzling out neoliberals' zeal for markets and means testing—there didn't seem to be much else on offer. Nowhere has this imaginative sclerosis been clearer than in the Democratic Party's establishment. The old guard isn't enthusiastically extolling New Deal means through neoliberal ends as it did in its Clinton-era heyday. It's defending smaller and smaller plots of political turf for reasons that are increasingly unclear to everyone involved. We may not be living in the rubble of neoliberalism now so much as its muck, like stubborn New York City snow piles that linger, collecting grime and trash long after people have started enjoying warm spring days in the park.

In positing all of human existence as an endless striving toward market society, neoliberals had to erase not just the possibility of a future but all memory of a past when humans managed to organize themselves in other ways. The kinds of tools needed to navigate out of the climate crisis—things like public ownership, full employment, or even just tough regulations—have receded into memory. The stories we're told about our warming world have suffered as a result. For about as long as it has existed in the public consciousness, the climate crisis has suffered from a dearth of scenarios. Down one path is a parched wasteland. Down another is a techno-utopian fantasy, with climate engineering solutions deployed through some combination of genius tech bros and bipartisan compromise. Mostly, establishment politicians just do their best to convince everyone that modest tweaks are the best we can hope for, be that carbon pricing, token investments in negative emissions technologies, or a new efficiency standard.

Unsurprisingly, corporate planning gurus won't be the ones to envision and make possible a future beyond apocalypse.

THE YEAR IS 2028, and the decade of the Green New Deal is underway.

The main goal of past economies had been to grow GDP as much and as quickly as possible, hoping ballooning profits would create widespread prosperity. Today's leaders now hold themselves responsible for other goals: keeping warming below catastrophic levels while ensuring that all people are employed and as many as possible can access a decent quality of life, whatever the weather brings. Although they have hit their share of snags, the new models seem to be working. What a few years earlier had seemed impossible—a carbon-free power sector, a federal green job guarantee, binding climate agreements—is now run of the mill.

A new financial crisis sharpened anger against the ultrawealthy and created an opening to talk about what and who the economy was for. Less than a week after the stock market tanked, Superstorm Julia barreled up the East Coast, hammering New York City with catastrophic floods, as the hills around Los Angeles burned. A poorly funded, controversial jobs program, modeled on the New Deal's Civilian Conservation Corps (CCC), passed through Congress in the weeks afterward and put people to work almost immediately. Unimaginatively, Congress called it the Climate Conservation Corps. It won out over another jobs measure pushed by Republicans, which had essentially been a generous tax credit for companies to hire more workers. One of the heaviest blows from the crash was dealt to independent fossil fuel companies in the Permian Basin, and the drill jacks that once bobbed along in West Texas sat still.

The recovery in West Texas became a kind of showcase. Rather than bailing out the fossil fuel companies again, the federal government, through the Oilfield Communities Act—introduced by the junior Democratic senator from Texas—bought up struggling oil and gas firms in the region and guaranteed five years of wage parity for workers while letting those over fifty retire on full pension and benefits, as they phased out drilling. National rent control alleviated housing costs in Midland and Odessa. Union training programs coordinated across the state's Central Labor Councils were expanded to reskill pipefitters to build and maintain utility-scale solar projects and wind turbines and install charging stations for public buses, as federal energy efficiency and clean energy standards created a new boom in Texas and around the country for building retrofits and electrification. But incentives were good enough to encourage many to switch sectors entirely and work as nurses and teachers, whose unions had joined with those at the refineries to raise wages across the board and push for sectoral bargaining for all clean energy jobs. Together, building trades and public sector unions pressured the Federal Reserve to purchase state and local government debt, too, keeping public budgets solvent as revenue streams shifted. Those who made the old minimum wage as gas station clerks and busboys at roadside diners serving oilmen now found better paid work through the CCC, administered at any one of the dozens of American Jobs Centers around the region. They converted reclaimed drill sites into hunting lodges and game preserves, cared for the elderly, and painted murals honoring Texas's indigenous history. The thirty dollars an hour wage and four-day weeks baked into the program meant people had time to go out and enjoy the parks and public concert venues sprouting up around them, funded by public infrastructure investments. This all softened the blow of ending fossil fuel exports. That was done, as much as anything, to ward off punishing sanctions from the EU and China. Lawmakers across the political spectrum sold it as a "new model of American energy independence."

This may not be the 2028 that comes to pass, but it's one that could be built by a Green New Deal. Sci-fi dystopias about a climate-changed future make the same mistakes Pierre Wack cautioned against in business forecasting: projecting business as usual out indefinitely. In the long sweep of human history, though, capitalism and its Great Acceleration has been a blip, and neoliberal dominance an even smaller one. There's nothing inevitable about them. One of the most damaging facets of that era's politics in the US, especially, has been to squash working people's ability to imagine a better future for themselves not determined by how many widgets they can collect. Utopias were for rubes, or enemies. Only corporate titans got to dream, and the worth of their ideas would be measured by the price their products could fetch or the billions they made. The irony of all this is that their visions have left us with a world where business as usual is becoming physically impossible. Ideology can endure, of course. But with more genuine disruptions—recessions and pandemics, fires and floods—has come more competition over who gets to shape the future and what ideas are in play. So far, this book has looked mostly at what the reigning ideologies have built and the consequences. The remainder of the book will look at what might come next: old ideas that cling on, new ones that crop up, and buried ones that reemerge. The roughly 1.1 degrees Celsius of warming already locked in won't establish a new normal of worsening weather so much as make normal irrelevant. It's a daunting prospect but not one without possibilities. If nothing else is set in stone, why should politics and economics—much less fossil fuel profits—be any different?

To be clear, egalitarian claims on the future are not winning the day; the monsters today are more powerful than the plucky heroes trying to slay them. With Biden in the White House and what—as of writing this—looks to be at least a few more years of narrow Democratic Senate control, the Green New Deal or any sweeping climate measures aren't imminent. My contention, in spite of all the evidence for pessimism, is that life in the next century can be better, even as it gets hotter. If that better, warmer world comes to pass, it'll owe the Green New Deal and the theory of change embedded within it some credit.

By the time it was popularized in 2018, the Green New Deal wasn't a new idea. Climate and environmental justice groups and academics had been proposing some version of it since the 1990s, and calls had long come from the Global South demanding a Marshall Plan for the earth, with large-scale investments from the North.[16] "The current neoliberal economics consensus that holds in America really isn't equipped to deal with a problem" like the climate crisis, Saikat Chakrabarti, one of the many people who've worked to bring the Green New Deal into the climate policymaking conversation, told me. "There's a whole new generation of economists and groups who have been working toward this for the last decade or two," he said, listing off a who's who of organizing outfits and economists who stood outside their profession's orthodoxy. "The Green New Deal is how you bring all those folks together to create a comprehensive plan." That plan, he said, "touches everything—it's basically a massive system upgrade for the economy."[17]

It wasn't brilliant policy design that pushed the Green New Deal into the spotlight at a particular moment, though. I happened to be talking to Chakrabarti that day in November 2018 for two reasons. First, he was about to become the chief of staff to the freshly elected congresswoman Alexandria Ocasio-Cortez. And just days after that election, Ocasio-Cortez had joined an internet-breaking sit-in at then aspiring House Speaker Nancy Pelosi's office with some two hundred young people in the Sunrise Movement, demanding Congress develop a comprehensive plan to transition the United States away from fossil fuels. After years in the wilderness, the Green New Deal—an open-ended call for a thorough, state-led transformation of society—had gone mainstream. Within a few months, that framework would be on the lips and websites of nearly every Democratic presidential candidate, inform the platforms of parties running for office abroad, and be a guiding, albeit mostly rhetorical principle behind European Union dictats. In short order, it became the standard against which nearly every large-scale emissions reduction plan was judged.

Embarking on their own kind of scenario planning, Sunrise had, leading up to that sit-in, spent well over a year figuring out what a climate movement that was up to its task could look like. Members read everything they could about the previous thirty years of climate policy-making, historic periods of realignment in American party politics, and sea changes on everything from civil rights to trade policy. The members of that initial group—almost all of them under the age of thirty—had gotten to know one another in part by working to divest colleges around the country from fossil fuels, either as student activists or as staff at the various green groups who supported their campaigns. After the 2016 election, they had a hard time imagining how to transition a movement built to pressure reasonably sympathetic Democrats—with few long-lasting wins to its name—into one that could reckon with Trump, all in time to effect the kind of transformative changes climate scientists were demanding and with enough force to do battle with the most powerful industry humanity has ever known.

"We had had a plan to focus on building popular support for climate change, anticipating that a Democratic president would be pushed to take action on climate if a majority of Americans wanted it and made enough noise," Sunrise's Executive Director Varshini Prakash said of the time just after Trump's election. "But our world had turned upside down, and there was no way that popular opinion and action would push a Republican trifecta government to care about climate change. When the dust settled, we came to an important realization: we have to figure out how to win elections."[18]

As Sunrisers came to see it, climate politics suffered from an enthusiasm gap. Majorities of Americans across the political spectrum reported believing in, caring about, and even being worried about the climate crisis. Many supported the idea of scaling up renewables and expanding research and development funding for clean energy, at least in theory. Seldom, though, was that what they were most worried about at the polls; for Democratic and Republican voters, climate change reliably ranked at the bottom of a list of priorities topped by jobs and the economy. And most people certainly weren't haranguing their elected officials about rising temperatures, however worried they claimed to be about them to pollsters. The GOP, meanwhile—following the Tea Party mobilization that had helped kill cap-and-trade—were all in on denial. Most climate plans on offer from Democrats were almost laughably out of step with the kind of wartime footing needed and never managed to pick up momentum. For the most part, climate change just wasn't talked about much at all. It got a whopping eighty-two seconds of airtime during general election debates in 2016 and barely came up in the midterms.

As detailed in previous chapters, the thesis that insider bargains and backroom deals would yield adequate climate policy had been tested and failed. To get anything done on climate, Sunrise surmised, there needed to be a real movement—one that lawmakers couldn't ignore, with more than a few of its own people in power on the inside to push it through. Like the Tea Party, with opposite politics, that movement would show up in district offices around the country to make them do the right thing and organize for more friendly faces to take their jobs if they refused.

The Sunrise Movement, if not the Democratic Party establishment, had learned the lesson of the cap-and-trade's failure in 2010, and it informed its theory of change. As Theda Skocpol wrote in her 2013 postmortem, "The political tide can be turned over the next decade only by the creation of a climate-change politics that includes broad popular mobilization on the center left. That is what it will take to counter the recently jelled combination of free-market elite opposition and right-wing popular mobilization against global warming remedies."

Skocpol added that "there must be changes in the course of national politics—changes to render GOP extremists less effective in setting national agendas, and changes to mobilize popular support for new legislation."

Sunrise started small, at least compared to its lofty ambition to build an "army of young people to stop climate change, and create millions of good jobs in the process." Mostly under the radar in the summer of 2017, the movement started building hubs, or chapters, in ten states. Sunrisers—including middle schoolers on up through elders in their late twenties—pressured Democrats to sign onto the No Fossil Fuel Money pledge, swearing off donations of more than $200 from fossil fuel industry employees and executives. They sat in at congressional district offices and bird-dogged candidates. Within a few months, they started canvassing for progressive Democratic challengers for congressional and state house seats, often alongside other progressive groups and outfits like the Democratic Socialists of America. The strategy wasn't complicated. As Prakash summarized: "disrupt, vote, disrupt, vote."

To get anything done on climate, they reasoned, Democrats needed to take power. And if Hillary Clinton's defeat in 2016 had been any indication, middling establishment centrism as usual wasn't the path toward it, nor did it hold the answers to keeping temperatures well below 2 degrees Celsius. The fight against the climate crisis was a fight for the soul of the Democratic Party.

Months before we talked about the Green New Deal, I had interviewed Chakrabarti in his then capacity as executive director of Justice Democrats, a political action committee, which, while less focused on climate, had reached similar conclusions. With a background in software engineering, Chakrabarti had helped found the payment processing app Stripe before becoming the director of Organizing Technology for Bernie Sanders's 2016 presidential primary campaign, where he developed its software for mobilizing volunteers and small donors.

Small donations were so baked into Sanders's campaign that their average size became a popular call-and-response chant at rallies: $27. Though it lent a grassroots feel to the whole operation, this was as much a pragmatic choice as a statement of the senator's politics. A grouchy democratic socialist who chided millionaires and billionaires wasn't exactly a draw to rich donors used to having their egos stroked, and he didn't want their money or the strings that came with it anyway. With big money came big asks, and no shortage of Democratic politicians were more loyal to their donors than to their constituents. But presidential races are expensive, so Sanders would have to find some other way to mount a campaign. As former Bernie organizers Becky Bond and Zack

Exley detail in their book *Rules for Revolutionaries*—part memoir of Sanders's 2016 race and part how-to guide for progressive insurgency— a campaign with grassroots fundraising means having "a base that wants to support you. If you don't have that base, you face two options: seek large donations from rich people and foundations, or build a base so you can seek small donor donations."[19]

Bond and Exley write extensively about what they call big organizing. The small-donor-fueled big organizing the Sanders campaign brought to presidential politics relied on being big in two ways: amassing a huge organizing budget and committed volunteer army and giving people a vision inspiring enough to make them invested in beating such crazy odds. "The messages were that Bernie was a long shot," Bond and Exley write, "and moneyed interests might be too powerful to overcome. That what we were talking about was nothing less than a political revolution and it would work only if millions of people joined us."

Justice Democrats translated that idea into House races, asking supporters to get behind a series of underdog candidates with big ideas about how to transform the party and the country. The Tea Party, after all, hadn't taken control of the GOP by triangulating around the Obama administration's first term; it went after its own first, enforcing a new party line and punishing those who strayed. "Electoral strategy wise, we're doing what the Tea Party is doing on the right. The Tea Party has an easier job," Chakrabarti told me, "in that their message is one of obstruction and destruction. It's easier to talk about that. I think it's harder on the left because we're trying to message creation, of a bigger and more inclusive future."

In a detail that didn't seem important enough to include in the piece I was writing at the time, Chakrabarti mentioned that among those they'd recruited was a "candidate from the Bronx in a super blue district, where DSA is very strong" to run against Joe Crowley, one of the most powerful Democrats in the House. He hadn't faced a primary challenger since 2004.

Alexandria Ocasio-Cortez's story is, by now, well known: while working as a bartender in Manhattan, she got interested in politics by volunteering on the Sanders campaign in 2016. Just after Trump's election she went to Standing Rock, to join the encampment there against the Dakota

Access Pipeline. Having already been nominated by her brother through online recruitment forms at the websites of Justice Democrats and Brand New Congress, another political action committee, she decided then and there on North Dakota's windy plains to give it a shot. With a grassroots fundraising operation and a dedicated battalion of volunteers going door to door for months, many of them from the DSA, she won the primary, and sent shockwaves through a political establishment that hadn't known her name a week earlier.

Like several other Justice Democrats, Ocasio-Cortez had run on a Green New Deal, though it wasn't the driving focus of her campaign. Sunrise had endorsed her and others who had sworn off fossil fuel donations and were running on ambitious climate platforms, several of whom—including Michigan's Rashida Tlaib and Minnesota's Ilhan Omar—won. Having just gotten through a successful vote cycle of their "vote, disrupt, vote" strategy, it was time to disrupt. Sunrise had planned for weeks on delivering its climate demands to the newly elected Congress with a big sit-in on the new lawmakers' first day of orientation, putting forth three demands: "1) champion a Green Jobs for All platform and 2) mandate that any Democrat in leadership must take the No Fossil Fuel Money pledge." The Friday after Ocasio-Cortez won the general election, Chakrabarti got a text asking if AOC would retweet news of the action Sunrise was planning to her millions of followers the next Tuesday.

Momentum had already been building and not only thanks to Sunrise's organizing. Just after Democrats retook the House, Pelosi announced she would resurrect a toothless House committee she had created in the lead-up to the cap-and-trade fight, the Select Committee on Energy Independence and Global Warming, established in 2007. Its mandate would be (inspiringly) to "educate the public about the impact of more frequent extreme weather events," per a *New York Times* report on the idea. Progressive policy in the new Democratic Congress, she said, would emerge from a "bipartisan marketplace of ideas."[20] For climate-conscious onlookers, including Sunrise, this seemed painfully out of touch. Wildfires had engulfed Northern California, and the IPCC had less than a month earlier released its report stating that capping warming at 1.5 degrees Celsius would demand sweeping changes in every sector of the economy, changes that would be "unprecedented in terms of scale." It also seemed like Democratic Party politics as usual—a repeat of strategies already tried and failed; the policy, or lack thereof, was as bad as the politics.

Like Sunrise, Chakrabarti and Zack Exley—who worked together first on the Sanders campaign and then through Justice Democrats—had been doing their own thinking about what an economic mobilization to bring down global emissions might look like, emphasizing nuts and bolts policy detail. For months they had been meeting with founders of the by then defunct Green New Deal Group in the UK like Ann Pettifor and other academics, thinkers, and movements trying to tease out what a reasonable, science-based response to the crisis could actually look like. Necessarily, this meant throwing out the economic stories and dogmas that had guided both parties for thirty-plus years. New Consensus, Exely and organizer Demond Drummer's new policy shop, was to be a kind of clearinghouse for this new, well, consensus on economy and society, intended to provide the intellectual scaffolding behind the political realignment that groups like Justice Democrats and Sunrise were working to bring about.

Having been a part of these conversations with Justice Democrats and her fledgling staff, Ocasio-Cortez would do one better than retweet the Sunrise action. After her first night of freshman orientation on the Hill, she and fellow freshman representative Rashida Tlaib joined a no-frills Sunrise training in a DC church on November 12, addressing those preparing for the next day's action. And Ocasio-Cortez, on her second day in Congress, would join the action. Rather than just calling out Pelosi's committee plan as insufficient with a big protest, Ocasio-Cortez decided to come armed with a counteroffer: a Select Committee for a Green New Deal. Empowered with the authority to convene experts, subpoena witnesses, and draft legislation, the committee was to spend a year bringing the brightest minds and movements in the country together to hash out an economic mobilization at a scale not seen since World War II, or maybe ever. The idea was that when Democrats took back the White House and, hopefully, the Senate in 2020, they could hit the ground running. But it all had to come together in a flash, funneling months and years of research on industrial policy and wartime economics into the wonky formality of a congressional resolution—essentially, another language to the DC newcomers who drafted it. As Sunrise's Evan Weber told me, the "New Consensus team, with input and feedback from AOC and us, wrote [the Select Committee resolution] within twenty-four hours and then we put all the pieces together over the weekend. Then it was Tuesday."

The Green New Deal itself had to that point been a "working title," as Ocasio-Cortez said later, "but it kept like leaking and catching and people just started writing articles calling it a Green New Deal before we even said anything or called it that ourselves."[21] She, her team, and Sunrise were well-acquainted with the baggage of the original New Deal—the fact that it created redlining, for instance, and purposefully excluded women and people of color from many of its programs. It wasn't really new in US politics; none other than *New York Times* columnist Thomas Friedman had coined the phrase in 2007, before Van Jones threw it around during his brief tenure as Obama's green jobs czar. People had used the label to describe the clean energy–related parts of stimulus, too. In spite of all this, the Green New Deal label stuck. Yet it hadn't even been a part of the branding for the sit-in, when Sunrisers flanked the halls of the Canon Office Building, donning T-shirts declaring that "We Have the Right to Good Jobs and a Livable Future."

The scene that day was raucous and tightly choreographed. In Pelosi's chambers, protesters huddled in. For one hundred feet of the long hallway approaching her office, the walls were lined with young people in color-coordinated shirts and signs, forming a sea of yellow, white, and black. Sunrisers took turns giving testimony about why the climate crisis mattered to them and led the crowd in chants and songs. Things really picked up when Ocasio-Cortez arrived, standing in the center of the office and high-fiving those sitting in, fifty-one of whom would be arrested in the next hour. Among them was eighteen-year-old Jeremy Ornstein who, cheeks flushed red and voice strained, recalled sitting on the phone with his dad earlier that year, who cried as he told him about the Tree of Life massacre, where a gunman killed eleven people during Shabbat services during the high holy days, citing the Hebrew Immigrant Aid Society's work assisting refugees. His grandparents had fled Hungary in the 1940s, Ornstein said, and he recalled that the first time he had to grow up was learning at temple how the Nazis massacred the Jews. He's had to grow up many times since, he said, and the climate crisis was forcing him and so many other young people now to grow up before their time.

"Because we have endured bullets and storms and fires. Because we have had to grow up one too many times. Speaker Pelosi, Democratic leadership, we are asking you to grow up," Ornstein shouted, gesturing toward her office behind him. "Speaker Pelosi, when will you confront the roots of division and hatred? When will you come up with a plan to stop the climate crisis and defend the homes of millions of would-be climate refugees?" Among the most numerous props that day, fittingly, were manila document envelopes with copies of the Select Committee proposal inside. "Dear Democrats" was screen printed in bold type on one side. The other posed a blunt question: "What's Your Plan?"

They didn't have one. Neither were establishment Democrats accustomed to thinking about politics in such terms. In decades past, the party had sold itself as a vehicle for big ideas that could transform society. Now—especially under Donald Trump—it just wanted to keep from losing more ground, having long ago subbed out ambitious projects like the New Deal or War on Poverty for lowering prescription drug prices and leveraging private investment to build infrastructure. They were still in some loose technical sense the party of Roosevelt, but decades of swallowing neoliberal dogma had chipped away at the idea that Democrats could get big things done or even wanted to. The Select Committee Pelosi proposed after the 2018 midterms would gather evidence about the crisis that would, at some point long down the road, inform a bill to make modest progress toward lowering emissions. As a demand, then, even a Select Committee for the Green New Deal—never mind the Green New Deal itself—was no less a moonshot than putting a democratic socialist in the White House had been for Chakrabarti and Exley.

Its vision was even bigger. The eleven-page document envisioned a fifteen-member, bipartisan committee, with six appointments from the minority leader and the chair picked by the Speaker of the House.[22] In essence, it was a plan to have a plan, or—per the text—to "develop a detailed national, industrial, economic mobilization plan for the transition of the United States economy to become greenhouse gas emissions neutral" over the course of a decade. Though much of the criticism lobbed at the Green New Deal since has taken issue with its lack of detail, what it proscribed were goals and guidelines for a yearlong process driven by the federal government to fill those details in. It recognized the climate crisis as a "historic opportunity to virtually eliminate poverty in the United States and to make prosperity, wealth and economic security available to everyone participating in the transformation," pledging to redress historic racial, regional, and wealth inequality and give labor unions a leadership role. The committee would look to provide opportunities for high-income work, as well as public and cooperative ownership in the wide array of sectors it would set out to decarbonize.

"The way things are done has not been getting results," Ocasio-Cortez told an interviewer that day. "We have to try new methods." In that spirit, none other than Ed Markey—cognizant of how much had changed in the last decade—introduced S.Res. 59, a companion resolution in the Senate to Ocasio-Cortez's follow-up resolution in the House, H.Res. 109. These measures, after they were unveiled in February 2019, prompted considerable backlash from the right, which attacked the Green New Deal as a big government plan to take away Americans' hard-earned hamburgers and plane travel. The right also found alignment on some points with the centrist pundits, who wrote the Green New Deal off as a bloated socialist wish list too loaded up with other priorities—like Medicare for All and a federal job guarantee—and not laser-focused enough on reducing emissions, which, they argued, could be accomplished more discreetly.[23]

The ambition was the point. Like the Sanders campaign, the Green New Deal was an ask big enough to inspire people to make it happen—and well in line with confronting a problem that touches every policy field imaginable. The Green New Deal, says Rhiana Gunn-Wright, former New Consensus policy director and architect behind the framework's earliest details, "opens an opportunity to renegotiate power relationships between the public sector, the private sector, and the people. . . . We are interested in solutions that create more democratic structures in our economy."

The Green New Deal, of course, is not the law of the land. Launched during the Trump administration with Mitch McConnell helming the Senate, establishment Democrats doubled down on defending their turf against it, too. Though it had a more urgent title, the Select Committee on the Climate Crisis that Pelosi ended up creating was weaker than its predecessor body formed in 2007. At the urging of then Democratic minority whip Steny Hoyer, the new committee was stripped of the power to issue subpoenas. Its chair, Kathy Castor, questioned the constitutionality of barring members who'd taken donations from fossil fuel companies. Frank Pallone—incoming chair of the House Energy and Commerce Committee—fought the idea of a Green New Deal committee for fear that it would undermine the authority of his prized new seat, although the plain text of the measure made clear it would do no such thing.

Even so, it has shaped climate politics more than its nonbinding congressional resolution would suggest. Talk of the Green New Deal dominated the Democratic presidential primary in 2020, and climate activists pushed Joe Biden, of all people, to embrace a $2 trillion climate plan, including large-scale investments in environmental justice. It lent a name, if not some spirit, to Europe's fraught Green Deal and recovery from COVID-19 and plans from Spain to South Korea, where center-left politicians had run successfully on their own versions of a Green New Deal. Perhaps most importantly in the US, it bridged the enthusiasm gap. Climate is now something people here are voting about: where warming had always ranked near or at the bottom of voters' priorities, it's become a major concern since the Green New Deal burst onto the scene. In the summer of 2020, the climate crisis was the third-most important issue to voters overall and ranked first among Democrats' concerns at the ballot box.[24] Seventy percent of registered voters that year wanted government action to address the climate crisis, and three-quarters wanted the country to generate all its power from renewable sources by 2035. Though it had been years in the making, Sunrise and Ocasio-Cortez jolted climate politics into a new world. For those involved, they made one seem possible.

IN INTERVIEWS WITH young activists over the years, I have heard again and again how hard it is for them to imagine a future marked by anything but catastrophe—or indeed any future at all. The American Psychological Association first defined eco-anxiety in 2017 as a "chronic fear of environmental doom."[25] Fifty-seven percent of American teenagers in 2019 reported feeling scared about climate change; 52 percent felt angry about it. Four out of five Australian students, according to one study, feel somewhat or very anxious about climate change. "It's like, the ice caps are melting and my hypothetical children will never see them," as one teen told the *Washington Post*, "but also I have a calculus test tomorrow." That the world is going to end can seem like a foregone conclusion reached before young people were born, reinforced by apocalyptic headlines, intransigent politicians, and a drumbeat of decidedly unnatural disasters.[26]

The nineties were supposed to herald in the end of history, when liberal democracies that extolled the heroic virtues of free markets and unfettered globalization would dominate and alternatives would cease to exist. Its flipside—some twenty-plus years on, with a recession, pandemic, and lots of scary science—is the end of the future. The late Mark Fisher described this as capitalist realism: less an excitement for neoliberalism than a cynical moralizing that anything else would be unrealistic, "analogous," he wrote, "to the deflationary perspective of a depressive who believes that any positive state, any hope, is a dangerous illusion." As Nancy Pelosi called it, a "Green Dream or whatever."[27]

An older, richer, and whiter generation has repeated a version of TINA for the climate crisis, within and without the Democratic establishment. Novelist Jonathan Franzen broadcast his own bleak climate model in the *New Yorker*, pairing his personal reflections on human nature with a survey of the climate science literature for something more earnestly nihilistic than the Democratic establishment's capitalist realism. "Call me a pessimist or call me a humanist, but I don't see human nature fundamentally changing anytime soon. I can run ten thousand scenarios through my model, and in not one of them do I see the two-degree target being met," he writes, vaporizing responsibility for the climate crisis out onto the masses. Its source can't be seen or grasped. Whether they like it or not, everyone will have to breathe the spores in eventually, so seek salvation where you can amid inevitable disaster. For Franzen, that meant harvesting strawberries and kale in a community garden.

Sunrise and climate strikers rallying behind a Green New Deal have offered something like an alternative, both to a sense of powerlessness and to all the doom and gloom. Karla Stephan, then fifteen, worked to organize climate actions first near her home in the Washington, DC, suburbs and then nationwide with the million-plus Youth Climate Strike in 2019. "We're striking from school for a day so we can fight for the rest of our lives," she told me. "I don't want to say Generation Z. We want

to be known as Generation GND"—referring to the Green New Deal—
"because we don't want to be the last generation."

For many getting involved in the youth climate movement, groups
like Sunrise and the climate strikes provide an outlet to channel all that
anger and anxiety into fighting for a vision of a fairer, more habitable
planet, while also processing the heavy reality of the climate crisis with
others. "Knowing that even in the best-case scenario we win and are able
to establish ecosocialism, the future still will be pretty tough. And that's
best-case scenario," said Matthew Fleming. When we spoke, Fleming was
a high school senior in rural Acton, Massachusetts. He had recently gotten involved in Sunrise's Acton hub and the Young Democratic Socialists
of America, after Twitter and left-leaning YouTubers like Contrapoints
turned him on to left politics. "I spent a lot of the summer pretty in the
dumps about climate change and the future in general. Sunrise has lifted
me out of that, and the organizers at the Acton hub have gotten me out of
that spiral and helped me realize that I could be a part of the change." Another member of the Acton hub, Sila Inanoglu, agreed: "Once the climate
crisis becomes super real to you, it becomes so awful if you don't have a
community to be with."

As Sunrisers and climate strikers sort through their anxieties via collective action and connection with fellow organizers, others have taken a
different, brutal route. Before gunning down twenty-two people in an El
Paso Walmart, twenty-one-year-old Patrick Crusius published a manifesto to the online forum 8chan, a virtual meeting place for neo-Nazis and
other far-right radicals. He described his monstrous act as "a response
to the Hispanic invasion of Texas." Its title ("The Inconvenient Truth")
was borrowed from Al Gore's 2006 documentary, which introduced millions to the imminent threat of global warming. Crusius, like most other
young people, takes that threat of the climate crisis seriously. "My whole
life," he writes, "I have been preparing for a future that currently doesn't
exist." Crusius despairs that the "environment is getting worse by the
year. Most of y'all are just too stubborn to change your lifestyle. So the
next logical step is to decrease the number of people in America using
resources. If we can get rid of enough people, then our way of life can become more sustainable." The twenty-eight-year-old man alleged to have
killed seventy-four people at a mosque and civic center in Christchurch,

New Zealand, months earlier—an inspiration to Crusius—voiced similar concerns at length in his own premassacre treatise, saying he intended to repel migrants displaced by climate change, who threatened a "Great Replacement" of the white race. The two killers spent time in the same 8chan discussion board with other young, alienated men, as much as eight hours a day in Crusius's case, and aligned themselves with the same obscure ideology now enjoying a troubling revival: eco-fascism.

The anxiety and anger the climate crisis inspires in young people is a blank slate to be painted with whatever politics and ideas are lying around. In times of upheaval, fascism and its derivatives have offered stories about the past—a kind of maximalist small-c conservatism or "reactionary futurism," as the *New York Times*'s Ross Douthat has called it. Whether it takes the form of vigilante ethnic cleansing or closing borders through public policy, the appeal of such backward-facing nostalgia will grow as temperatures rise, for fantasies of dryer and whiter times. In that sense, the bitter partisan politics that have surrounded climate politics in the United States have held a silver lining. It's a blessing, of sorts, that as white nationalism has come more fully out into the open, climate politics have been associated largely with Democrats, progressives, and even the radical left, who—for all the flaws of their varied approaches thus far on climate—have largely not described it as a race war over scarce resources.

That stark partisan divide probably won't last. Environmental concerns among young conservatives are growing nearly as fast as they are among young people in general. The veteran Republican strategist and pollster Frank Luntz noted in a 2019 memo that 58 percent of GOP voters under forty are more worried about climate change than they were a year ago.[28] Sixty-nine percent of Republicans, he adds, fear that the GOP's climate stance is harming its chances with younger voters, and Republicans are beginning to trickle over the rubicon. If it were possible to flip a switch and make today's Republican Party take the climate crisis seriously, though, what would be the result?

Europe, where climate denial isn't viable, might hold a preview, with its far-right politicians spouting ecological consciousness and some green social democrats, in turn, warning about an influx of climate refugees. The policies that flow from that zero-sum approach are the ones we're already seeing stateside: more militarized borders, terrorized immigrant communities, and atrocities such as El Paso and Christchurch—ginned up by all that heated rhetoric. The eco-fascist shooters in those slaughters trained their guns on the people they saw as eating up the scarce resources to which they felt uniquely entitled. In government, a xenophobic right awakened to the climate crisis will do the same at a horrifying scale.

Whether old-school denial leaves the stage of US politics abruptly a month from now or fades away over five years, the climate crisis will be less of a weapon in culture wars between left and right than the ground on which they're fought, with rising temperatures becoming the rationale for everything from vigilante ethnic cleansing to hardened borders in the name of maintaining minority rule. Without a tangible, abundant alternative future on offer, the right's utopian ethno-nationalism and strongmen's protection will become more attractive prospects than dead-end centrism. For now, those pushing for that alternative—to grow democratic majorities and wield them in the interest of people and planet— remain a small if growing force in Congress. The Squad (Ocasio-Cortez, Omar, Tlaib, and Ayanna Pressley) welcomed Jamaal Bowman, Cori Bush, and Mondaire Jones into its ranks in 2020. Progressives still face near constant pushback from the Democratic Party's establishment wing. If more centrist types want to avert electoral and planetary oblivion from a force more fearsome than Trump, they might do well to adopt their colleagues' advice: organize to expand the Democratic base, do everything possible to improve peoples' lives, and give them real hope that a better world is within reach.

The twenty-first century's defining fight on climate change, in any case, won't be over whether it exists or not. It will be over how societies choose to respond, and who gets to live and live well in a warming world. There are now, as the Shell Scenarios team might say, many possible futures before us.

PART 2

GREEN DREAMS VERSUS ECO-APARTHEID

CHAPTER 6

PICK GOOD! BE SMART!

WHEN AUBREY MCCLENDON gave Billy McFarland half a million dollars, he probably saw a bit of himself in the enterprising twenty-something. At twenty-three, McClendon had founded a business that bought and bundled oil- and gas-rich land out to drillers in package deals. He and partner Tom Ward eventually spun that into Chesapeake Energy, a venture with its own drilling operations through which McClendon pitched the still niche business of fracking to investors the world over. After going public with a $25 million valuation in 1993, Chesapeake would in the 2000s help transform the energy sector and become one of Wall Street's best-performing stocks. At twenty-two, McFarland had founded a company called Magnises, "Latin," he has said, "for absolutely nothing."[1] The business created sleek, black metal dupes for its members' charge cards that granted entrance into members-only events and "club-houses" in New York and Chicago, furnished by $3.1 million in venture capital funding. His next project—the one he's best known for, to which McClendon gave his cash and blessing—was a "media company" (app) that revolved around booking celebrities to show up at parties and that in 2017 was to host an influencer-stacked music festival in the Bahamas, on an island reportedly once owned by the drug kingpin Pablo Escobar. In all, McFarland would raise $26 million for the biggest party that never happened. In festivals and fossil fuels, McClendon and McFarland were playing similar games.

The Fyre Festival was one in a summer rich with scams, less a season than a microstage of capitalism. These viral grifts—ranging from multibillion-dollar medtech companies to lavish New York lifestyles—had a lot in common: idiosyncratic, invariably millennial protagonists; seemingly visionary business models; apps. They all had more ostensibly legit validators lending funds and credibility. McFarlane had McClendon and (more famously) Ja Rule. Ousted WeWork founder Adam Neumann had the hearty backing of SoftBank's Masayoshi Son and his $100 billion Vision Fund. With that, he pumped his ever-expanding coworking empire up to a $47 billion valuation before its implosion. At one point, he called JPMorgan Chase CEO Jamie Dimon his "personal banker." Elizabeth Holmes, founder of the ill-fated blood testing firm Theranos, swooned former defense secretary James Mattis, Henry Kissinger, and the coauthor of the Climate Leadership Council's carbon-pricing plan discussed earlier in these pages, former treasury secretary George Shultz. At one point, Shultz sicced lawyers on his own grandson for blowing the whistle on Holmes, hailed by *Forbes* as the youngest-ever self-made woman billionaire.[2] These founders weren't as unique as they seemed. Along a time line marked by low interest rates and cheap, easy debt, fossil fuel entrepreneurs struggling to make their way in the world used many of the same tricks to attract billions of dollars worth of financing, change laws, and transform the world's energy landscape. Before too long, their grift would come crashing down, too—but not before enlisting high-profile supporters across the political spectrum to help keep it going.

Like McFarland, McClendon lived large and was one hell of a salesman. As Bethany McLean details exhaustively in *Saudi America: The Truth About Fracking and How It's Changed America,* his great success in selling fracking to the world wasn't thanks to any great technical skill. The breakthrough use of hydraulic fracturing and horizontal drilling to unearth difficult-to-access fuels had been pioneered years earlier by Texan George P. Mitchell. Fracking itself had been around since the 1940s, invented to solve a geology problem.

Drilling in places like Saudi Arabia's expansive Ghawar oil field is like sticking a straw in a soda. Oil comes up cheaply and easily there, with costs of production as low as $7.50 a gallon, sometimes lower. By contrast, in the Permian Basin in West Texas and New Mexico, drillers need prices of at least $50 per barrel just to break even. Unlike the Persian Gulf's vast reserves, shale oil of the sort found in West Texas is stored in dense rocks, requiring the extensive use of sand, water, and chemicals to blast (frack) gas and oil free and coax it to the surface. Even then yields are relatively low. Horizontal drilling can expand them. Whereas straight down vertical drilling might let a company access one hundred feet or so of buried gas and oil, drilling across a shale formation—for up to three miles, in some cases—multiplies the amount that can be taken out of each drill site by expanding surface area being fracked. All this takes big up-front investments and a near constant injection of cash into drilling companies to keep production going. Even on good wells, production declines sharply after about a year.[3]

What McClendon brought to this business in the new millennium, critically, was cash. "As oxygen is to life, capital is to the oil and gas business," as one analyst told McLean, no less so than for unconventional drilling. And no one was better at raising capital than Aubrey McClendon.[4] His sales pitch was straightforward: through technology and the power of innovation, fracking had transformed oil and gas from a gamble into a sure bet, more akin to manufacturing, he said, than the usual game in conventional oil production of go fish, where new wells may turn up dry.

Still, investors and oil majors like ExxonMobil had discounted unconventional drilling through most of the 2000s, arguing that high costs of production would only make sense if oil prices surged. Beyond those costs was another problem: most American refineries weren't set up to process the lighter oil fracked domestically, having been built for the heavier crude imported from places like Venezuela and Canada. Retrofitting refineries would require sizable investments that companies seemed unwilling to make. Nixon-era laws meant to boost US fossil fuel production and supplies in the wake of the oil crisis greatly curtailed the amount that could be exported abroad, barring Canada and a couple of other exceptions. So even if all that new oil could be drilled profitably—a big if—only so much could be sold.

McClendon set out to prove skeptics wrong, making the case to investors wary of fossil fuels that there was a new normal and that stable, high fuel prices, which made fracking economically appealing, were here to stay. In the early 2000s, he argued this new technology was already disrupting business as usual; further investment could only make it more efficient and profitable. Funds could either get in early or get left behind. As a sweetener, McClendon would also hand banks exorbitant fees to underwrite the debt Chesapeake needed to snap up as much land as it could, as quickly as possible. They happily obliged, and McClendon burned through cash. As he told *Natural Gas Intelligence* in 2005, "Asking me what to do with extra cash is like asking a fraternity boy what to do with the beer."[5]

The backdrop to McClendon's success was the biggest energy panic since the 1970s, in which fears that oil would run out translated into soaring prices from 2002 on, climbing to over $3.00 per gallon by 2007. Demand from China and India was skyrocketing and projected to keep growing apace. Politicians feared easily accessible oil could run out and that the US was particularly vulnerable to any supply shocks given that it imported some 60 percent of its oil. At the start of George W. Bush's second term, the Department of Energy published a report noting that even under optimistic forecasts, "world oil peaking would occur in less than 25 years," leaving "the US and the world with an unprecedented risk management problem."[6] Adding to American officials' troubles was the fact that much of the remaining, easily accessible ("conventional") oil was in countries with which it had fraught relationships, like Iraq; even a war there hadn't managed to allay supply concerns, despite the Bush administration's best efforts to divide up the spoils of their desert conquest for its good friends in Houston.[7] If anything, inflaming tensions in the Middle East had made things worse.

The panic was great news for Chesapeake, whose less volatile, mostly gas business started looking more attractive; the company's market value surged to $35 billion by the summer of 2008. McClendon could now sell shale fracking as a path to energy independence. Chesapeake's Oklahoma City headquarters at the time was lavish, including a café staffed by professional chefs, mirroring the sprawling Silicon Valley campuses Google and Facebook would become known for. McClendon—one of the richest men in America, for a time—binged on houses from Maui to Minneapolis and was said to have had one of the world's finest wine collections.

McClendon's early success notwithstanding, unconventional fuels— including fracked gas and oil but also tar sands—were still too costly to pursue at scale, much less provide an answer to what ailed oil. Without an easy fix, economists began to worry that skyrocketing fuel prices would spell trouble for the economy as a whole, as people spending more to fill up tanks getting to school and work conserved on spending elsewhere.

To help remedy this problem through the 2000s, central, retail, and investment banks encouraged households to take on more debt to cover the cost of bigger expenses like cars and homes, leaving more discretionary cash on hand for consumer spending that would drive growth. The question of whether borrowers could actually pay all this new debt off—or keep their homes in the long run—was mostly irrelevant; between 2000 and 2007, US household debt jumped by an astounding $7 trillion.[8] Investment banks bundled that reckless new borrowing into novel financial products called mortgage-backed securities to be traded and gambled on for profit, with little attention paid to the risk housed within them. With bankers riding high, consumer borrowing continued to grow through the first quarter of 2008, after the start of the economic downturn.

Demand for fuel started waning as consumers felt crunched and became wary about the future. Long-term projections for fuel demand grew more pessimistic. Oil prices eventually crashed with the rest of the global economy and played a not insignificant role in making that happen. The world had entered its worst economic downturn since the Great Depression. One product perfectly embodied the boom and the bust: at their peak in 2006, as the War on Terror was revving up, the United States sold 71,524 Hummers, those hulking, luxury-ish, gas-guzzling military-style vehicles that got twenty miles per gallon on the highway; in 2010, Americans bought just 3,812.

LIKE MOST COMPANIES across the world, Chesapeake Energy took a hit in 2008 too. Collapsing fuel prices saw its value drop to just $16 a share that fall, and its seemingly dwindling fortunes cost shareholders some $30 billion. McClendon, meanwhile, was awarded a $75 million bonus. His best years, though, were still ahead of him. Like McFarland and Neumann, he'd have the crash—and the government's response to it—to thank.

For decades, the Federal Reserve had relied primarily on one thing to accomplish its dual mandate of price stability and full employment: targeting short-term interest rates through the federal funds rate, which determines how much banks charge one another to lend out Federal Reserve funds overnight. Roughly speaking, higher rates raise the cost of borrowing money and discourage investment. Lower rates, then, make borrowing and debt cheaper for businesses and consumers, spurring new investment and hiring and spending with it. The Fed began cutting the federal funds rates to target lower interest rates in the fall of 2007 to account for stagflation. This didn't happen fast enough to relieve pressure on consumers and the financial system. Yet by the time the recession set in and the stock market crashed in September 2008, the federal funds rate was already at a low of 2 percent; collapsing it down to virtually zero didn't carry the punch of stimulus monetary policymakers needed. Having exhausted what had for over a decade been considered its sole policy tool, the Fed unveiled a range of asset-purchasing programs and new debt-buying facilities to prevent a depression and keep the country's biggest banks from going bust.

Over three rounds of quantitative easing (QE), the Federal Reserve quadrupled the size of its balance sheet, bringing trillions worth of banks' toxic assets and long-term Treasury securities onto its balance sheet and pumping $700 billion into the financial system to bail out the banks that had engineered the crisis. "Policy steps that would have seemed exceptional, and that would have commanded many hours of debate in January 2008," economist Matthew Berg wrote later, "barely seemed to merit a second thought in the tumult of September and October."[9] While estimates vary, researchers, accounting for cumulative support offered to the world's biggest financial institutions from 2007 to 2012, found that the total dollar footprint of the Fed's postcrash interventions amounts to a whopping $29 trillion.[10]

The low interest rates that followed the crash meant the capital needed for unconventional oil and gas drilling was now easier to come by than ever. Between 2006 and 2014, syndicated loans to the oil and gas sector—those provided by two or more lenders—ballooned from $600 billion to $1.6 trillion.[11] In buying up Treasury bonds, central banks' QE further raised the price and depressed the yields that big institutional investors like pension funds had traditionally relied on to pay out to retirees, sending them on the hunt for higher yields in often riskier places like corporate bonds and private equity funds. With few other options, pensions throughout the US and Europe began investing—among other places—in oil and gas companies, which tripled the amount of outstanding bonds they issued between 2006 and 2014. For a while fuel prices remained low, keeping more widespread unconventional drilling off the table. But they recovered roughly on pace with the rest of the economy, stabilizing above $90 a barrel between 2011 and the middle of 2014.

Frackers had other help, too. The American Clean Skies Foundation—founded by McClendon—pitched the already industry-funded MIT Energy Initiative on writing up a study they would sponsor. *The Future of Natural Gas*, released in 2011, made the case for gas as a critical bridge to a low-carbon future. The cochair of the study was MIT Energy Initiative cofounder Ernest Moniz. His fellow cochair, Anthony Meggs, would go on to work for the gas company Talismen Energy before the report was released. Moniz would become Obama's secretary of energy in 2013, where he moved "expeditiously" on approvals for liquified natural gas export permitting.[12]

For oil and gas companies, this all came together in a perfect storm. Low interest rates made the kinds of debt needed to finance new shale drilling cheap. QE sent big institutional investors looking for new places to put their cash, and now high oil prices made unconventional drilling attractive, particularly as conventional drilling remained stagnant. And policymakers were happy to see the country's oil and gas sector grow, lessening dependence on imports and bolstering economic recovery. Enter: the fracking boom.

It took off first in the Bakken Field stretching through North Dakota and Montana. Production there—nearly all of it through horizontal drilling—doubled between 2012 and 2014, producing millions of barrels of oil a month.[13] From producing almost no unconventional oil in 2008, American wells overall produced 2 million barrels per day by 2013 and 3.5 million per day by the next year; there was a 62 percent increase in oil production between 2010 and 2014, when loans to shale oil companies reached almost $250 billion. Refineries retooled to process lighter shale oil to keep up with demand.[14] Sleepy towns like

Williston, North Dakota, exploded with man camps, temporary housing for the mostly young men who flocked to the oil fields for well-paid work roughnecking and driving. Housing prices and just about everything skyrocketed, and at one point Williston residents were renting out their walk-in closets for $1,000 a month. Similar dynamics visited other shale formations—the Appalachian Basin and Marcellus Shale in the Northeast and the Permian Basin in West Texas.

But the oil fields weren't the only boomtowns. The same brave new world of postcrash economics that birthed the fracking boom injected Uber, WeWork, Airbnb, Theranos, and countless other new firms with tens of billions of dollars. Tech workers, famously, swarmed to the San Francisco Bay Area on the promise of being part of the next big initial public offering (IPO), start their own thing and maybe get bought out by a Facebook or Google. The chill of the dot-com boom had thawed; venture capital funds and angel investors had money to spend and were eager to fund the next Bill Gates or Larry Ellison. Rents in San Francisco's working-class and bohemian neighborhoods shot up, as did its displacement and homelessness crisis. Businesses in the city had trouble attracting entry-level service workers at fifteen dollars an hour since it was impossible for them to afford to live within a one- or two-hour commute.[15] Black-and-white charter buses zoomed up and down previously sleepy streets, shuttling workers forty-five miles to and from Silicon Valley using public bus stops and lanes.

With so much cash floating around the economy, some of the biggest companies to emerge after the financial crisis encouraged a more shaky relationship to profit-making than their old economy peers, making flashy pitches to investors eager to stow cash somewhere.[16] At the peak of the dot-com boom in 2000, 81 percent of newly public companies were unprofitable; in 2017, 76 percent of companies that went public were unprofitable, the highest number since just before the dot-com bust.

Fracking has operated on a similar footing, the idea behind both being that big initial investments would provide the necessary start-up capital for companies to scale up fast and work out the kinks on their way toward profitability. For years, Bethany McLean writes, "the value the public market was willing to accord a fracker was based not on a multiple of profits, which is a standard way of valuing a company, but rather as a multiple of the acreage a company owns. It was a bit like the old dotcom days, when internet companies were valued on the number of eyeballs."[17]

Pressure to grow has given fracking a strange, vicious cycle of a business model. Since well yields drop off quickly after initial ground is broken, more had to be drilled to show growth. There was plenty of money on offer from Wall Street to fund that, but that meant companies had to burn through what they had in the short term to look like a worthwhile investment, acquiring new land and producing more with what they had to prove that the business was growing. Fracking companies drilled like crazy, making slim if any profits, promising that technological breakthroughs would improve their balance sheets. That didn't happen, at least at the scale promised, and before too long companies began working against their own and their investors' interests.

Global demand for oil, at the same time, was starting to fall as the rapid growth in countries like China and India began to slow. The oil not being imported to America, rebuffed by domestic production, went elsewhere and drove down prices.

OPEC, whose members benefit greatly from high prices, faced a choice: It could cut supply to drive prices up but continue to lose market share by effectively subsidizing the small American producers who needed higher prices to stay afloat. Or it could weather lower prices for a while longer. Already drowning in debt, hundreds of small American firms that had cropped up in the boom were vulnerable to even slight fluctuations in the price of oil. Independent wildcatters with costly extraction methods would go under well before an Aramco or Pemex, if such a thing was even plausible, given their state backing. With low production prices, virtually unlimited reserves, and a massive sovereign wealth fund, OPEC opted to stick it out and keep supply steady.

The gamble—if it can be called one—paid off. Oil prices dropped, sliding from highs of over $100 per barrel to $26 per barrel by February 2016. Bigger companies like Equinor and Shell wrote down billions of dollars of shale investments as fast as they could. Smaller companies lacking the cash flow to pay off their mounting debts squeezed workers and operations as much as they could. By 2014, fracking companies' net debt exceeded $175 billion. By the second quarter of 2015, oil producers were spending 83 percent of the cash they had on hand to pay down debt. In all, one hundred oil and gas companies filed for bankruptcy in 2015 and 2016. Some that went under were absorbed by bigger companies. Private equity firms smelled blood, swooping in to load drillers up with even more debt and collect lucrative fees for the privilege of restructuring them. By the end of the bust, it's estimated that nearly half a million workers had lost their jobs. The promise of fracking had been energy independence. Ironically, it now looked like that "independence" could be undermined by a handful of technocratic managers spread across state-run oil companies across the globe.

Having been ousted from Chesapeake on bad terms after a string of ethics and financial management concerns, McClendon weathered the crash at American Energy Partners, his new umbrella venture containing several different companies. He made out all right for a while, raising $15 billion the year after stepping down as Chesapeake's CEO. Things went south after fuel prices crashed in 2015, but McClendon kept spending and growing, approaching vulture funds for a lifeline and mortgaging hundreds of millions of dollars worth of his own property. Chesapeake's IPO way back when had given him a personal stake in its wells on the condition he also covered a portion of the drilling costs. With low prices, profits from those wells dried up, and keeping up with costs got more expensive. Amid the bust and climbing debt, McClendon's math stopped working; his backers started walking. To make matters worse, he was under criminal investigation after having allegedly fixed land lease prices in 2010.

On March 1, 2016, McClendon was indicted by a federal grand jury. The next morning, he was killed after crashing his Chevy Tahoe into a concrete wall going ninety miles an hour.

AUBREY McCLENDON'S LIFE may have ended in 2016, but fracking had a long future ahead of it. A few months before he died, the federal government extended a hand to the flailing industry that would not only spur its recovery from the oil price crash but transform the world's energy landscape. If you weren't following closely, you might have missed it.

As early as 2013—as companies started to run up against the limits of US refinery capacity—oil and gas companies started organizing their lobbyists to lift the ban on oil exports, put in place by the Energy Policy and Conservation Act of 1975. At one point in 2013, oil interests had just sixteen registered lobbyists in Washington working on oil exports.[18] By the end of 2015—in the final push to get the export ban lifted—they had three hundred lobbyists on the Hill pushing for it and spent $38 million pushing the rule change in the third quarter alone, according to the Center for Responsive Politics.[19] Their main opponents were both greens concerned about the carbon implications of new exports and refinery companies and adjacent unions, who stood to lose business and jobs, respectively, if drillers shipped crude to be processed abroad.

Although the GOP controlled the House and Senate—giving them a relatively friendly audience in Congress—the measure would still need White House support to get through. So the industry reached out to a who's who of Obama-era officials to ask them to speak out in support of lifting the ban. As McLean writes, Democrats from resource-rich states, like North Dakota's Heidi Heitkamp, who pushed actively in the Senate to lift the ban, were an easy draw. But former treasury secretary Larry Summers got in on the act too. At an event at the Brookings Institution, he argued that the merits of repealing the ban were "as clear as the merits with respect to any significant policy issue I have encountered."[20]

Among the most effective ammunition in support of the change was a January 2015 paper, coauthored by Jason Bordoff and Trevor Houser, backing plans to lift the ban. Bordoff's opinion carried weight. He had been a special assistant to Obama and senior director for Energy and Climate Change on the staff of the National Security Council before leaving the administration to direct Columbia University's Global Center on Energy Policy in 2013. Houser, meanwhile, would go on to become Hillary Clinton's top adviser on climate and energy issues during her 2016 run.

In its previous iteration, the center (then called the Center for Energy, Marine Transportation and Public Policy) had gotten funding from a number of major energy companies, including ExxonMobil. Now, as the export ban fight heated up, energy companies with major interests in fracked oil and gas—including several members of American Producers for Crude Oil Exports—doubled down on their support to Columbia's School of International and Public Affairs (SIPA), which houses the center.[21] Annual reports show that the Louisiana-based liquified natural gas (LNG) company Cheniere donated at least $500,000 when the Center for Global Energy Policy's launched in the 2012–2013 school year—launch events that featured then New York City mayor Michael Bloomberg, acting secretary of energy Daniel Poneman, and ConocoPhillips chairman and CEO Ryan Lance. One of its two founding research projects would be, that year's report states, a study of energy exports that "will make policy recommendations for how government should reform trade restrictions."[22]

SIPA's annual reports don't specify which of its several programs ultimately received Cheniere's money, although no others are explicitly energy related. And the money kept coming. In 2013–2014, the company gave at least another $500,000 to SIPA.[23] ConocoPhillips and Statoil each gave between $100,000 and $499,000 that year, and Laredo Petroleum and Pioneer Natural Resources pitched in somewhere between $25,000 and $99,000. Exxon gave $25,000 directly to the Center on Global Energy Policy's markets program, according to company disclosures. The following year, Cheniere founder Cherif Souki donated at least $1 million to SIPA and Cheniere reupped its $500,000-plus gift. Statoil and ConocoPhillips gave in the same range they had, joined by the Tokyo Gas Company. ExxonMobil, Pioneer Natural Resources, and a newcomer—Continental Resources—all gave between $10,000 and $99,000.[24]

Bordoff and Houser's paper cited, among other considerations, national security benefits and the potential to boost the US economy by bringing down gas prices, writing that the "original rationale for crude export restrictions no longer applies." Just three of the report's eighty pages are devoted to a discussion of potential climate and environmental impacts. "We can support domestic production while still meeting our climate change objectives," they conclude, "but that requires new policy to reduce US oil consumption and production-related GHG emissions, as well as action in other sectors." This assessment differed markedly from a paper released that same year by the Center for American Progress, which found that lifting the ban could unleash emissions equivalent to the lifetime pollution of 135 coal-fired plants.[25]

Oil lobbyists trying to sway Obama to lift the export ban weren't starting from scratch. The Obama White House had already been peddling US natural gas around the world and supported its ramped-up use domestically as a cleaner alternative to coal, the only fossil fuel the administration made a point of moving away from. "Natural gas isn't just appearing magically," he said in a 2012 debate with Mitt Romney. "We're encouraging it and working with the industry." Obama declared before Congress in 2013 that the "natural gas boom has led to cleaner power and greater energy independence. That's why my administration will keep cutting red tape and speeding up new oil and gas permits." Shortly after taking office that year, Energy Secretary Ernest Moniz—who had cochaired a 2011 MIT study that recommended the US remove barriers to gas imports or exports—had streamlined the process for approving gas exports, surmounting concerns about potential trade law violations. Even green groups like the Sierra Club had painted fracking as a "bridge fuel," and the green side of Obama's energy strategy focused on boosting support for renewable fuel and "clean" coal and gas, as he had through the American Recovery and Reinvestment Act. With the notable exception of coal, that meant not going after polluting fuels directly. Often, it meant actively encouraging them.[26]

Tyson Slocum, the director of the progressive advocacy group Public Citizen's Energy Program, had been working on the Hill to corral members of Congress to stand against any revisions to the language of the Energy Policy and Conservation Act (EPCA) and strengthen protections against gas exports. Obama himself had threatened to veto a stand-alone bill allowing oil exports when it first went through the House before faltering in the Senate. But the industry's choice to squeeze it into a must-pass spending bill ultimately won it the needed support. In exchange, the White House negotiated two consolation prizes: a temporary extension of clean energy tax credits, which were set to expire abruptly that year, and agreement from Republicans not to block $500,000 for the Green Climate Fund. "We had actually cobbled together a decent critical mass of members of Congress to stand firm and build a wall to preserve the existing language that significantly limited the ability to export," Slocum said. "And then all of a sudden, folks on Obama's negotiating team just signed it all away by agreeing to a very bad legislative deal."[27]

The tax credit extension would last for five years, whereas the export ban would be indefinite. "Any way you look at it objectively, it was a really bad deal," Slocum said. The spending bill sailed through Congress with a broad margin of bipartisan support, and Obama signed it into law on December 18, just days after the Paris Agreement was settled in France.

"If you were wondering how seriously world leaders took the obligations they imposed on themselves in Paris over the weekend, the early returns would indicate: not very," writer and 350.org cofounder Bill McKibben wrote afterward. "Barely 48 hours after all the back-patting at the climate conference had ended, word leaked out in Washington that the administration and Congress were preparing to lift the 40-year ban on oil exports, a major gift to the oil industry."[28]

The White House's agreement to lift the ban marked a new beginning for oil and gas companies. The US government had saved them from an oversupply crisis of their own making, though it took several months for its effects to be fully felt. At the start of 2020, roughly one in four barrels of oil extracted in the US was sent abroad, up from just 4.75 percent in December 2014. Total fuel exports grew more than 750 percent, from 400,000 barrels per day (bpd) in 2015 to 3.4 million bpd in October 2019. The US surpassed both Russia and Saudi Arabia to become the world's largest producer of both natural gas and crude oil. A 2018 analysis by Rystad Energy found that projected new growth in oil production was on track to double over 2019 levels by 2032, expansion premised almost entirely on exports.[29]

Such enthusiastic projections wouldn't last. Few companies managed to start turning a profit on their fracking operations despite technological improvements, making the export ban look like more of a lifeline than a permanent fix for what ailed the sector. By 2018, Wall Street banks were starting to lose patience with oil and gas producers and began cutting off their finances. Prices stayed stubbornly low. Thanks to a persistent supply glut, disruptions to oil markets that once would have spiked prices—attacks on oil tankers in the Middle East, potential war in Iran—barely registered, with prices stabilizing after modest jumps over the course of days, not weeks or months. Throughout 2019, forty-two exploration and production (E&P) companies filed for bankruptcy, carrying a total debt load double the amount tied up in the previous year's bankruptcies.[30] Bankruptcies doubled among oil services companies, too. "The significant uptick in bankruptcy filings in the E&P and oil services sectors in 2019 illustrates the increasingly speculative character of the industry as its financial rationale of high risk and high reward deteriorates," the Institute for Energy Economics and Financial Analysis (IEEFA) wrote in its brief on the 2019 bankruptcies and debt overhang. "High risk is now producing chronic value destruction."

Companies in the Appalachian Basin were the hardest hit. Chevron put its assets there up for sale, and the biggest company operating in the region—EQT—laid off 25 percent of its workforce. Just eight firms with major operations there faced a combined $26 billion long-term debt burden and just $12.4 billion in market capitalization; nearly all that debt is due to be paid back before 2027. Chesapeake Energy led the pack on debt that year, with nearly $10 billion. Office vacancy rates in Houston—the epicenter of the American oil and gas business—climbed 24 percent in the third quarter of 2019. Overall, oil and gas companies lost $400 billion in value between 2016 and 2020, as Goldman Sachs recommended selling off traditionally valuable stock in giants like ExxonMobil. As of 2018, 35 percent of all horizontal drilling was done by privately backed companies.[31] "In this environment," IEEFA concluded, "it is difficult to see a financial pathway forward for oil and gas producers."

The worst, as we'll see in Chapter 9, was yet to come.

REPUBLICANS AND EVEN some Democrats have warned climate activists against the dangers of governments picking winners and losers in the private sector, a charge often lobbed at clean energy subsidies. Provide firms a level playing field to compete, the thinking goes, and the magic of the market will do the rest. As the Tea Party governor of Texas in 2011, Rick Perry railed against energy regulations and subsidies. "Get rid of the tax loopholes, get rid of all of the subsidies," he said in a 2011 interview. "Let the energy industry get out there and find—the market will find the right energy for us to be using in this country."[32]

But whether through cheap credit, preferential land leasing, lax regulatory enforcement, repairs to the public roads destroyed by boomtown traffic, trade policy, foreign wars, or diplomats bent on creating markets for American oil and gas companies abroad or any other manner of state support, the US government and many others have time and again picked one clear winner: the fossil fuel industry.

From 2009 to 2014, the value of direct and indirect subsidies given out to the fossil fuel industry ballooned by 45 percent, reaching $18.5 billion in 2013.[33] Despite Obama's frequent calls to end subsidies to the fossil fuel industry, those handouts exploded throughout the course of his administration. By the time he left office in 2017, the federal government and states spent over $20.6 billion padding fossil fuel companies' bottom lines. That many of these subsidies offer more support to new drilling than existing operations has been an especially valuable boon to fracking, where yields drop off rapidly in the first several years and new wells are needed frequently.

This is all on top of a corporate tax rate that, thanks to the Trump tax cuts, now sits at a historic low. Oil and gas companies enjoyed an effective tax rate of just 3.6 percent in 2018. Chevron, in fact, paid a negative 4 percent federal tax rate, receiving $181 million back in rebates. Chesapeake, EOG Resources, Halliburton, Dominion Resources, Kinder Morgan, and Occidental Petroleum—all companies with considerable business in unconventional drilling—paid no federal income taxes in 2018, joined in that by Amazon, Netflix, and eighty-two other companies. G20 countries' export credit agencies further provided some $31 billion in financing to fossil fuel activities, outranking those institutions' funding of renewables twelve to one.[34] Worldwide, accounting for tax breaks and the unpriced cost of fossil fuels' health and environmental impacts, among several other factors, the IMF pinned the annual cost of fossil fuel subsidies the world over at $5.3 trillion.

None of that, of course, takes into account the vast diplomatic and military resources the US spends opening up new markets for fossil fuel companies abroad, rivaled only in scale by its commitment to boosting sales for US-based arms' manufacturers. Ironically, Hillary Clinton's State Department was more aggressive and effective in promoting US fossil fuels overseas than Rex Tilleron's. Unveiled in April 2010, her Global Shale Gas Initiative worked closely with US-based drillers to encourage foreign countries to tap into their shale reserves, turning so-called energy diplomacy into a top departmental priority.[35] Embassies hosted conferences on fracking around the world and deployed US experts to work with government officials abroad in developing fossil gas, priming the pump for governments to grant American companies major concessions on their fuel reserves. Unsurprisingly, similar efforts kept up through the next administration. Documents unearthed as a part of the impeachment proceedings against Trump, for instance, show then energy secretary Rick Perry eagerly pitching American LNG companies to Ukrainian dignitaries. "Clearly," Perry was advised to communicate in one trip to the EU, "as new U.S. LNG export terminals come on line, U.S. exports will become even more attractive and cost competitive in the European gas market. Europe will benefit from these supplies," winking that countries could get their gas from the US instead of Russia. He boasted about the growing export capacity of Sabine Pass, Dominion Cove Point, and Corpus Christi export facilities as proof.[36]

It's fitting that, a few years into his tenure as energy secretary, Perry changed his tune on energy markets. During a press briefing at the National Renewable Energy Laboratory, he was more circumspect about the state's role in the economy than he had been when he railed against tax loopholes and subsidies. "The government's been picking winners and losers since government was created," he said. "We do it by tax policy, we do it by regulation, we do it by permits. Pick good! Be smart!"

IT'S GOOD ADVICE, which previous generations of Democratic politicians have followed toward more egalitarian ends. As Franklin D. Roosevelt's administration rolled out its alphabet soup of New Deal jobs and relief projects in the wake of the Great Depression, political opponents decried several of the projects under the Federal Emergency Relief Administration (FERA) and Works Progress Administration (WPA) as "boondoggles"—wastes of time, money, and effort and evidence that the government should stay out of the business of putting people to work.

Attracting particular ire were jobs apportioned out to white collar workers, a less-remembered part of the Depression-era New Deal projects than the men who built majestic dams and repaired roads. As Nick Taylor details in *American Made: The Enduring Legacy of the WPA*, FERA funded a number of research projects at New York City universities, including "the compilation of a standard Jewish encyclopedia, a study of the making of safety pins, and sociological investigations into matters such as the non-professional interests of nursery school, kindergarten, and first-grade teachers."[37]

Lloyd Paul Stryker, a criminal attorney overseeing hearings for New York City–area FERA projects, called them "high-spun theoretical bunk." Other critics brought up dog shelters, elaborate building façades, and ski lodges as examples of government excess. In response to these jabs, Roosevelt quipped, "If we can boondoggle our way out of this Depression, that word is going to be enshrined in the hearts of the people for years to come." WPA director Harry Hopkins was blunter: "They are damn good projects—excellent projects. That goes for all the projects up there. You know some people make fun of people who speak a foreign language, and dumb people criticize something they do not understand, and that is what is going on up there—God damn it!"

Conventional wisdom now holds that the private sector—and its visionary founders, in particular—are better at coming up with world-changing ideas than the public sector, which is allegedly bloated and allergic to outside-the-box thinking. Corporations' hunt for profits and lack of bureaucratic constraints, it is said, compel cutting-edge research and development in a way that the government is simply incapable of. Supportive venture capital and angel investors help them realize their vision. If founders get rich, it's proof their ideas are meeting some pressing societal need that serves the collective good.

Attuned to investors' search for the next tech wunderkind, Adam Neumann's and Elizabeth Holmes's wealth and their companies' multibillion-dollar valuations were evidence that there was something behind all their lofty promises to transform the world for the better—not that they had talked a few incredibly wealthy people into believing they could actually make it happen. When companies run up big debts, they're taking heroic risks to go against the grain. Mismanagement and mistakes come with the territory of doing such heady and important work at such a rapid pace; eccentric personalities are encouraged because they create big ideas. In the public sector, meanwhile, even minor missteps are evidence of bureaucratic bloat and inefficiency. If a government spending project is making headlines, it's rarely for a good reason.

To Billy McFarland, Adam Neumann, and even Aubrey McClendon's credit, there's a certain kind of genius required to convince investors with cash to burn that a company deserves a multibillion-dollar valuation. It's just not the kind that lends itself to building a decent society.

This phenomenon didn't start with either the tech or fracking booms, of course. Defending his and other tycoons' sacred right to accumulate virtually unlimited wealth, steel magnate and philanthropist Andrew Carnegie in 1889 described an "ideal state," where the wealthy would administer their surplus "for the common good," where it can be a "much more potent force for the elevation of our race than if it had been distributed in small sums to the people themselves," that is, through taxation. "Thus is the problem of Rich and Poor to be solved. The laws of accumulation will be left free; the laws of distribution free. Individualism will continue, but the millionaire will be but a trustee for the poor," Carnegie concluded, leaving the rich rather than democratic processes to apportion societal wealth "for the community far better than it could or would have done for itself."[38]

Decades later, neoliberal attacks on the New Deal order worked to cement a like-minded consensus in the public imagination and the heads of policymakers: that companies rationally seeking to maximize their profits are best suited to make most of the higher-order planning decisions about how a society's vast resources should be distributed. Who gets a job? What kinds of infrastructure should be built? Who should have health care? How should emissions be reduced? That the state played no role in these matters—and should simply clean up corporations' mistakes—was always a story told with a wink and a nod by neoliberals eager to leverage government toward specific ends, be that subsidies or trade policy. But broadcasting it out helped spread myths about a dynamic and innovative private sector versus profligate government, however much the former relied on the latter.

It was in acting on this belief in the private sector's planning prowess that the Treasury poured trillions of dollars into propping up the banking, insurance, and auto sectors after the 2008 financial crisis. While this lifeline effectively nationalized many of the country's largest financial institutions, the government didn't demand much say at all in how those institutions were run or what they did with that extra cash. The 500 million shares in GM the government bought up in March 2009, for instance—60.8 percent of its total market capitalization—gave the White House all the leverage it needed to mandate any number of changes, from fuel efficiency to higher wages. Instead, Obama vowed that "the federal government will refrain from exercising its rights as a shareholder in all but the most fundamental corporate decisions."[39] This all seemed odd to Swedish finance minister Bo Lundgren: "For me," he said at the time, "that is a problem. If you go in with capital, you should have full voting rights."[40]

What did the White House get in return for its generous extension of cash and goodwill to the sectors it trusted to be responsible stewards of the nation's financial health? Banks that were already too big to fail in 2008 are bigger than ever and funneled much of that money right back into the same kinds of rapid-fire speculation that created the crisis in the first place, all the while lobbying to get the modest restrictions placed on them after 2008 wiped off the books. Aside from financing all manner of private sector boondoggles, banks poured prodigious amounts of money into fossil fuels. Between 2016 and 2019 alone, JPMorgan Chase—which received $25 billion in Troubled Asset Relief Program (TARP) funds, meant to save too-big-to-fail banks—financed $196 billion worth of coal, oil, and gas projects around the world. Altogether, major banks furnished fossil fuels with $1.9 trillion over the same period.[41]

Climate policy tends to be thought of primarily as an additive project, but there's plenty of emissions that can be abated by simply peeling back the ample support already offered to the world's biggest polluters. Researchers with the Stockholm Environmental Institute have found that up to half of all oil developed globally through 2050 would be unprofitable if not for state subsidies.[42]

That support can shift its focus, too. The American Recovery and Reinvestment Act (ARRA) in 2009—while disastrously small and lacking sufficient force to take on either the economic or climate crisis—offered a preview in miniature for what active support for a low-carbon economy could look like. Among other things, the Recovery Act enabled tens of billions of dollars' worth of investment in climate-related infrastructure as well as loan guarantees and cash grants to clean-energy companies. It was a turning point in making wind and solar cost competitive. The stimulus program invested $90 billion in these technologies, and renewable power generation doubled over the course of Obama's first term.[43] But Obama kept most of the stimulus's greatest accomplishments quiet. As a result, its public face became defined by the right. Fox News, Rush Limbaugh, and company hammered one of the few supported projects to go bust: thin-film solar cell manufacturer Solyndra, which defaulted on a $535 million loan guarantee from the Department of Energy in 2011. Solyndra was an outlier; within three years, the same Energy Department program—initiated under George W. Bush and plumped by the stimulus—was turning a $5 billion profit, bringing a higher rate of return than most venture capital funds. Congress, in fact, had at the start of the program set aside $10 billion to cover any losses, considered a natural part of funding innovative, precommercial research.[44]

The Department of Energy's Solyndra loss also pales in comparison to those that followed in the private sector. Theranos investors—including George Shultz and Trump education secretary Betsy DeVos—lost some $600 million, and SoftBank forfeited $6.4 billion in its pricey gamble on WeWork—the largest wipeout of shareholder value since Enron's collapse.[45] Chevron wrote down $11 billion worth of its shale investments in early 2019, in line with other oil majors' write-downs on unconventional fuels around the same time.[46] And over the course of just one week that same year, Chesapeake Energy investors lost $1 billion.

Innovation and breakthrough technologies will be key to tackling the climate crisis. And risk is a necessary part of that. But public sector investment is held to impossibly high standards—particularly when it shoulders the risk of cutting-edge projects that venture capitalists and angel investors are too sheepish to take on. As economist Mariana Mazzucato has written, private investors haven't done much to earn their reputation as risk takers. Corporations and venture capitalists often adopt conservative thinking and fall into "path dependency." They're reluctant to invest in important early stage research that won't necessarily turn a profit in the short run. This kind of research is inherently risky, and the vast majority of this kind of protean R&D (research and development) fails. For every internet—birthed in the Defense Department—there are well over a dozen Solyndras, but it's virtually impossible to have one without the other. Yet for all its patient investments, the government gets neither credit nor a cut when the successful innovations it helped spur on take off.

From occasional bailouts to annual subsidies, government intervention in the US economy is *already* rampant and expensive—and destructive. The question of how to pay for a Green New Deal or its constituent parts—from large-scale electrification to a federal job guarantee to Medicare for All—is asked constantly. The question of how much it'll cost us to maintain business as usual is hardly raised. That sum includes not just the billions of dollars worth of funds spent annually for the fossil fuel industry's benefit—propping up companies that wouldn't be profitable otherwise—but the cost of the climate crisis itself in lives and dollars.

Traditional climate economy models, as noted in Chapter 2, have tended to grossly understate the costs of climate impacts, projecting what we've been able to observe far out into the future and ignoring the compounding effects of climate change we haven't yet observed. A group of scientists and economists at the London School of Economics and Political Science, the Potsdam Institute for Climate Impact Research, and the Earth Institute at Columbia University set out to correct for that, noting the many results of rising temperatures that can't be neatly quantified into cost-benefit analyses—massive risks that tend to get omitted from climate models altogether.[47] We do, though, have some lowball estimates: Moody's Analytics pegs the cost of climate damages at $54 trillion by 2100 if the world caps warming to just 1.5 degrees Celsius and $69 trillion if the world warms by 2 degrees Celsius, warning that warming past that point could risk triggering tipping points that could lead to irreversible warming feedback loops.[48] "It's not a shock to the economy," Moody's Analytics chief economist Mark Zandi told the *Washington Post*. "It's more like a corrosive."

The question, Rick Perry rightly points out, isn't whether the government should pick winners and losers. It's who ends up on which side. There's no exact science for arriving at the valuation for a company like WeWork or Chesapeake Energy. Neither is there any easy way to put a price on the end of the world. Both are the product of assumptions about what matters in the economy and who holds the power to make that call. Left to the private sector, those calculations have tended to rank the planet—and most of the people on it—at the bottom. What the booms and busts of the last decade help prove is that there's no shortage of cash floating around the economy, ready to be unleashed to get off fossil fuels instead of propping up the companies that get rich off them. Will we trust the Jamie Dimons, Adam Neumanns, and Aubrey McClendons of the world to channel that toward the transition to a low-carbon society? Or will we make a plan to build one?

CHAPTER 7

PLANNING FOR A GOOD CRISIS

IN THE EARLY spring of 2020, a diminutive strand of RNA known as SARS-CoV-2 seemed to rip open half a century of economic orthodoxy. First detected around a wet market in Wuhan, China, the respiratory infection quickly spread out from that city of 11.1 million to other parts of Asia and then to Europe and the United States before moving on to the rest of the world. Reported cases in the US topped 100,000 just a few weeks after the World Health Organization declared a pandemic; fatalities there caught up to China's days later as the virus began to move south, and the richest country in the world would soon be home to the deadliest outbreak in the world. The about-face from Global North leaders talking about COVID-19 as a matter of foreign policy to, suddenly, pressing domestic crises seemed to happen overnight; their problem was now ours. On February 25, Trump's top economic adviser, Larry Kudlow, told CNBC that US containment of the coronavirus was "pretty close to airtight." The next day, Trump himself said his administration had "done a great job in keeping it down to a minimum."[1] By March 29, after weeks of rejecting warnings from public health officials and comparing the disease to the annual flu, Trump suggested in a press conference that 100,000 people dying of COVID-19 would be evidence of his administration having done a "very good job." By October, 200,000 people were dead. No one had a plan.

Unlike the climate crisis, wealthy countries were some of the first and worst hit by COVID-19. Much like it, though, casualties mapped neatly onto existing inequalities. In addition to the elderly, those who were on the losing end of the US health care system and who had endured decades of structural oppression suffered most, prone as they were to a host of preexisting conditions that tended to exacerbate the fierce respiratory illness. Within the first month of the outbreak in the US, Chicago's Black residents made up 72 percent of the disease's victims but just 30 percent of the city's population. Similarly outsized figures were reported in Louisiana, Michigan, and New Jersey; relatedly, a 2018 study conducted by EPA scientists found that Black communities nationwide face a 54 percent higher health burden from air pollution compared to the overall population.[2] About the worst preexisting condition a person could have in the face of COVID-19 was being poor and Black, which often meant living in a zip code long considered an expendable dumping ground for toxic waste.

The consequences of decades spent starving the public sphere—apportioning care out based on ability to pay, leaving critical planning decisions up to for-profit firms—were put on full display, with hundreds of thousands left dead in the wreckage. American hospitals were dangerously ill-equipped to handle an influx of cases, with a pool of health care professionals simply too small to cope and a persistent misallocation of ventilators and basic personal protective equipment, or PPE. Those left to treat the crisis took to social media with pleas for support, sometimes through crowdfunding campaigns. Freezer trucks pulled up to hospitals to be loaded up with the dead who wouldn't fit in morgues surging past capacity. Prisoners at Rikers Island in New York, who, like many other inmates around the country, reported being "left to die," were offered six dollars an hour and PPE to dig mass graves.[3]

To contain the spread of the disease, economies the world over ground to a halt. Local, state, and in some cases national governments ordered nonessential businesses to shut down and people to stay in their homes. Those deemed essential workers—health care professionals, grocery store cashiers and stockers, sanitation workers, Amazon warehouse employees, and more—put themselves at grave risk to prevent societal breakdown, often for little pay and with scant protective gear. "I have been coming in sick because I'm worried that I'll lose my job or just be punished if I call out," UPS package handler Angel Duarte told the *New York Times*. "I am 23, and I have no savings, and I have a 4-month-old son."[4] In some ways, those forced to choose between their health and a paycheck were lucky. Prior to the coronavirus crisis, 40 percent of US residents reported they wouldn't be able to handle a $400 emergency. Over the course of just two weeks in March and April, ten million people, many in low-wage sectors like retail and hospitality, filed for unemployment, erasing five years' worth of jobs gains and breaking new records for jumps in joblessness several times over.[5] Millions, in turn, lost access to their employer-based health care. Within a few days of the novel coronavirus taking hold on American shores, a recession—if not depression—looked imminent; in the second quarter of 2020, ending in July, US GDP shrank by 32.9 percent, a contraction four times greater than the worst quarter of the Great Depression. Those lucky enough to avoid or survive the virus may never recover from the economic fallout that followed it.

The pandemic wasn't unrelated to the climate crisis. Biologist and IPCC author Colin Carlson told me before the worst of it had settled over North America that it "would be difficult to make a case for climate involvement in this outbreak," although the "rate at which things like this happen is increasing because of climate change." Warming climes are more welcoming environments for the mosquitoes that have historically tended to spread around ailments like malaria and yellow fever closer to the equator. And ecosystem destruction—be it by warming weather or industrial development—is increasingly pushing species up against one another, where viral and bacterial loads can mutate and hop from creature to creature in events known as spillovers. Aside from pouring prodigious amounts of greenhouse gases into the air, factory farming and the habitat destruction involved in it create petri dishes for these sorts of transmissions to happen. For years, scientists had warned that a warming world was one more prone to pandemics. It's possible COVID-19 will be one of the first in a long line of outbreaks now endemic to our planet. What seems all but certain is that the rest of this century will be one filled with crises that, much like the coronavirus, will crash into the ones already festering. The historic heat waves, derecho, wildfires, and hurricanes that rolled across our disease-riddled country that summer were a kind of dress rehearsal.

Even the Trump administration's deadly, fumbling response showed the scale, if not the quality, of government action that was still possible.

Within weeks of former vice president Joe Biden having been declared the presumptive Democratic nominee, mainstream liberals—who had spent that primary season chiding that Bernie Sanders's moderately social democratic agenda was too divisive and unrealistic—called for nationalizing whole supply chains to produce sorely needed protective equipment and ventilators by order of the state. The Fed cut its interest rate targets down to virtually nothing and unveiled a dizzying spread of tools to keep the global economy from imploding outright, enlisting the gargantuan asset manager Blackrock to buy up corporate debt. Under pressure from social movements and tenant organizers, cities and states placed moratoriums on evictions and utility shutoffs. Trump even instructed the Department of Housing and Urban Development to place a stay on evictions in public housing. While New York cut local Medicaid funding, it also effectively brought its health care system under state control and coordination. An early fiscal stimulus measure passed through Congress doubled the figure that then president-elect Barack Obama's top economic adviser Larry Summers cautioned *against* proposing in his recovery package for the financial crisis of 2008, fearing it would be too big and controversial, even as Democrats controlled Congress and the White House and experts recommended something even bigger. In late March of 2020, Congress quickly passed a $2 trillion stimulus package. The Coronoavirus Aid, Relief, and Economic Security (CARES) Act included a $450 billion slush fund for Treasury Secretary Steve Mnuchin to hand out to whichever corporations he wanted and a generous tax break that, as it turned out, would mostly benefit millionaires.[6] The White House declared that the deficit simply didn't matter in a crisis, broadcasting its willingness to spend whatever it took to get the economy—by which Trump mostly meant the stock market—back on track. It was a job too important to be left entirely up to a clever set of market-based mechanisms or even for the private sector to tackle on its own.

Yet however much money Trump wanted to throw at the stock market, however much liquidity the Federal Reserve poured into the financial system, it wasn't enough to drain fluid-filled lungs or train up a new class of doctors, nurses, and medical technicians. With some anecdotal exceptions, it also hadn't compelled companies to manufacture sorely needed equipment out of some spirit of corporate social responsibility. Trump

eventually invoked the Korean War–era Defense Production Act to compel General Motors to make ventilators, but he used it sparingly. Bodies
kept piling up. In a line that might sound familiar by this point in the
book, right-wing economists Art Laffer and Stephen Moore complained
that the cure might be worse than the disease, as the *Wall Street Journal*'s
editorial board argued similarly that a government-enforced shutdown
of the economy or hasty, overreaching response measures could do more
harm than simply unleashing everyone to spread the virus among themselves. With casualties topping twenty-five thousand and a curve not
beginning to flatten, Trump—weary as ever about his prospects for reelection in November—suggested he might start to reopen the economy
in May. He appointed Laffer, Moore, and the Heritage Foundation to figure out how and cheered on Koch-funded protests against state shutdown
measures.[7] The right's preferred solution, in other words, was to sacrifice
as many lives as necessary at the altar of the market.

There were alternatives. Every other Asian and European country
that the virus spread to first had a flatter curve of mortality than the
United States, which by that April had the deadliest outbreak of any
country on earth. The Indian state of Kerala, for instance, had invested
in a strong public sphere through thirty years of communist governance and provided a humane alternative to Prime Minister Narendra
Modi's crisis authoritarianism. Primed by an ugly Nipah outbreak two
years earlier, state officials implemented a rigorous testing regime just
after the first case was diagnosed, while ensuring people's basic needs
were looked out for amid social-distancing measures.[8] Defying bleak
projections, the African Union helped roll out rapid testing, masks, and
ventilators across countries pursuing a range of strategies, helping cap
continent-wide death rates at numbers orders of magnitude below those
in the US alone.[9] New Zealand imposed strict shutdowns before the first
death and flattened its curve of cases and infections dramatically. South
Korea contained its early outbreak through the rapid deployment of
tried-and-true methods: fast testing, tracing, and isolation for suspected
cases.[10] After brutally suppressing news of the virus's initial outbreak—
suppression that could have bought precious time, maybe even prevented
a pandemic—China's spread of the disease was contained largely to one
province through strict, surveillance-aided shutdowns, social-distancing

enforcement, and an ability to rapidly bring on new infrastructure; in a matter of days, the Chinese Communist Party constructed two new hospitals with 2,600 beds.[11]

The US could do no such thing, or at least wouldn't. Conference centers and sports stadiums in the US were converted into makeshift hospitals as governments scrambled for more space. Meanwhile in Philadelphia, the recently shuttered Hahnemann University Hospital—home to 496 beds—was being held hostage by its new owner, investment banker Joel Freedman. He demanded the city pay him roughly $29,000 a night for use of the facility he was reportedly angling to sell off to luxury developers.[12]

Vultures trying to turn a profit in the crisis abounded. A shell company with the right to two Theranos testing patents—yes, *that* Theranos—sued the testmaker BioFire for alleged intellectual property theft before backing down under public pressure.[13] And the drugmaker Gilead temporarily restricted access to an experimental treatment called remdesivir amid overwhelming demand, leaving doctors and patients in a lurch. The company promised only that its treatments for COVID-19 would be "affordable" should it pass through clinical trials, not free, as a public health emergency killing hundreds a day might seem to demand; it was only after a public prodding by freshman representative Katie Porter that the CDC director committed to eventually making them free.[14] For its part, the Trump administration ensured that drug companies' intellectual property right to charge patients exorbitant fees would be protected as they developed vaccines and pandemic treatments with public money.[15] It engaged, too, in what might accurately be called crisis protectionism: trying to convince European drug companies to create treatments that would only be available to the US and seizing planes of medical equipment bound for other countries on the tarmac and rerouting them to American shores.[16]

The Trump administration wasn't about to let this crisis go to waste. While the public's and lawmakers' attention was trained on the pandemic, the EPA pushed to relax virtually all of its own enforcement authorities.[17] At the urging of the oil and gas industry, for example, it unveiled a new set of fuel efficiency standards giving car manufacturers a license to pollute that many of them didn't even want.[18] Another proposal looked to privatize oil-rich indigenous territory, and auctions began to lease out vast swaths of public lands for drilling as cheap as two dollars per acre.[19] The

Federal Aviation Administration proposed reviving commercial supersonic jets like the Concord, because why not. Fossil fuel companies got generous bailouts as the United States Postal Service drowned. The National Labor Relations Board greenlit a rule allowing a minority of workers to decertify a union election, and Texas effectively banned abortions, using the virus as a half-baked excuse.[20] With the prospect of primaries and even November elections looking shaky, the GOP railed against expanding vote by mail nationally, in a not so subtle indication that they just didn't want everyone who could vote to do so. At a March press conference, Trump invited up the CEOs of CVS, Walgreens, Walmart, and other health-care-adjacent companies to advertise corporate America's commitment to the cause outside the White House, shaking each of their hands as he introduced them. America's response to COVID-19, he seemed to be suggesting, would be the greatest public-private partnership the nation has ever seen.

Altogether, the Trump administration's response to the coronavirus crisis was, well, pretty Trumpian: a big, shambolic, murderous display of administrative incompetence that heaped praise on the private sector and scorn at scientific expertise and international institutions. He left plenty of room for whichever White House friends in industry saw an opportunity, and racism coursed through every piece of it. Besides COVID-19's disproportionate blow to communities of color, the White House insisted on calling it the Chinese or Wuhan virus—a point Trump was so emphatic about that administrative representatives torpedoed cooperation with G7 countries when they refused to adopt the same nomenclature.[21] Much as he's helped further neoliberal goals, though, Trump himself was never much of a free-market ideologue, the results of which showed in his pandemic response. Like most of the supposed market fundamentalists, he was happy to call on government when the moment suited him; in a pandemic, there was no avoiding it. "Just as there are no atheists in foxholes," Peter Nicholas wrote contemporaneously for the *Atlantic*, "in a national emergency, there's no truly laissez-faire government."[22]

We may be living in the America neoliberalism built, but the architects are dead, the foundations are sinking, and the dwellers are pissed off and dying of whatever toxic asbestos concoction was injected into the walls decades ago. What had always been a contradictory bundle of

PR, sociopathy, and earnest political philosophy has given way to people rattling off ideas no one much cares to defend anymore. That was true for senior Democrats, too; Joe Biden redoubled his opposition to Medicare for All, waited for months to go on an offensive against Trump over his handling of the virus, and reiterated that he would follow the science. Under pressure, the party's congressional leadership pushed to make Republican stimulus proposals more expansive and egalitarian by degree, winning key protections for airline workers and historic unemployment insurance extensions, for example. Although Pelosi and Schumer had promised ensuing rounds of stimulus would correct for the mistakes of the CARES Act, all too predictable Republican obstruction made that elusive; a second, $900 billion package passed at the end of December with modest improvements. Like Biden, the Democrats' congressional leadership offered no clear vision of its own for what a country that could deal successfully with a pandemic might look like. Even the editorial board of the *Financial Times*—that reliable mouthpiece of capital—mustered more fiery ambition, forced to admit drastic change was necessary:

> Radical reforms—reversing the prevailing policy direction of the last four decades—will need to be put on the table. Governments will have to accept a more active role in the economy. They must see public services as investments rather than liabilities, and look for ways to make labour markets less insecure. Redistribution will again be on the agenda; the privileges of the elderly and wealthy in question. Policies until recently considered eccentric, such as basic income and wealth taxes, will have to be in the mix.
>
> The taboo-breaking measures governments are taking to sustain businesses and incomes during the lockdown are rightly compared to the sort of wartime economy western countries have not experienced for seven decades. The analogy goes still further.
>
> The leaders who won the war did not wait for victory to plan for what would follow. Franklin D Roosevelt and Winston Churchill issued the Atlantic Charter, setting the course for the United Nations, in 1941. The UK published the Beveridge Report, its commitment to a universal welfare state, in 1942. In 1944, the Bretton Woods conference forged the postwar financial architecture. That same kind of foresight is

needed today. Beyond the public health war, true leaders will mobilise now to win the peace.[23]

The coronavirus didn't wash away the hallmarks of neoliberalism so much as make them look more grotesque than ever when hauled out into the sun. Still, the sheer size and scope of the state seemed to expand beyond liberals' wildest dreams, as the Federal Reserve took up new powers and government spending crept up to 13.2 percent of GDP in the US, a figure that would have been virtually unimaginable just a few months prior.[24]

COVID-19 and its fallout seemed to bring a well-loved, almost cliché quote among the left to life, from Antonio Gramsci: "The crisis consists precisely in the fact that the old is dying and the new cannot be born; in this interregnum a great variety of morbid symptoms appear."

WHEN THE GREEN New Deal reemerged into headlines in November 2018, unemployment in the US sat at 3.7 percent. Even supporters of the program voiced warranted skepticism about its viability. Sure, the climate crisis is important, but the government hardly ever spends huge sums on big social programs anymore—least of all when the economy appears to be doing relatively well, by conventional accounting. The window for a massive stimulus opens when there's a recession, and we weren't in one. Times have obviously changed since then, although the path to an ambitious climate response remains far from certain.

Joe Biden will be the president by the time this book is released, having won decisively in an election that should have been by all accounts—given the blood on Trump's hands—a blowout. Instead, Trump collected ten million more votes than in 2016, Democrats lost seats where they were expected to gain them. After run-off elections in Georgia, the party managed to win back control of the Senate, held by the narrowest of margins. Biden was pushed by movements during his campaign to adopt a climate platform more ambitious than the one he ran on in the primary. But his administration will be hard-pressed to get any of that through Congress, left mainly to find creative uses for the executive branch—that is, if he decides to treat his $2 trillion commitment to a green-tinted stimulus as anything more than lip service to progressives and isn't completely shut down by the 6–3 right-wing majority on the Supreme Court. Democrats' underwhelming performance in 2020, moreover, doesn't bode well for winning back more power in upcoming elections. If anything, there's much more to be lost.

Understanding what the road toward anything like a Green New Deal looks like now, when all manner of crises are boiling over, means taking its namesake seriously. The New Deal—in all its deep flaws and contradictions—was more than just a big spending package that helped to drag the US out of the Great Depression. It reimagined what the US government could do, what it was for, and who it served. To effect such a drastic sea change in this country's politics, it did something climate policy in the US has historically struggled with: it made millions of people's lives demonstrably better than they would have been otherwise. That, in turn, helped solved the other big dilemma facing a Green New Deal and just about any major progressive legislative priority: the tangible mark New Deal programs left in nearly every county in the US helped to build a sturdy Democratic electoral coalition that could bat off challenges from the right and endure for decades. Even as many of its gains have been clawed back by a revanchist right, hallmarks like Social Security remain so broadly popular that even the GOP has stopped trying to go after them. A Green New Deal should aim even higher.

Like today, the bar for successful leadership some ninety years ago was pretty low. A very rich man with even richer friends, Herbert Hoover was mostly blind to the effects of the Great Depression on working people and for a while denied there was any unemployment problem at all. Before becoming president, Hoover had made his fortune in mining, transforming himself from poor Quaker boy to lowly engineer to magnate. He gave away large chunks of his fortune to charity and fancied himself both a man of the people and a magnanimous captain of industry. Hoover assumed his fellow businessmen were philanthropic types, too. As he would find out in the waning days of his administration, America's businessmen might fund libraries and museums, but they had neither the will nor the ability to fix the problem they had helped create. The Depression defined and destroyed his administration and nearly took down the whole concept of liberal democracy with it.

In May 1930—with an unemployment rate screeching past 20 percent—Hoover assured the US Chamber of Commerce that "I am convinced we have now passed the worst. . . . The depression is over."[25] That
December, his State of the Union address promised that "the fundamental strength of the Nation's economic life is unimpaired," blaming the
Depression on "outside forces" and urging against government action.[26]

"Economic depression," he said then, "can not be cured by legislative action or executive pronouncement. Economic wounds must be
healed by the action of the cells of the economic body—the producers
and consumers themselves." Ideologically opposed to the idea of state
intervention in business, Hoover that year had convened a compromise:
the Emergency Committee for Employment, to gently nudge the private sector into putting 2.5 million people back to work through local
citizens' relief committees, comprised mostly of local officials and business executives. After several months it hadn't worked; members of the
committee could point to no evidence that it had created any jobs at all.
Committee head Arthur Woods petitioned the White House to create
a public works program with federal funding instead. Hoover refused,
and the committee withered away shortly afterward as unemployment
continued to skyrocket. Its replacement was an advertising campaign
coaxing individuals to give to charity. Announcing the plan via radio
address, Hoover bellowed that "no governmental action, no economic
doctrine, no economic plan or project can replace that God-imposed responsibility of the individual man and woman to their neighbors."[27] Just
before the 1932 election, Hoover warned that a New Deal—what Franklin Roosevelt was campaigning on—would "destroy the very foundations
of our American system" through the "tyranny of government expanded
into business activities."[28]

Hoover had a relatively successful career up until the crash, with a
well-regarded run as secretary of commerce that included his successful management of the Great Mississippi Flood of 1927 by marshalling
public and private resources toward recovery. That Hoover is widely remembered as a loser is thanks mostly to who and what he lost to. Roosevelt's blowout victory in the 1932 election—where he won forty-two
of forty-eight states—ushered in a profound change in American life.
With it came fourteen years of uninterrupted, one-party control over the

White House and both chambers of Congress, secured not by the kinds of authoritarianism that were common through that era, and which well-heeled American elites mused might be needed, but by democratically elected Democratic majorities. Accounting for two brief interruptions just after the end of World War II, Democratic control would extend on for a total of forty-four years in the Senate and fifty-eight years in the House.

Until he left office, Hoover refused to budge on his overall approach, as he would through the rest of his life. He pleaded with Roosevelt to denounce the agenda he had just run on, which included such things as widespread unemployment insurance, a job guarantee for the unemployed, tackling soil erosion, and putting private electric utilities into public hands. As the financial system collapsed, the unemployment rate floated around 25 percent, and fascism was on the march in Europe, Hoover did nothing. Federal Reserve chairman Eugene Meyer begged him to reconsider and declare the bank holiday he knew that Roosevelt was already planning as president-elect. "You are the only one with the power to act. We are fiddling while Rome burns," he told Hoover. The president was unmoored: "I have been fiddled at enough and I can do some fiddling myself."

Hours after taking the oath of office, Roosevelt and his top advisers embarked on a marathon session to save and restore faith in a banking system on the verge of collapse. Within thirty-six hours, the administration declared a nationwide bank holiday. Before it ended, on the afternoon of March 9, Roosevelt spent two hours presenting one of the earliest New Deal programs to his closest advisers. It would be a jobs program, he explained, that would "take a vast army of these unemployed out to healthful surroundings," doing the "simple work" of forestry, soil conservation, and food control. By that evening, the program's final report explains, the proposal was drafted "into legal form" and placed on the president's desk. At ten, he convened with congressional leaders who brought it to Congress on March 21. It was signed into law on March 31, and the first recruits of the Civilian Conservation Corps (CCC) were taking physicals by April 7 before being bussed from their homes in New York City to Westchester County, freshly issued clothes in hand.[29]

By July, the program had established 1,300 camps for its 275,000 en-rollees. Between 1933 and its end in 1942, the CCC's workers—average age 18.5, serving between six months and two years—built 125,000 miles of road, 46,854 bridges, and more than 300,000 dams; they strung 89,000 miles of telephone wire and planted three billion trees.[30] Among the most expansive and maligned of New Deal programs, the Works Prog-ress Administration—derided as full of boondoggles and government waste—built 650,000 miles of roads, 78,000 bridges, and 125,000 civilian and military buildings; WPA workers served 900 million hot lunches to schoolchildren, ran 1,500 nursery schools, and put on 225,000 con-certs. They produced 475,000 works of art and wrote at least 276 full-length books. From 1932 to 1939, the size of the federal civil service grew from 572,000 to 920,000.[31] The WPA's predecessor, the Civil Works Administration, created 4.2 million federal jobs over the course of a single winter. Much of that work was in construction, but the program also employed 50,000 teachers so that rural schools could remain open, rewilded the Kodiak Islands with snowshoe rabbits, and excavated pre-historic mounds, the results of which ended up in the Smithsonian.[32] In the first year of its operation, 1939, the Civil Aeronautics Board built three hundred airports. They did it all without so much as a cell phone or computer.

Like the original, a Green New Deal won't—if it's successful—be a discrete set of policies so much as an era and style of governance. It will be the basis of a new social contract that sets novel terms for the relation-ship between the public and private sector and what it is that a govern-ment owes its people. Likewise, the New Deal was designed—learning as it went—to solve a problem the United States had no blueprint for: creating a welfare state capable of supporting millions of people essen-tially from scratch and with a wary eye toward those countries abroad that were handling a catastrophic economic meltdown in very different, far crueler ways. The New Deal might be best described by a spirit of what Roosevelt referred to as "bold, persistent experimentation": flawed, con-tradictory, ever-evolving, and very, almost impossibly big. "It is common sense," he said in the same speech, "to take a method and try it: if it fails, admit it frankly and try another. But above all, try something." More than giving bureaucrats carte blanche to move fast and break things, the New Deal crafted a container in which innovation and experimentation could take place, providing a combination of ample public funds and rigorous standards, all to be overseen by a set of dogged administrators. As Paul Krugman would write some seventy-five years later, the "New Deal made almost a fetish out of policing its own programs against potential corruption," well aware of the hostility its new order would face from those invested in continuing on with business as usual.[33]

The New Deal's spending programs depended on its public relations efforts, both making a show of its distaste for graft and corruption and showcasing the benefits of the New Deal in every county in the country. It also paid plenty of attention to optics. Posters created through the Federal Arts Program advertised the successes of New Deal programs in striking detail, and public works programs were designed with presentation in mind. "At Hoover Dam," architecture critic Frederick Gutheim would write of the Tennessee Valley Authority's flagship project, "one was impressed by the sheer size. But at a TVA dam one was reminded of humanistic values, of power serving man . . . of the virtues of public ownership of hydroelectric power."[34] In constructing observation decks and breathtaking approach roads to New Deal dams, Roland Wank—the TVA's socialist, Hungarian émigré lead architect—is remembered as having seen to it that his projects "were approached as one would the Acropolis."[35]

Compare that boldness to the Obama administration's response to the Great Recession eighty years later, seemingly designed to be as uncharismatic as possible. Obama had wanted to use the recovery, in part, to transform and decarbonize the electric sector by building out an array of transmission lines that would bring the country's grid fully into the twenty-first century. But like health care, electricity—his top economic adviser Larry Summers counseled—was primarily the domain of the private sector.[36] When it came to transmission lines, he said, "the government's job is to remove regulatory obstacles." Any reforms would have to wait for a legislative push in the form of a cap-and-trade bill. The goal now, according to Summers, was to get people spending, preserve the financial system, and not rattle the markets by placing too many limits on corporations or by causing too big an increase in the federal deficit. Most of Obama's top economic advisers harbored a basic weariness about government spending too much or mucking around in the private sector, even if they conceded the basic Keynesian point that governments should spend their way out of recession. Summers famously advocated that spending to be "timely, targeted, and temporary," so as to not overextend the welcome or footprint of big government to Americans who allegedly hated it. And transformative as it might have been, the 10 percent of the ARRA devoted to clean energy investments took the form of loan guarantees for companies and public sector research. The upsides were nearly incomprehensible to the general public. By 2010, just 6 percent of Americans believed the stimulus had created *any* jobs.[37]

And yet for all of the Democrats' attempts to hide their spending footprint, the GOP attacked the ARRA as full of boondoggles anyway. Obama responded by pivoting back to belt-tightening austerity in 2010 with a stark State of the Union address. "Like any cash-strapped family," he pledged, "we will work within a budget to invest in what we need and sacrifice what we don't. And if I have to enforce this discipline by veto, I will."[38] A New New Deal the Obama stimulus was not.

Campaigning for reelection in 1936, FDR was, by contrast, unapologetic. He told a crowd at Madison Square Garden that year about his administration's struggles against "the old enemies of peace—business and financial monopoly, speculation, reckless banking, class antagonism, sectionalism, war profiteering." These forces, he said, "had begun to consider the Government of the United States as a mere appendage to their own affairs. We know now that Government by organized money is just as dangerous as Government by organized mob."

"Never before in all our history," he bragged, "have these forces been so united against one candidate as they stand today. They are unanimous in their hate for me—and I welcome their hatred." In stacking top cabinet posts with Wall Street allies like Summers and Timothy Geithner, the Obama administration chose instead to welcome the "enemies of peace" into its ranks. While they controlled Congress and the White House, Democrats squandered not just their opportunity to pass climate policy in Obama's first term, but also to maintain and build on the enthusiasm that had elected him into a durable coalition. Obama's too-small stimulus and subsequent turn toward austerity helped cost the party not just the House, Senate, and White House over the next eight years, but also statehouses around the country, where GOP majorities have proceeded to gerrymander away democracy and ossify minority rule.

Much as voters have longed for some return to normalcy through the Trump years, resurrecting the politics of the Obama era threatens to conjure worse monsters down the line. During his campaign, Joe Biden obliquely promised a "Rooseveltian" presidency.[39] If he intends to follow through on that, or even just secure a future for the party in Congress, 2021 will need to look more like 1933 than 2009. By extension, Democrats should be pushing for the midterm elections in 2022 to resemble those of 1934—not 2010.

THERE ARE MORE aspects of 1934 than not, of course, that should be avoided. The New Deal's achievements aren't easily parsed out from its uglier elements. As historian Ira Katznelson has argued, New Deal–era policies "both rescued and distorted American democracy."[40] Core to what kept the United States from going the way of Europe was a prodigious expansion of the military budget and, perhaps most damningly, a compromise with segregationist Southern Democrats. Many of the New Deal programs entrenched Jim Crow statutes, which had the effect of codifying segregation into American life in ways we still live with today. Southerners aren't solely to blame, either. It was northern technocrats—namely, Treasury Secretary Henry Morgenthau Jr.—who led the push to exclude largely nonwhite and female agricultural and domestic workers from the landmark Social Security Act of 1935, arguing it would make the program easier to administer.[41] The New Deal was the United States' non-communist alternative to fascism, not a program for liberation. Greater than the sum of its parts, it created a new America and a new conception of what a government was supposed to do. As Katznelson writes, the result was a compromise, an enduring two-sided state "characterized by democratic advantages yet marked by antidemocratic pathologies."[42]

The New Deal's accommodation of Southern Democrats and its technocratic embrace of Jim Crow norms shouldn't, though, obscure either its transformative effects—including for many people of color—or the threat it posed to the powers that be in corporate America. Thanks to the work of Roosevelt's "Black Cabinet" (the Federal Council of Negro Affairs) and advisers like Robert Weaver and Mary McLeod Bethune, Black employment at federal agencies tripled throughout the course of the New Deal. Universal programs and some targeted specifically at historically oppressed peoples, including African Americans and Native Americans, improved infrastructure, education, and living standards broadly, even as deep inequities remained intact. The foundations of the modern right were crafted in retaliation both to the New Deal and the war mobilization that followed it, and for good reason. The New Deal empowered organized labor to a degree previously unheard of, reined in corporate greed, and created an expansive network of public services against the wishes of industry. It popularized the idea that the government could play a positive role in the lives of its constituents, providing not just needs like food and electricity but wants as well, from concert halls to affordable theater performances to rose gardens. To fight the Axis Powers during World War II, the federal government effectively created a centrally planned economy that subsumed profits to the public interest. Federal agencies controlled prices and wages and placed strict rules on corporations, built whole supply chains practically from scratch, and rationed goods ranging from oil to sugar. Companies that failed to comply with production demands faced a federal takeover.

Economist Rexford Tugwell was a member of FDR's Brain Trust, a group of academics who helped develop policy recommendations leading to the New Deal. For Tugwell, the New Deal stood between the US and "those militant nationalist movements with which the world has had too much experience lately," as he told a crowd of reporters in 1934. "The New Deal is not something which can establish itself in the mind of a dictator or a small governing group. That was the fatal theory of the system from which we are turning away. Its base has to be as broad as the economy which has to be brought under control and as deep as the minds and hearts of the people whom it affects."[43] Despite his reputation as "Rex the Red," Tugwell—like almost all of the New Dealers in government—can't neatly be described as an enemy of private ownership and corporations. In *Time* the same year, he called outright government control of industry "expensive and repressive."[44]

As Tugwell saw it, the New Deal's throughline wasn't socialism or even big government, but a thoroughly democratic political economy.

"The essence of the New Deal," he noted, "is that it recognizes and gives expression to the people whose wants are going unsatisfied because of the failure of the industrial and political institutions which they have established in the hope of satisfying those wants," he said. What was demanded by those majorities of voters who backed the New Deal, Tugwell argued,

> is the making over of the institutions controlled by and operated for the benefit of the few, so that regardless of their control they shall be operated for the benefit of the many. In all this there is no thought or need to change the individual so that he may conform to some pattern or be fitted to some industrial scheme about to be created. The reverse is true: that the industrial scheme shall be made over to fit the individual and supply his wants. What the Old Order describes as "rugged individualism" meant the regimentation of the many for the benefit of the few. The social mission of the New Deal has a somewhat higher standard of individualism—it believes in freeing the many from the regimentation of the few.

Roosevelt would make the case more memorably at a low point for the New Deal and a high one for fascism abroad. In 1941, he promised Four Freedoms: of speech and worship, and from want and fear. The great irony of seeing the New Deal as a democratic bulwark against fascism is that the US couldn't earnestly, at the time, be described as a democracy. Black southerners lived under a regime of racist terror that had served as inspiration for the Nazis themselves, wherein hundreds of thousands of people were effectively barred from the democratic process and forced to live under minority rule. By leaving that system intact, and in some ways bolstering it, the New Deal would ensure that its proponents' most ambitious, egalitarian dreams could never be realized. As is discussed further in chapters to come, any Green New Deal that leaves white supremacy in place is similarly doomed.

What is remembered as the history of a great man (or several great men) in the White House was in reality the product of bitter fights within a cabinet that pitted deficit hawks like Henry Morgenthau Jr. against economic planners sympathetic to left ideas like Rex Tugwell, and on the floor of Congress. Unions vied for a greater say over how the economy was run and WPA workers railed against the administrators who ran it—all as conservatives attempted to scale it back. It was only after a new class of bullish New Dealers was elected in the 1934 midterms that a more ambitious Second New Deal—including Social Security and the National Labor Relations Act—could take shape.

Battles over the New Deal were often waged in the workplace; strikes—as Raj Patel and Jim Goodman have noted—increased year-on-year from 1930 to 1937 as workers joined together into unions that enjoyed new and unprecedented legal protections. There were 1,856 work stoppages in 1934, the most, at that point, since World War I. The biggest and most disruptive among them were led by avowed communists and socialists like longshoreman Harry Bridges, who, like much of the era's left, was a staunch critic of New Deal policies as he pushed them to go further. The three-day general strike he led in San Francisco led to the unionization of ports up and down the West Coast. The Wagner Act, which guaranteed workers' right to organize, was passed, and the National Labor Relations Board was formed the next year.[45] There were more than 4,500 strikes in 1937, in part, as a reaction to Roosevelt's widely unpopular attempt to balance the federal budget.[46]

By expanding the ranks of organized labor, the New Deal helped in turn to build institutions that would protect and build on its gains for decades to come. Crucial to that were African American voters, whose support many Democrats now take for granted. In 1932, most Black Americans—those that were able to vote, that is—voted Republican, which many still considered the party of Lincoln. By 1936, they abandoned the party en masse to vote for FDR, with many having moved North to escape Jim Crow regimes and pursue work outside the South, where life and New Deal jobs programs alike remained segregated. They joined unions in the process. By 1938, the majority of African Americans registered to vote were registered Democrats.[47] As much as any individual program, the New Deal Coalition—of union members, African Americans, farmers, urban voters, and more—would be among the New Deal's most important, enduring contributions to American politics.

That many of the New Deal's relief programs had been designed as emergency relief meant, however, that *every* annual budget represented a new fight for funding, especially by programs under more political scrutiny like the WPA. The New Deal as a whole wouldn't have been possible without Roosevelt's election in 1932. But it had to be won continuously over a troubled decade, through battles behind closed doors in the White House, in the courts, in the counties it touched, on the floor of Congress, and on shop floors and picket lines. A Green New Deal won't be any different. And like the original, it should work to build a coalition that will keep fighting to protect its wins decades down the line.

DAYS AFTER ROOSEVELT'S inauguration in 1933, three and a half years into the crash, and in the midst of the bank holiday, Walter Lippmann wrote, "Every crisis breaks a deadlock and sets events in motion. It is either a disaster or an opportunity. A bad crisis is one in which no one has the power to make good use of the opportunity and therefore it ends in disaster. A good crisis is one in which the power and the will to seize the opportunity are in being. Out of such a crisis come solutions."[48]

The Depression, Lippmann determined in that column, was a good crisis. It's not clear whether the same can be said of either the coronavirus or climate crisis. Not long after shutdown measures were announced in March 2020, UK-based economist Christine Berry wrote a word of caution to leftists eager about the right's embrace of expansive spending policies, in some cases resembling what the Labour Party had pushed under the left-wing leadership of Jeremy Corbyn. If the enemy had changed, the opposition now risked shadowboxing a threat that had changed form. Progressives, Berry warned, were "no longer competing with small-state neoliberalism. Across the world, it is competing with a new breed of right-wing nationalism—sometimes coupled with neoliberalism, to be sure, but sometimes quite happy to countenance state intervention if it helps cement their electoral coalitions. If this crisis may be creating some of the conditions for socialism, it is also accelerating many of the conditions for fascism. . . . Fundamentally, we must remember that *we are not in charge* of the course of events."[49]

Expansions of state power and the scope of the public sphere, Berry was arguing, aren't an unalloyed good. And not all turns away from neoliberal propaganda are created equal. Who holds the keys in these moments means everything. There were plenty of bold ideas put forward in the months after the coronavirus crisis for how to rebuild a better economy after the pandemic was under control. Those plans matter, but the best of them is worthless without the power to put them into action. Empowered by legislation like the Wagner Act, strikes and union power both escalated throughout the 1930s, pushing the New Deal to go further in expanding the safety net and challenging big business. In the decades afterward, corporations and the right became more organized and managed to break the unions and the institutional power they held both among ordinary people and in the White House. The right didn't have the public on its side, but had plenty of money to make up for it. Consequently, the vision it pushed was expressly antidemocratic, channeling not just as much wealth but as much control over society as possible into as few hands as possible. The neoliberal settlement between public and private that crystallized in the 1980s—and which we're still living under today—was the result of that changed balance of class forces.

No one could rightfully accuse Donald Trump or the Republican Party that has congealed around him of adhering to any kind of strict neoliberal orthodoxy; today's actually existing neoliberals are mostly skilled European technocrats. But the fact that he picked up a slightly different set of tools than his predecessors doesn't mean he wasn't serving the same interests. Spending prolific amounts of public money to prop up the economy's worst actors isn't something to be celebrated; state support has always flowed freely to the fossil fuel industry, after all. As we'll see in Chapter 9, that support only grew in the wake of the pandemic. Big government isn't necessarily good government so long as democracy is muzzled.

There's a real risk, too, that whatever the neoliberalism of the last few decades morphs into will be much worse for working people. Economist Laurie MacFarlane, in openDemocracy, a UK-based political website, argued that what might well emerge in the place of that would be an authoritarian capitalism, comfortable with large amounts of state ownership and bolstered by a fearsome surveillance state that constrains both wages and privacy. As even the fossil fuel industry pays lip service to climate action, any emergent authoritarian capitalism could easily be tinted green: think new resource wars between state-backed enterprises over control of the technology minerals needed to power an ever-expanding fleet of electric vehicles or the vast amounts of data generated from energy efficient smart homes being traded at a profit.

It's worth being specific, then, about the *kind* of public sphere a Green New Deal is looking to build and expand, ensuring it's designed in service of a fairer and more deeply democratic society. There's no path to keeping emissions below 2 degrees that doesn't run through sustained democratic—and in the case of the US, Democratic—majorities, who will need to fight for and defend rapid decarbonization against some of the most powerful industries the world has ever known. Building and keeping those democratic majorities—and giving their members a meaningful stake in and say over the energy transition—will be as key to successful climate policies as the details of any clean tech manufacturing loan guarantee or renewable portfolio standard.

There's only so far that lofty messaging about the promise a Green New Deal can go if no one believes it can deliver. The New Deal built its coalition by actively improving lives and legally empowering democratic institutions, namely unions. The trouble now is that, for good reason, that coalition has less and less reason to believe the Democratic Party will do either. As the party has shrunk from its belief in good government, it's taken its historic base for granted, never mind trying to expand it beyond a small slice of mostly white suburban swing voters. Consultants have poll-tested Democratic messaging into oblivion and become wholly reactive to the GOP while neglecting even basic questions of institution building, abandoning both workhorse state parties and the flashier person-to-person, hope-and-change-inflected electoral organizing that first brought Obama to the White House.

For the party to win in 2022, it seems, and to keep winning, the burden of proof now rests on the Biden administration to show that Democrats can govern for the many and not the few. With an uphill battle in the Senate, it's hard to imagine that happening absent pressure from within and without the halls of power. Deep organizing outside Washington must be done to start cohering what might tentatively be called a Green New Deal coalition for a multiracial democracy that builds on the one created by its namesake. That coalition might include essential workers, from Amazon warehouse workers to delivery drivers in precarious, poorly paid jobs bound up in the on-demand fossil fuel economy;

Latino voters, who were critical in flipping Arizona blue in 2020 after organizing against brutal immigration crackdowns and within unions like UNITE HERE; care workers, including educators and nurses in trusted, low-carbon professions, already unionized, who have struck to defend the public sphere against privatization and austerity in recent years; clean energy workers, who were badly hit by the pandemic recession but stand to grow rapidly and could swell unions' ranks and electoral might; and Gen Z and millennials, now the largest and most diverse age bloc in the country, acutely worried about their economic and ecological futures. It's hard to imagine any of the above cohering around the same politics the Democratic Party has been projecting.

The basic goals that mattered to Tugwell and the most ambitious New Dealers are what should matter to Green New Dealers: expanding democracy and tearing down barriers to collective freedom. Among other things, a Green New Deal should ensure our freedom to breathe clear and unpolluted air, to find a new home when ours floods or catches fire, to experience joy and contentment, and to live on a habitable planet. Neoliberalism promised freedom too, if only for the 1 percent: the freedom to break up unions and pollute unimpeded, from regulations, and, perhaps most importantly, from democratic oversight. As Stephen Moore—charged by Trump to reopen the economy after COVID-19 shutdowns—said in an interview, "Capitalism is a lot more important than democracy. I'm not even a big believer in democracy." The right's fear of Red Vienna in the 1920s extended to a New Deal that expanded the influence of labor unions and working people in American politics, to the Black freedom movements that James Buchanan and Milton Friedman opposed, and to calls for a democratic world order in the 1970s that the West suppressed. If certain neoliberal nostrums have been thrown out in whatever new age of right-wing politics and governance we're entering post-COVID-19 and mid–climate crisis, white supremacy and a contempt for democratic control at each level of governance will be steady themes among reactionaries.

For decades, the right has maintained a creative and strategic edge. Its leading lights have proposed a slate of bold new ideas about how to design states to suit private interests. But it has had a destructive bent, too, and worked diligently to take certain allegedly dangerous ideas for how to build a better world off the table entirely. Business conservatives' broader victory—and hindrance to effective climate policy—was to take decision-making power over the economy out of democratic control, leaving higher-order economic planning up to markets that the state was tasked with protecting. Since before even Hoover's presidency, the bête noire of these antidemocratic forces, predictably, has been public ownership. As ever, the fight over it has been a fight about power.

CHAPTER 8
POWER TO THE PEOPLE

HOW MANY PEOPLE has PG&E, Northern California's monopoly electric utility, killed? And how many people got rich off their deaths?

Thanks to a measure known as inverse condemnation, the company was responsible for 1,500 fires between 2014 and 2017. It plead guilty to eighty-four counts of involuntary manslaughter for those killed in the 2018 Camp Fire it helped spark, which torched the California hills of Butte County and burnt whole towns—most famously, Paradise—to the ground. Its old, poorly maintained high-voltage transmission lines and the neglected brush surrounding them proved a deadly combination. Facing at least $30 billion of legal liability from the state and lawsuits from wildfire victims, they filed for Chapter 11 bankruptcy in 2019—an event the *Wall Street Journal* called the "first major corporate casualty of climate change." It reemerged after seventeen months with access to a $21 billion pool of state funds. As part of the settlement, PG&E agreed to pay out $13.5 million to seventy thousand wildfire victims—half of that in nonvoting PG&E stock, the value of which had plummeted by 80 percent since the Camp Fire.[1] PG&E had killed before, albeit without such a clear verdict: the 2010 explosion of a PG&E pipeline claimed eight lives; fires sparked by its equipment in 2015 and 2017 killed twenty-four people altogether.[2] As I write this, it's under investigation for sparking a fire in 2020 that killed four.[3] In all likelihood, PG&E will kill again.

Assessing just how much responsibility PG&E bears for those deaths depends on how far back you want to look. There's no easy explanation for what causes wildfires, and inverse condemnation is a blunt diagnostic tool. The utility, though, bore plenty of blame. Its top brass poured millions into salary bumps for top executives and billions into payouts for the utility's wealthy shareholders, all the while knowing its equipment was dangerously outdated. As federal judge William Alsup put it in a 2018 hearing, "PG&E pumped $4.5 billion in dividends and let the tree budget wither." During the 2017–2018 state legislative session, it also spent $11.8 million on lobbying—much of that to shrink down what it owes and scrap inverse condemnation off the books. As journalist Lee Fang pointed out in The Intercept, the company shelled out $2.1 million wooing politicians in 2019 and spent millions more over the years on consultants to clean up its public image. They've gotten a decent return on investment. California State Assembly Bill 1054, rapidly signed into law as an emergency measure in the summer of 2019, limits the amount of money the state's utilities can be made to pay out for wildfire damages, allocating $21 billion of liquidity to help companies mitigate those costs—much of that to be provided by ratepayers.

Yet understanding what's going on in California requires a wider view. The state's housing crisis pushed low- and middle-income residents out into more affordable dwellings in its fire-prone wildland-urban interface, and wealthy homeowners and developers have fought back against land-use planning that could have restricted building in places where fires have been a regular occurrence for centuries. Firefighting tactics that look to eliminate all fires rather than control those that happen naturally have made devastating blazes more likely, failing to burn off brush that otherwise builds up into combustible piles. Through droughts, global heating dries that kindling out, and warmer wind gusts carry sparks toward it. Institutionally, there's plenty of blame to go around: the California Public Utility Commission (CPUC), the fossil fuel producers driving climate change, the US Forest Service, even the tech companies that have driven up rent and property costs in the Bay Area, and, zooming out further still, Western society's relationship to nature.

Californians more broadly have rightly questioned whether private utilities should exist at all and whether a right as basic as electricity should be left to the mercy of a for-profit firm. In a poll conducted by Change Research, 62 percent of state residents said it's a bad idea to have utilities' shares traded on Wall Street.[4] Seventy-seven percent agreed that it's "always a bad idea to have investor owned public utilities. They're more concerned about shareholders' returns than creating a safe infrastructure for the public." Another poll found that less than one in eight respondents supported allowing PG&E to fix its own problems and maintain its existing structure.[5] Facing backlash and few viable alternatives, California governor Gavin Newsom in 2019 seemed open to the possibility of a state takeover. Campaigners throughout California had, by the time that poll was taken, been speaking out against the bailouts for well over a year and called to bring the utility under public ownership. Politicians joined them, putting forward their own plans to take PG&E into Californians' hands.

"How are we benefiting from having private ownership? I see no benefit," Representative Ro Khanna, whose district lies within PG&E service territory, told me in 2019, just after calling for public ownership. "We have public schools. We have firehouses. We have public police stations. There's no reason we shouldn't have public electricity."

As Khanna and other public ownership advocates readily admit, it's no silver bullet for what ails PG&E or the electricity sector as a whole, now in dire need of rapid decarbonization. The energy sector is responsible for approximately two-thirds of global carbon dioxide emissions, and transportation and electricity together now account for about half of US emissions. Renewables remain a relatively small, if growing, slice of the power pie in the US. Greening the power sector, then—considered one of the lowest-hanging fruits of decarbonization—means tackling several problems at once: bringing a massive amount of activity onto the electric grid and converting the source of that electricity over to zero-carbon fuels, all the while scaling back the amount of energy we use overall and making that system resilient against the wildfires, hurricanes, extreme heat, and the many other climate impacts already accelerating in a world warmed, as it is now, by at least 1 degree Celsius.

If all goes well, that means moving much more of our lives onto a grid that needs to roughly double in size: not just fueling up cars with roadside electric chargers instead of at gas stations, but also having fewer cars overall as electrified mass transit helps more people get to and from work, school, nights out, and weekends at the beach. Homes will be warmed with heat pumps instead of gas furnaces, and meals will be prepared on induction rather than gas cooktops. To cope with upticks in demand throughout the day—when whole coasts wake up to brew their morning coffee and come home and make dinner—appliances from washers to refrigerators will get smarter, cleaning your clothes while you're away at work at peak sunlight and collecting all manner of data about when the sun shines on yours and your neighbors' roofs and when you and your neighbors are likely to binge-watch Netflix.

At any given moment, we probably aren't thinking too much about the for-profit companies, cities, or co-ops that we in the US pay each month to keep the lights turned on; if you are thinking about your power provider, you might be on the bad end of a shutoff or bill dispute. Despite having engaged in much of the same ugly lobbying and climate denial and, in PG&E's case, the deaths of scores of people, these electric utilities—now poised to amass an incredible amount of power and information in the coming years—have largely avoided the same kind of scrutiny as fossil fuel producers. Their business model, cloaked as it is in layers of complexity and inaccessible technical jargon, has helped shield them from public ire. Today's utilities, including the ones that are sparking fires, have a lot of work to do toward decarbonization. Most of the charges lobbed at the fossil fuel producers—that they've poisoned politics, misled the public about the existence of the climate crisis, and poured vast sums into fighting anything that might challenge them—land on utilities, too. Using the same playbook, they've spent decades stamping out anything like an alternative to power for profit. An equitable transition to a low-carbon world should revive those alternatives for the twenty-first century.

For a time when electricity was taking off, overlapping sets of wires erected by competing power companies left some cities littered with power lines and some places—mostly rural areas—with none at all. That didn't last long. At the height of the Gilded Age, early electric companies began to rival the scale of giant railroad conglomerates, as magnates like Thomas Edison protégé and General Electric founder Samuel Insull absorbed competitors into their growing empires. The burgeoning social movements of the day didn't differentiate much between robber barons and electricity bosses in charge of massive, vertically integrated firms known as holding companies. Encompassing several different power providers, they became as reviled as their counterparts in steel and finance.

Sensing some kind of reform on the horizon, it was investor-owned utilities (IOUs) themselves that over a century ago first pushed to establish statewide utility commissions as a grand bargain. These new public utility commissions (PUCs), which now exist by some name in every state, were a means of cementing their monopolies, appeasing progressive reformers, and heading off a much less savory alternative for them: publicly owned electricity. At the time, the concept was gaining rapid ground at home and abroad. Some seven hundred public power systems were created between 1895 and 1906. In 1907, Insull's National Electric Light Association (NELA)—a trade association for privately held utilities, including PG&E—drafted model legislation for several states submitting themselves to regulators, who in exchange granted them a captive, reliable market in the form of designated service areas.[6]

The commissions did provide some safeguards for consumers, reining in prices and making service more reliable. They ensure for-profit utilities fulfill their basic mandate of providing affordable and reliable services to state residents, setting the rate of return power providers can recoup from customers for new infrastructure projects. And having a single entity serving an area made more logistical sense than any return to competing wires replicating service. It's why electricity, like railroads, is commonly referred to as a natural monopoly, wherein market competition is a less efficient means of service delivery. Simultaneously, though, the creation of these PUCs prompted widespread consolidation within and across regulated service areas and emboldened company executives with powers normally reserved for government actors, like eminent domain. PG&E alone swallowed up some five hundred companies throughout California during this period. And by 1929, Insull—having left GE years earlier—made a profit off of nearly 10 percent of the country's grid through his various chairmanships and presidencies. By 1930, ten utility holding companies like his controlled 75 percent of the electric industry.

Without competitors, power companies could focus on buying off the regulators tasked with curbing them, whether by throwing money behind candidates they supported or lobbying governors and lawmakers to make favorable votes and appointments. For as long as they've existed, regulatory capture was effectively baked into the business model of America's investor-owned electric utilities, who can spend modest amounts of money to have friendly regulators elected or appointed and swing the outcomes of votes in often sleepy and sparsely attended meetings loaded with technical jargon.

Despite their relative success in establishing monopolies, private utilities continued waging war on their publicly owned counterparts. A seven-year investigation by the Federal Trade Commission (FTC), begun in 1928, unearthed private utility executives' elaborate public relations push to warn of the "dangers to the American Way of Life that would come if the utilities were ever allowed to slip from private enterprise to public control," complete with an annual advertising budget of up to $30 million—$450 million in today's dollars. As with most utility expenses, this was funded largely by ratepayers. NELA planted "public information committees" in every state. The association's member private utilities rebranded themselves as "public utilities" or "public service companies." They bought off professors, labor leaders, and newspapers from the *Brooklyn Daily Eagle* to the *State Journal-Register*, in Springfield, Illinois, nixing critical coverage and filling the press with proutility editorials sometimes bylined by judges and governors.[7] Utility employees, Sharon Beder writes in *Power Play: The Fight to Control the World's Electricity*, "were trained in public speaking and given courses in public relations. The recurring message was not only that municipal ownership had been a failure everywhere it had been tried, but also that it threatened American democracy."[8]

Among the most pernicious techniques of this PR blitz was encouraging ratepayers to buy up stock in private utility holding companies, intended to spread the idea that IOUs were a collective endeavor along the lines of their publicly owned counterparts. The kinds of "customer ownership" ratepayers were encouraged to purchase, though—bonds, securities, and other nonvoting stock—didn't come with any actual ownership control or voting rights. As NELA pointed out, the main goal wasn't to raise more money from ratepayers but to win them over ideologically. The group's customer ownership committee boasted about the creation of a "stalwart army of sound-thinking owners of private property" that would be "the nation's greatest defense against socialism or communism—and every step toward public ownership is a step toward communism."

Through the Depression and the investigation, Insull promoted stock in his own utility holding company as a more secure investment than government bonds. Finding it difficult to raise cash after the crash, his Middle West Utilities Company, spanning thirty-nine states, eventually collapsed in 1932, costing some six hundred thousand stockholders—including its own now-jobless employees—$4 billion, or $75 billion in today's dollars. The conclusion of the FTC investigation would lead to the breakup of NELA in 1932, although it reformed into the Edison Electric Institute (EEI) within twenty-four hours. It still operates today under the same name and a broadly similar mission.

Though Herbert Hoover was a reliable friend to private utilities, Roosevelt campaigned against them, capitalizing on the fact that Insull was being pinpointed as a chief cause of the country's economic troubles. "Where . . . a city or county or a district is not satisfied with the service rendered or the rates charged by the private utility," he urged during a 1932 stump speech in Portland, Oregon, "it has the undeniable basic right, as one of its functions of Government, one of its functions of home rule, to set up, after a fair referendum to its voters has been had, its own governmentally owned and operated service."[9] After winning, he and the New Dealers would become enemies of the IOUs.

Roosevelt was no socialist, but like many at the time, he considered public ownership a necessary tool for filling gaps left by the private sector. And private utilities at the time left plenty of them. The New Deal expanded public power accordingly. Though IOUs had snapped up monopolies in population centers, some 90 percent of rural homes still lacked electricity at the start of the Depression, limiting not just their access to electricity but to sorely needed economic development more broadly. For private utilities, extending power lines to customers spread out over tens or hundreds of miles just wasn't worth the cost—especially considering that the vast majority of those potential customers happened to be poor.[10] When service was available, it usually meant making a costly up-front investment for lines, meters, and wiring. A study from the Mississippi Valley Committee of the Public Works Administration found at the time that "unless the Federal Government assumes an active leadership, assisted in particular instances by State and local agencies, only a negligible part of this task can be accomplished within any reasonable time."

Founded to remedy this problem in 1933, the Tennessee Valley Authority (TVA) was envisioned by Roosevelt as "one great comprehensive plan" for not just electricity but also for reforestation, flood control, national defense, agricultural and industrial development, and land use.[11] It would be, as he put it, "a corporation clothed with the power of Government but possessed of the flexibility and initiative of a private enterprise," intended as the pilot project of a model to be scaled out through the rest of the country. The TVA operated on the principle of "administration at the grassroots," with federal administrators working closely with farmers and other locals to help craft plans for power lines, economic development, and resource conservation. Its flagship project—the Norris Dam— seemed to embody those principles. The TVA's chief architect, Roland Wank, ensured the expansive project came with breathtaking views as well as a visitors' center and observation deck to attract tourists to the region. He designed housing, as well, for those displaced by its construction. "TVA's job," one administrator for the program told the documentarian George C. Stoney in 1940, "is to save the land and the water not alone for the people who happen to live in the Valley at the present time, but for people in cities, people in other parts of the country, people not yet born, all of whose lives will be affected eventually."

A complementary agency, the Rural Electrification Administration, or REA, was initially created in 1935 to help finance grants to private utilities to electrify rural areas throughout the country. The IOUs refused to cooperate with the programs' aims, wanting to use federal funds to extend service only to more populous and well-off rural areas and charge exorbitant rates; effectively, the $100 million of initially allocated REA funds was treated by IOUs as a subsidy for what they were already doing. Private utilities eventually refused to cooperate altogether as more and more of their applications for federal funds were denied. Modeled on the successful cooperatives created to distribute TVA-generated power, the REA's ten-year extension doubled down on public, nonprofit power, with a mandate to erect power lines as well as install plumbing and string telephone lines, with below-market-rate loans available for rural residents to buy household appliances, as well.

The REA made it possible for rural communities to electrify themselves and string wires via cooperatives owned and operated by their members. Having established a broad set of criteria for which communities were eligible to apply for REA loans, administrators let rural residents decide whether they wanted power lines and organize to have them built. The program then provided them the necessary financing and technical assistance to carry out their vision. As REA director Morris L. Cooke explained in 1938, New Deal–era advisers at the federal level would offer guidance "as to organization, methods of accounting, home demonstration projects [and] engineering practices." But nearly just as central to the actual provision of power were the added benefits of well-organized communities. "The immediate and tangible results will be to bring electricity to a large proportion of American farms, to stimulate employment and manufacturing, and to raise living standards in rural communities," Cooke wrote. "The intangible values—building self-reliance and training leaders in every community—should prove no less satisfying."

Both the TVA and REA attracted powerful enemies. Investor-owned utilities began erecting "spite lines" to try to undercut the agency by extending service to wealthier rural residents, since TVA legislation prevented the agency from duplicating service already provided by the private sector. Wendell Willkie, head of Commonwealth & Southern Corporation, a utility holding company, and EEI locked horns with the TVA in a long-running legal and PR skirmish that turned the full weight of the private utility industry's machine against the New Deal. In its propaganda, private utilities assured rural residents that they were looking out for their best interests. In private, the profit motive ruled. "Only in the imagination," a Philadelphia utility manager told an EEI conference in 1935, "does there exist any widespread demand for electricity on the farm or any general willingness, or ability, to pay for it. . . . The possibilities of the market are vastly exaggerated."

After winning a legal battle against the EEI, the White House–backed Public Utility Holding Company Act of 1935 (PUHCA) broke up trusts like Insull's, vastly limiting the scope of what IOUs could do and own. Enforced by the Securities and Exchange Commission, PUHCA confined them to provide basic electrical services within a defined geographic area and kept nonutility companies—say, oil companies or investment banks—from owning and making a profit off the utility business. In a show of how closely linked they were to IOUs, forty-six of the country's forty-eight PUCs came out against it.

Despite opposition, in just ten years, with a $100 million annual appropriation from Congress—$1.8 billion a year, in 2019 dollars—the REA had financed the construction of 380,000 miles of transmission lines, 42 percent of the transmission lines ever built in the United States. Today, over nine hundred rural electric cooperatives (RECs)—owned and operated by their members—stretch through forty-seven states, serving 42 million ratepayers and 11 percent of the country's demand for electricity. They also serve 93 percent of the country's "persistent poverty counties," 85 percent of which lie in nonmetropolitan areas. REC service areas encompass everything from isolated farmhouses to mountain hollers to small cities, with the highest concentrations in the South, the Midwest, and the Great Plains. Overall, publicly owned and cooperative utilities serve 49 million customers in forty-nine states, compared to the 110 million served by IOUs, according to the American Public Power Association and Department of Energy. RECs alone service a whopping 75 percent of the country's total land area.

While most co-ops obtained their power from private companies, municipalities, or bodies like the TVA, starting in the 1960s and '70s they began to construct generation and transmission (G&T) cooperatives via low-interest loans from the federal government to build transmission lines and coal plants, which politicians were especially eager to back following the oil crisis. The federal government would continue to expand the Power Marketing Administration (PMA), unveiling its fifth and final major project—the Western Area Power Administration—in 1977.

By the 1960s, 90 percent of rural homes could turn their lights on thanks to injections of federal cash; the public sector had done what the private sector wouldn't.

THE NEW DEAL's public power programs were a microcosm of its larger contradictions. Black households enjoyed relatively few gains from rural electrification compared to their white counterparts. Landlords who rented to African Americans often chose not to pass power bill savings down to their tenants. The interests of Black tenant farmers in rural development programs were subsumed to those of white landowners, and African Americans were barred from the best-paid positions within the TVA and consigned to its most menial jobs. John Rankin, a Mississippi House member who pushed hard for the TVA and one of the country's most dogged advocates for public power, fought for rural electrification as doggedly as he fought against antilynching bills. Southern Democrats broadly supported the TVA for the benefits and job creation it would bring to a region long neglected by the utility giants. For all its transformation of life in the South and its built environment, the REA and TVA—like much of the New Deal—self-consciously did nothing to disturb Jim Crow.

Their foundations in segregation extend well into the present. Many co-ops are still unrepresentative of the areas they serve. In southern states, many that serve majority Black service areas have for years maintained all-white co-op boards and employed majority white staffers and contractors; some co-ops haven't had proper elections in several decades. In the Black Belt, where the civil rights movement swept majority Black local governments into power, RECs continue to lag behind, often featuring little to no minority representation.

Nsombi Lambright is the executive of One Voice, an affiliate of Mississippi's NAACP, which trains up African American co-op member-owners to run for boards on nine of the state's twelve co-ops, many of which serve majority African American counties. Lambright knew of only three Black board members in the co-ops they've been targeting. While Mississippi is 37 percent Black and nearly half its residents get their power from co-ops, its co-op boards are 91 percent white and 96 percent male. Rates can be well above what's offered in adjacent IOUs, with some co-op members—many of them living in persistent poverty counties—paying as much as 40 percent of their income on monthly bills. "The only thing that has pretty much been common among the co-ops is that the members of the board tend to be the existing power structure in the community," Lambright told me.

"What we're dealing with here is that racism is still alive in America. These RECs were started in the 1930s and '40s. In Alabama at that time, there was no justice, no equal rights," John Zippert, of Epes, Alabama, told me. With the Federation of Southern Cooperatives Land Assistance Fund (FSC)—a group supporting Black cooperative development in the rural South, founded in 1967—Zippert, a civil rights movement veteran, has worked with member-owners of the Black Warrior Electric Membership Corporation to make it more democratic. While Zippert estimates that as much as 60 percent of Black Warrior's 26,000-plus membership is Black, "up to this point," he said, "they have not had a Black board member."

Looking to change this, the federation brought around one hundred people to Black Warrior's annual meeting in Choctaw County in the summer of 2016—only to have it immediately called to a close. "They had sixty seats put out for 1,300 people," he said, referencing the co-op's stated quorum requirement. "They insisted that this wasn't a real meeting, and they didn't have a quorum. Some people asked questions about the election procedure. They didn't answer and didn't want to answer." Like many co-ops, Black Warrior requires that 5 percent of its members be present at its meetings in order to hold them. With member-owners spread out over eleven counties in rural West Alabama, there hasn't been a proper meeting—by Zippert's estimation—in fifty years.

Despite the fact that member-owners legally own the means of energy production, today's publicly owned utilities are hardly the socialist covens NELA feared they would be. In the century since the rural electric cooperatives were created, the democratic and cooperative principles that once governed them have devolved, in many places, into old boys' networks eager to maintain business as usual—including massive amounts of coal power. That wasn't thanks to some backwardness supposedly inherent to rural democratic institutions so much as decades of deliberate erosion. Republican politicians since the 1950s have targeted the rural electric cooperatives in their broader pushback against the New Deal. Dwight Eisenhower attempted to raise the interest rate on loans co-ops got from the federal government, and Richard Nixon and Ronald Reagan each tried to scrap some loans entirely, leading co-ops to seek alternative financing from Wall Street and raise rates as a result. Reagan attempted to sell off the PMAs altogether in the 1980s, citing that the low-cost power they provided wasn't generating adequate returns for the Treasury. Ultimately it was the Democratic Clinton administration that finally delivered the heaviest blow, eliminating the REA's low-interest loan program altogether and privatizing the Alaska Power Administration. Barack Obama considered privatizing the TVA entirely.

Starving the public sector of funds and then blaming it for inefficiency and mismanagement has been a core strategy to privatize public goods and services, or nix them altogether. Though the National Rural Electric Cooperative Association (NRECA), the trade association for electric cooperatives, now spends much of its political capital defending fossil fuels, its considerable lobbying might was developed largely to fight off persistent rounds of cuts—not to advocate for revanchist energy policy.

That's not much of a consolation. NRECA enthusiastically battled the Obama administration's Clean Power Plan, which sought to phase out coal-fired power plants, and joined with right-wing groups to back many of the Trump administration's environmental rollbacks, lobbying extensively in support of fossil fuels and the politicians who support them. When One Voice attended the NRECA Annual Meeting in New Orleans in early 2020, Lambright reported a "sea of older white men" of the sort found at annual meetings in Mississippi co-ops. Several One Voice members had the cops called on them. "Some of the workshops were informative for folks, but they didn't feel welcome," she said.

While rules vary from state to state, many co-ops aren't required to submit rate increases to regulators or even follow energy efficiency and renewable portfolio standards. Public power was exempt from the early bargain struck between IOUs and PUCs, where they agreed to be regulated in exchange for their monopoly and new powers. Since the structure of public utilities was already *de jure* democratic—ratepayers or elected governments legally own them—the idea was that they could run themselves in their own best interest, outside the grasp of craven magnates like Insull. That turned out not to be the case.

Liz Veazey is the cofounder of a group called We Own It, which organizes rural electric cooperative members around the country to reassert democratic control over their utilities. They run reform candidates for co-op boards and push for better service and cleaner energy sourcing. She's well aware of the problems plaguing co-ops but cautions against seeing them as a monolith. "Some of that critique is relevant in that some co-ops are really not living up to the cooperative principles of member education and democratic participation," she said. "The flip side is that there are really good examples too. I don't think that model inherently leads to bad things."

Among the biggest problems We Own It members face are intransigent co-op leadership teams. It's not hard to figure out why longtime board members are eager to hang on to their seats: all in all, it's a pretty sweet gig. The average annual salary for co-op board members is $17,000, which Veazey estimates requires about five hours of work per week. Some also get health benefits and free trips to industry conferences in hot tourist destinations. Maintaining these perks has been a major disincentive against merging co-ops, maintaining what might otherwise be redundant salaries and benefits. That all contributes to rates being cheaper at some IOUs than in the co-ops that neighbor them.

Plenty of member-owners are fighting back. In 2007, member-owners of the Pedernales Electric Cooperative—the largest in the country, serving 300,000 people throughout Texas Hill Country—filed a $164 million class action lawsuit over utility leadership siphoning off ratepayer money for their own benefit.[12] With the offending parties convicted, Pedernales underwent extensive governance reforms and implemented policies friendly to renewable energy. Tapping into $86 million of funds from the USDA's Energy Efficiency Conservation and Loan Program, it built out its own 11 megawatt and growing solar farm, incentivized rooftop panels, and allowed member-owners to source 100 percent of their power from renewables for less than fifty cents extra per month. Southern Arkansas's Ouachita Electric Cooperative and Roanoke Electric Cooperative, in North Carolina, have taken advantage of grants from the USDA to offer energy efficiency programs and broadband at no up-front cost to members-owners. A few years back, Ouachita partnered with a private developer to open the largest solar farm in the state. Great River Energy—a G&T serving twenty-three co-ops in Minnesota—committed in 2018 to generating 50 percent of its power from renewables by 2030 while keeping rates flat through 2019 and price increases below the cost of inflation through the next decade.

There's work to transform co-ops happening in the heart of coal country, too. Started in 1976, the Mountain Association for Community Economic Development (MACED) has worked to build economic alternatives to coal in central Appalachia. Recently, much of that work has involved RECs. MACED has partnered with a network of six distribution cooperatives on a program called How$mart Kentucky. Through contracts between MACED and the RECs, the cooperative makes it known to its member-owners that they can save money on their monthly bills with a set of upgrades, such as weatherization to make buildings more energy efficient.

Chris Woolery, whose family hails from Eastern Kentucky, started working as an auditor with the How$mart program after he lost his home and contracting business during the fallout from the financial crisis. He now coordinates the program as a MACED staff member. "This program allowed me to transfer some of my skill set into a job that I love, that has meaning, helps people, and helps the region," he told me. "Bad housing stock is a resource," Woolery said, noting that How$mart upgrades could create hundreds and potentially thousands of jobs for Kentuckians over the next several years. One partner, the Jackson Energy Cooperative, has 51,000 members. Retrofits on all of the homes and businesses it services could take up to fifty years, Woolery estimates, by which point several thousand of the already serviced homes would need to be upgraded again.

As promising an engine of job creation as energy efficiency might be, MACED sees it as just one part of a potential transformation of the region, involving investments in everything from local food to health care to tourist attractions. On the renewables front, Woolery thinks of energy efficiency as a "gateway drug to clean energy." Indeed, members of their partner co-ops, he noted, have been asking about solar, largely out of interest for the savings it could offer them on their monthly bills. In response, one REC, the East Kentucky Power Cooperative, brought a sixty-acre, 32,300-panel solar farm online in 2017.

Not far away, in the Tennessee Valley, residents remember the jobs and electricity the TVA provided their families as well as the land it took from them to make way for hydropower.[13] The Tennessee Energy Democracy Movement is working toward a TVA fit for the twenty-first century; in 2020, they drafted a "people's vision" out of thirteen listening sessions throughout its service area involving hundreds of people. As in most co-ops, those who get power from the TVA have little access to renewables, and there are few good avenues for public engagement. "Clearly, the public power model in the Tennessee Valley is broken; more likely, it was never fully realized," the campaign writes. Participants outline wanting more renewable energy, high-quality jobs, and lower rates, but they also want more democratic input into how the TVA is run.[14] Brianna Knisley, who organizes TVA members through the nonprofit Appalachian Voices, described a careful balance between local input and federal authority. "People want actual mechanisms that give communities and residents some real negotiating authority over the decisions being made," she said, "and it can't all just be up to organized communities to take all this on. There have to be other structures in place that are playing a regulatory role," pointing to the need for the Tennessee Public Utility Commission to actually regulate the municipal power companies and co-ops that distribute TVA power. "There aren't any perfect answers. We're talking about deep democracy. What's always been missing is real democracy, and real community control over what's happening with the TVA."

America's public power can take different forms, too. Liz Veazey, who lives in Nebraska, gets her electricity from the Omaha Public Power District (OPPD), one of three such utilities that comprise the state's entirely public grid. The Husker State is no eco-socialist paradise: 63 percent of its power comes from coal, and its congressional delegation is exclusively Republicans.[15] But its public power is also a far cry from more staid co-ops and municipally owned utilities, let alone the most recalcitrant IOUs.

Set up by George Norris, who championed both the TVA and REA in the Senate, the system is divided between three public power districts. A statewide Power Review Board (PRB) can determine whether rates throughout the state are fair, reasonable, and nondiscriminatory, although PRD directors don't need to seek permission from the board to change rates. Nebraska's Public Service Commission also has some limited authority over matters like transmission lines, but in general the power districts self-regulate. As in Nebraska's unicameral legislature—another of Norris's ideas—power district director elections are nonpartisan. Thanks to pressure from member-owners and some newly elected directors, the OPPD has recently committed to reaching net zero carbon emissions by 2050 and to bring online 600 MW of solar—what could be one of the largest solar arrays in the region.

"If we don't carry out maintenance there is no return of financial benefits to any outside stakeholders," OPPD board member Eric Williams, Veazey's neighbor, told me. "If we don't do the maintenance then the customer-owners see lower quality of service, and they would ask us to maintain the high quality of service we've been maintaining. We do not have a private profit motivation." If customer-owners don't like what they see, they can vote the directors out or run themselves. "Every now and then there are discussions about privatization in Nebraska. In general," he added, "those have not gone very far because the results that public power has delivered have been great." OPPD consistently ranks among the Midwest's top utilities for customer satisfaction.

THOSE CALLING TO put a state takeover of PG&E on the table in California are keenly aware of how public ownership can go wrong. "Obviously there are public utilities that are run in ways we wouldn't want them run," said Keith Brower Brown, a member of East Bay Democratic Socialists of America (DSA) and organizer on the Let's Own PG&E Campaign. "We're not naïve to the fact that public ownership is no panacea. We're democratic socialists because we don't just want more government control but more democracy."

With active campaigns in Providence, Chicago, Boston, and New York, DSA's work to switch off IOUs nationwide is open to various alternative models, sticking to broader principles—decarbonize, decommodify, democratize, and decolonize—than any particular existing model. All the anger being directed at PG&E over the fires led eight DSA chapters in its service area to identify an organizing opportunity. "More than anything, it showed the incredibly pressing need, in a state that's increasingly flammable, to get the profit motive out of the way between us and safety," Brower Brown said.

PG&E's bankruptcy settlement, struck in 2020, allowed it to maintain its basic structure, building in more stringent oversight from the CPUC and mandates for a change in leadership. But a stipulation of that settlement allows the CPUC to revoke the company's charter should it fail to meet safety requirements or burn down more homes. In that event, its assets would be transferred to a nonprofit body called Golden State Energy. A ratepayer-owned company regulated by the CPUC, the board of the utility would initially include five members appointed by the governor and four by the legislature. Eventually, it would transition to having six ratepayer-elected members and three legislative and gubernatorial appointees.

Mari Rose Taruc heads up the Utility Justice Campaign. The campaign sees worker and community control of energy as core to a just transition off fossil fuels. Above all, though, Taruc wants to be clear that the future of the state's grid—including a prospective Golden State Energy—is designed to keep the needs of California's most vulnerable communities front and center. Having opposed the bankruptcy settlement for being too lenient on PG&E and paltry on payouts for wildfire victims, climate justice advocates also wanted to see a better-defined trigger for transitioning PG&E to Golden State, ratepayer protections, and board representation for front-line communities, including those worst hit by both wildfires and fossil fuel power generation.[16] "This is all a creative process. I think that's what the climate crisis brings. There are old systems and infrastructure that are crumbling, and we have to reimagine something new," she said. "We have to look at how we're going to have a system that can make it through the climate crisis. Those conversations need to have the right people in them."

Public ownership and climate justice advocates have bristled at takeover plans that could splinter service, fearing that one-off municipalizations in wealthier cities with more resources for a takeover—like proposals from the mayors of San Jose and San Francisco—threaten to create a two-tiered energy system, leaving poorer areas with worse service and higher rates as rich customers splinter off. That's not just a problem for California. Homeowners who can afford renewables are already beginning to defect from their grids nationwide and generate most of their power from solar panels on their own rooftops. The physical properties of fossil fuels lend it to more centralized production, requiring dug-up coal, oil, or gas to be shipped to power plants that generate electricity that is then spread throughout the grid. Wind is more centralized, requiring industrial-scale investments. But solar power can be generated virtually anywhere by anyone with the right equipment. To provide no-carbon power at a mass scale, at least some degree of centralized renewable power (utility-scale solar) and transmission lines are still likely to be needed for years to come; not everyone can put solar panels on their homes. In the Golden State especially, however, expanding decentralized solar is a key part of protecting homes against climate impacts. Microgrids at the neighborhood level can be turned back on quickly after disasters strike and remain operating when old transmission lines need to be turned off to lessen wildfire risk or deal with increased demand from heat waves.

Public policy can help build these out. Pro-renewables policies in California—most notably including a mandate the state be carbon neutral by 2045—have made solar cheaper and more accessible than it is in many other parts of the country. PG&E has been made to comply and now has some of the most renewables on its grid of any IOU in the country. But without a well-thought-out plan for what an equitable, rapid transition away from centralized fossil fuel power looks like, California and states around the country run the risk of creating the kind of bifurcated electricity system campaigners fear: wealthier communities running on clean, resilient distributed power and a centralized, dirtier, blackout-prone grid serving everyone else, with the remaining polluting power plants still sited in communities of color. Taruc related the situation facing California's grid to a longtime mantra of the state's climate and environmental justice groups: "The transition is inevitable, justice is not."

A state takeover would be unprecedented and opens up a number of questions: Would all of PG&E's assets be part of the co-op or state-run provider, or would its gas operations and generation facilities—including its beleaguered Diablo Canyon nuclear plant—spin off as separate entities, or be sold off to the highest bidder? How will any publicly owned PG&E mesh with another quasi-public part of the state's energy system, Community Choice Aggregation, which allows smaller groups of ratepayers to collectively bargain for better rates and cleaner sourcing? And who's going to pay all that debt? Even with all the uncertainties, other options don't inspire much more hope—including business as usual. And this time around, distressed asset investors, better known as vulture funds, are sniffing around in search of a quick buck. Baupost Group, run by the billionaire hedge fund investor Seth Klarman, bought up $6.8 billion worth of an obscure financial product known as subrogation claims against PG&E, wherein an insurance company sells the right to sue to recoup the cost of damages borne by its policyholders. Effectively, Klarman bet that PG&E would have to pay out insurance claims for wildfires and was right. When the matter was settled in courts in 2020, Klarman, who had bought up claims for thirty-five cents on the dollar, likely netted $1 billion on blazes that killed dozens.[17]

Public ownership campaigners see action at the state level as essential to making sure a decarbonized energy system is in fact a more equitable one. New York City DSA is part of a coalition of groups pushing to bring that state's monopoly IOU—Consolidated Edison, or ConEd—under public ownership by expanding the scope of the New York Power Authority, already the country's largest public power provider, created by FDR during his time as governor. Governor Andrew Cuomo and several state legislators have expressed some openness to taking over both Con Ed and New York's main gas provider, National Grid, after both were embroiled in scandals over rates hikes and poor service provision in 2019.[18] Gustavo Gordillo—a member of New York City DSA working on its public power campaign—is clear that none of these changes will happen in a vacuum or without a major fight. Most public utilities, he explains, were set up "decades before decarbonization was known to be something we needed to do. Most of the publicly owned utilities were not set up with those goals in mind. If we set up new public utilities as part of a Green New Deal, that would be a built-in goal and make for a very different situation."

Given the high levels of union density in the utility sector—from generation to distribution—labor groups are nervous about what a poorly designed transition both off fossil fuels and into public ownership could look like. The International Brotherhood of Electrical Workers (IBEW) protested a state takeover proposal from San Francisco state senator Scott Wiener, citing concerns about job security and pension protection.[19] That's understandable. Within the private sector, utilities are still one of a dwindling number of union-dense industries. And the rooftop solar business now exploding in California and elsewhere is currently dominated by nonunion labor. Public ownership campaigns around both PG&E and ConEd are clear about the need to maintain wages and benefits for utility workers, as was Wiener's proposal; what's more, most utility work in general is unionized, including in public power. The Let's Own PG&E campaign is calling to go a step further, for majority public ownership and minority worker ownership of public power.

Even if changing ownership isn't a cheat code in and of itself for a more democratic, decarbonized energy system, the mere *threat* of public ownership can be a powerful motivator for IOUs. When Boulder, Colorado, passed a ballot initiative to municipalize its Xcel Energy assets, the utility kicked off a legal battle that's still tied up in the courts today. To save face and prevent more defections throughout its eight-state spanning service area, Xcel announced it was phasing out some of its coal power and committing to be carbon-free by 2050, cutting emissions by 80 percent by 2030. The announcement also came shortly after the election of Colorado governor Jared Polis, who campaigned on taking the state to 100 percent renewable energy by 2040. As the price of renewables has plummeted and climate concerns have increased, a wave of such commitments have sprouted up across even more traditionally conservative utilities, to the point of driving a wedge in the traditionally strong relationship between fossil fuel companies and utilities, who are quitting coal and even gas and skipping straight toward renewables. As with fossil fuel producers, it's not enough to take these pledges at face value, particularly when there are no regulations in place to make sure they follow through. Optimistically, they can be seen as evidence of a rare, happy coincidence of public pressure and economics traveling in the same direction.

The key factor in what makes for a responsive utility seems to be how much genuine, democratic oversight exists for its leadership. That's why campaigners for public ownership aren't proposing to scrap the CPUC or any other statewide regulators. By contrast, they want it to be less prone to the kind of regulatory capture that has defined its relationship to the state's IOUs and that has been built into its business models for one-hundred-plus years. Still, huge barriers remain. The kinds of oversight and investment needed to bring utilities and regulators both into the twenty-first century have been pushed out of political favor over the last several decades by the same interests that have fueled climate denial and the climate crisis itself—IOUs included.

UTILITIES HAVE BEEN active in spreading disinformation and blocking climate-friendly measures at just about every level of government, constraining not just how the public thought about climate change but also the set of tools policymakers had to respond to it. Having them run democratically and in the public interest could change that. "The most compelling argument for public ownership is how much money monopoly IOUs are spending on a daily basis to prevent climate solutions," David Pomerantz, director of the Energy and Policy Institute, said. "It hasn't gotten as much attention as companies like Exxon, Chevron, or Shell, but I think utilities are at least as pervasive in terms of the effects of the amount of money that they spend blocking solutions."[20]

For regulated utilities, regulatory capture is all in a day's work. "Compared to other kinds of companies that aren't regulated monopolies—say, Apple or Walmart—investor-owned utilities are operating in totally perverted markets. Those companies' ability to operate is about outcompeting competitors. Regulated utilities don't have to do that," said Pomerantz. "Their profits are literally prescribed by these three- or five-member commissions. If you're a regulated monopoly utility that's regulated by a PUC, that's the whole game. You operate as a company by how well you can get those regulators in your pocket."[21]

Even well-intentioned regulators who aren't either wining and dining with utility executives or former utility executives themselves face a steep path to thinking too far outside the box industry has created. IOUs, Pomerantz said, have "built up an information environment around those regulators where they produce nearly everything they read or consume," from trade publications to conferences on utility management to experts housed in campus energy centers whose boards and donor rolls are filled with industry interests. "If they want answers to certain technical questions, the only people that can answer them are the utilities themselves."

Unwitting customers often end up paying for their power providers to pollute politics.[22] While lobbying expenses have to be disclosed publicly and can't be paid for by ratepayers, utilities' membership dues to

EEI, for instance—which freely intervenes in the policymaking process—are in most states considered an operating expense that is allowed to be added to customers' monthly bills. Beyond their historic involvement in bodies like the Global Climate Coalition and funding the work of climate skeptics like Patrick Michaels, some of IOUs' most effective work in slowing climate action more recently has been at the state level. In 2015, the Energy and Policy Institute and the *Washington Post* reported on a concerted campaign by EEI to undermine clean energy policy around the country.[23] Since IOU profits come from the rate of return they get on massive infrastructure build-outs, the proliferation of rooftop solar in particular makes it harder for them to argue that new infrastructure and rate increases are justified. In December 2019, EEI ran a conference, called Campaign Institute, for high-level government affairs and communications staff at IOUs to learn how to defend themselves from "major categories of risk," from antipipeline protests to clean energy mandates to public ownership campaigns, with advertising and statewide ballot initiatives, among other tactics.[24]

State capitols are pelted with utility cash, handing IOUs there a win-win: defeating prorenewables measures and placing industry-friendly Republicans in office. In 2018, the parent company of APS (Arizona Public Service) in Arizona poured $37.9 million into a ballot measure that would have required power providers to source half their energy from renewables by 2030, the most expensive such fight in the state's history.[25] Between its PACs, CEOs, and lobbyists, Dominion Energy furnished Virginia politicians with $2.6 million worth of political contributions in the lead-up to statewide elections in 2019.[26] "It's not an envelope of cash under the table saying that if you vote for this bill we'll be good to you. They give campaign contributions to legislators that they know will be friendly to the types of projects they want to price gouge people on, and they host parties," said House of Delegates member Lee Carter. "They're a constant presence at every event in the state capitol. It's not just the assembly, the governors, or the State Corporation Commission," the state's public utility regulator whose three members are appointed by the legislature. "It's local politics. It's nonprofits. It's every aspect of civic life in Virginia."[27]

Through its cozy relationship to Ohio state legislators and $30 million in political spending, the utility FirstEnergy secured a controversial

$1.1 billion bailout for flailing coal and nuclear plants to continue operating, making sure to kneecap renewables as well.[28] Charges unveiled from an FBI investigation the following year found the bill, HB 6, had been the result of a $61 million conspiracy wherein the utility paid Republican Ohio Speaker Larry Householder handsomely for his work on the issue in a blatant pay-for-play setup—collecting $100,000 toward his Florida vacation home. FirstEnergy, moreover, had funded the push to make him House Speaker in the first place. Before the charges were announced, EEI had given FirstEnergy an award for its work in passing the bill.

EVEN IF PUBLIC ownership manages to neutralize IOUs' epic political spending, absent much larger changes utilities will still operate within a grid designed to serve for-profit interests. After the passage of PUHCA in the 1930s, most IOUs existed as vertically integrated entities: selling power through their transmission and distribution lines that was generated in their own power plants, with rates regulated by the PUCs. A little-noticed provision in the massive Public Utilities Regulatory Policy Act (PURPA) of 1978 opened up electricity generation to any company that could put its electrons on the grid at low cost, spawning a new class of so-called merchant generators. Deregulation in the 1990s aimed to transition entirely from monopoly power providers to a system where customers were free to choose where they got their energy from on an open market. The sudden introduction of several middlemen into the energy equation—from financial institutions to new bodies like independent system operators (ISOs) and regional transmission organizations (RTOs)—proved chaotic.

Deregulation's biggest disaster was in California. PG&E, SoCal Edison, and the CPUC, along with Enron, had all been active proponents of deregulating the state's energy system, with the CPUC arguing in a memo that it would bring much-needed "market discipline" to the electricity sector. It took a few years to set up the necessary infrastructure after deregulation became law in 1996. Almost as soon as the system was up and running, Enron devised elaborate and colorfully named schemes (Death Star, Jedi, etc.) to manipulate newly formed markets, cutting off power they could then flood back in once the price had gotten high enough. The state was hit by rolling blackouts as electricity prices soared. PG&E—now paying exorbitant costs for wholesale power—blamed the state's booming economy, hotter than usual temperatures, and the rise of personal electronics like computers and cell phones. The company filed for bankruptcy in 2001, and angry protests roiled around the state. Because California's experiment with deregulation was so disastrous, other states put the brakes on plans to follow their lead. Today's system remains a patchwork of regulated and deregulated energy markets.

The industry's real prize, though, was the Energy Policy Act of 2005. Though it threw in a number of sweeteners for clean energy companies, the sweeping bill also exempted fracking from clean water regulations (the Halliburton loophole), handed billions of dollars in tax breaks to fossil fuel companies and utilities, and, most importantly, repealed PUHCA. Since then, vertically integrated utility holding companies like AEP and Duke Energy have radically consolidated their inordinately coal-fueled business in ways that have begun to mirror Insull's pre-Depression empire.

The one-two punch of deregulation and repealing PUHCA was in some ways the culmination of a successful forty-year attack on the New Deal order, which had by that point achieved a certain level of bipartisan respectability. Signing a bill to deregulate the financial sector in 1999, Bill Clinton proclaimed that "the Glass-Steagall Law is no longer appropriate for the economy in which we live." Deregulation had always been a misleading name for this project, especially when it came to electricity. The goal was never really a smaller government or free markets, as such, but a set of rules that protect and expand profits. Enron has been treated as a bad actor, but it was in many ways a model for a Clintonite new economy in which energy transformed from a physical product into a financialized commodity, transforming fossil fuels from gas in a tank into numbers on a spreadsheet.[29]

Deregulation did have some major upsides. It took away power providers' inborn incentive to sell more electricity to their customers and thus use more fossil fuels. Introducing some level of market competition into monopoly-controlled power markets would help phase out that coal power and open the door for more renewables. In many states now, a growing number of investor-backed third-party solar providers will slap a few panels onto your roof for the right price; if the right state-level incentives are in place, that price will be cheaper.

But just as public ownership is no panacea for what ails utilities, breaking their duties up into ever-smaller chunks isn't either. Small, when it comes to energy, isn't necessarily beautiful, and the "wires-only" utility of the future that clean energy companies have dreamed of—wherein stagnant old utilities are left simply to maintain their lines—might open gaps that leave millions out. There's nothing inherent to local and diffusely owned energy generation that addresses the need for a managed, society-wide transition away from fossil fuels or the extension and upkeep of much-needed energy infrastructure to communities that, for any number of reasons, are either unable or reluctant to switch. That takes planning led by the state. Good planning doesn't preclude private investment; plenty of clean energy companies will make out well in an energy transition. But it ensures that the transition isn't held hostage by their ability to turn a profit.

It's not monopolies, scale, or even necessarily formal ownership structures that determine whether a utility is operating for the benefit of its members and the planet. For public power campaigners, energy democracy—as reformers have called it—is less about the legal designation of a transmission line than a theory of change about how utilities are run and who they're accountable to. Energy democracy requires genuine democratic oversight both from the public and from regulators that are not bought and sold by the firms they're meant to be regulating. As Sean Sweeney, the director of the group Trade Unions for Energy Democracy, put it to me, "You don't necessarily fetishize that it's big, and you don't fetishize small. You look at all the options and ask: What is the best way of delivering clean renewable power to ordinary people in order to meet basic needs?"

SO FAR, CALLS for converting IOUs away from public ownership have picked up steam in larger cities where coal makes up a relatively tiny portion of the energy mix. For a Green New Deal to create a grid that runs on renewables and in the public interest, it'll have to take on the even thornier politics of electricity where politicians aren't as likely as they are in California or New York to be pushing for clean energy mandates. Coal country needs to be on board for zero-carbon power, too.

Co-ops with longstanding agreements with coal-fired generation and transmission cooperatives or merchant generators are obligated to keep buying power from them until the contract is fulfilled. Functionally, this is a debt to be paid off through decades and billions of dollars worth of dirty energy. In providing billions of dollars worth of subsidies for new coal-fired power plants, the Energy Policy Act of 2005 only compounded the problem. Some RECs have contracts for coal power that extend as far out as 2075.[30] As coal continues to become more costly than natural gas and renewables, such agreements have become a ball and chain for utilities—and a massive disincentive against decarbonization. Among rural electric cooperatives, Veazey said, "the sense is that their interests are intertwined with coal and fossil fuels. The federal government has helped facilitate that by providing low-interest loans and grants to get co-ops to build fossil generation." Unsurprisingly, co-ops' strongest ties to coal are in places that have been on the losing end of deep cuts to social programs and industrial disinvestment, including from the coal industry itself. Even if they wanted to get off coal, for many co-ops there seem to be few easy or cheap ways out.

Coal's intransigence is a problem for IOU customers too. Where they exist, regional ISOs process bids from different distribution utilities to determine which generation plants will operate the next day and which won't, based on what source is cheapest at the time. A process known as self-scheduling allows coal generation to cut in that line, even when it's less cost effective. And because the job of an ISO is to make electricity demand equal to electricity supply at any given moment, this means that, so long as those coal plants are running within them, "they are pushing off some other cleaner and cheaper resource that would otherwise be operating," Joe Daniel, then senior energy analyst for the Union of Concerned Scientists, said.

In the Southeast, where there are no ISOs, massive utility holding companies with generation assets—namely Southern Company and Duke Energy—dominate. Since regulators have already approved those utilities to collect the cost of those assets back from customer bills, their ratepayers are left footing the tab for power they don't need to keep fossil fuel executives from swallowing the cost of bankrupt plants that it'd be cheaper to replace with renewables. Coal plants that IOUs have already gotten approved by PUCs they've helped to populate, Daniel told me, "get cost recovery regardless of the market clearing price. They have customers that have to pay them. They're not relying on markets for revenue. They have set rates by regulators. . . . It doesn't matter what the price is." Essentially, this is socializing the risks of uneconomic, polluting power while privatizing the rewards. As his research has found, running uneconomic coal plants is costing utility customers $1 billion a year.[31]

Navigating utilities away from coal isn't just about ownership or oversight. It also means finding some way to pay off debt for fossil fuel assets so they can stop running well before existing contracts say they should, a particular problem for co-ops that operate on a not-for-profit basis. A few have begun to come up with answers. After seeing its rates rise 106 percent to keep coal going, New Mexico's Kit Carson Electric Cooperative recently bought itself out of a long-term contract with the Tri-State Generation and Transmission Association in order to transition 100 percent of its daytime power sourcing over to solar by 2022. The move is expected to save member-owners $30 million over the course of a decade and inspired several other member-owner co-ops of Tri-State to try to spin off as well. After the Rocky Mountain Institute estimated that Tri-State—which sells power to forty-three co-ops and public power districts across four states—could save at least $600 million by using less coal and investing in renewables, state legislatures in Colorado and New Mexico, each served by Tri-State, passed clean energy targets in 2019 that made coal an even worse bargain. Under market and policy pressure, and with an increasingly agitated group of member-owners, Tri-State in early 2020 released a Responsible Energy Plan (REP) to phase out coal and replace it with renewables via six new solar farms and two wind farms.[32] Co-ops in Wisconsin and Indiana followed suit.

Absent federal intervention and funding, though, the ubiquity of co-ops' debt to coal producers will almost certainly continue to hamstring efforts to decarbonize the economy, as massively polluting plants keep running for decades to come.

A joint report from We Own It, CURE (Clean Up the River Environment) Minnesota, and the Center for Rural Affairs lays out potential paths forward. The USDA's Rural Utilities Services (RUS) could "take ownership of rural electric cooperative coal assets in exchange for forgiving the debt" and in turn require co-ops to make investments in energy efficiency programs and renewable energy. "Then," the report adds, the RUS "could work to quickly retire the use of these fossil fuel assets in the interest of the American people," not to mention co-op member-owners. Because the RUS has taken similar action before, this could happen without new legislation. The only limit would be the RUS budget. Thanks in part to a lack of transparency among electric cooperatives, exact figures on how much coal debt they hold has been difficult to come by. In reviewing what data was available, report authors estimated that a fifth of loan guarantees offered to co-ops by the RUS in 2010—$8.4 billion—is tied up with coal infrastructure, with potentially more coal debt financed by CoBank, the National Rural Utilities Cooperative Finance Corporation, and Wall Street banks.[33]

Whichever form it takes, what's needed for coal-heavy co-ops to make a rapid transition to clean energy is a coal debt jubilee. Alternatively, the RUS—through the annual farm bill—could help refinance co-ops' debt in exchange for requiring investment in renewables, energy efficiency upgrades, and modernized infrastructure. In any case, alleviating coal debt for co-ops would free member-owners up to invest in renewable energy that they themselves could own, operate, and reap the benefits of—not least of which would be cheaper rates. Public entities also aren't allowed to take advantage of the investment and production tax credits that at their peak allowed installers to deduct 30 percent of the cost of new solar and wind installations (respectively) from their federal taxes, giving third-party, for-profit installers a leg up over any generation and transmission cooperatives that might want to go green. An extension of those tax credits could change that.

There are other challenges. The coal report's authors note that the current makeup of co-op boards means they could take up coal industry talking points about the unreliability of renewables and reject them even when it's the cheapest option. Along similar lines, under immense pressure from fossil fuel interests, Tennessee passed a temporary ban on new wind development, and Oklahoma scrapped a statewide tax credit for renewables worth $500 million.

It's easy to overstate how effective NIMBY (Not In My Backyard) campaigns against clean energy infrastructure have been, and the idea of some kind of popular backlash against renewables is a convenient story for those invested in incumbent fuels; in reality, most of these fights are fueled by people who want to keep making money off incumbent fuels. And thanks to just how cheap renewables have gotten relative to fossil fuels, many red state power providers are switching to renewables for purely economic reasons, leapfrogging over the "bridge fuel" of natural gas. In total, twenty-nine states have enacted mostly modest renewable portfolio standards.[34] That doesn't mean popular opposition won't be a force to reckon with. The Tea Party employed a mix of grassroots and grasstops activism and managed to transform the Republican Party and American politics within a few short years. There's no guarantee that a disinformation-fueled grasstops revolt against wind and solar—of the kind already taking place in states across the country—won't still stymie progress at the national level.

THE PROBLEMS CALIFORNIA faces in a warming world are ultimately much bigger than PG&E, however big a role they've had in fueling them. The challenges facing the country's grid are bigger still. "Public ownership alone would be an enormous step," Brower Brown told me. "But without a Green New Deal, it wouldn't be enough. There's the huge problems of ecological restoration and fire prevention that PG&E or any other utility can't deal with on its own," he said.

Developers can't continue to build in the most fire-prone parts of the state, particularly if insurance companies balk at insuring repetitive loss properties. Massive amounts of new public housing could alleviate the housing crisis facing cities like San Francisco and keep Californians from having to seek cheaper homes in the urban-wildlife interface. The service area PG&E now serves will require massive amounts of labor just to trim back trees along its transition and distribution lines to make them less prone to sparks. Looking back to the original New Deal, this could be a job for a revitalized CCC that puts millions of people to work across the state; California already has its own poorly funded version. Bringing PG&E under state ownership won't stop fires from happening and can't ensure Californians are better protected when they are stopped.

More broadly, the United States currently has no plan to resettle the millions of people who will be displaced by fires, floods, and other climate impacts over the next century. A health care system that keeps people tied to their jobs and unable to take up new opportunities elsewhere isn't making matters any better. The question is whether we take on that spread of challenges with the painfully limited set of tools laid out a century ago by private utilities, fossil fuel companies, and the neoliberal think tanks they seeded: an environment where corporations must be gently coaxed into doing the right thing, all-powerful markets know best, and any kind of public ownership is an unthinkable step toward tyranny.

The blessing and the curse of transforming the grid today is that very little of that old model seems to be working. The ways that regulators and power companies have managed it since the nineteenth century, Gretchen Bakke writes in her expansive history of the grid, "are slowly but surely being relegated to the trash heap of history" by an oddball co-alition of, as Bakke describes it, Silicon Valley smart-guys, aging hippies, and multinational corporations. "The object of their ire," she writes, "is not so much the big grid but the habitual and increasingly ineffective and uninteresting ways that the electricity game has been run."[35]

The sheer abundance of renewable power presents challenges for a grid designed mainly to distribute electrons, not accept them from millions of microscale power providers spread out over hundreds of miles. And with new, massive amounts of renewable power coming online every day, the central task of the grid has shifted: instead of managing power supply to meet customer demand, the grid has to manage demand so that it can keep up with supply. Who will write the future of this new grid remains, in some sense, up for grabs. As of now, it's the multinational corporations and Silicon Valley smart-guys who look poised to win out. For-profit firms already dominate wind and solar generation, including fossil fuel companies like Equinor that have made hearty investments in wind power. Shell has ramped up its interest in utilities in recent years, too. With oil's future in flux, more unseemly investors may flood into the power sector in the years to come. Utilities have already begun turning to

Google and Amazon Web Services (AWS) to handle the superabundance of data now under its control and that is poised to grow as a new fleet of electric car charging stations and smart dishwashers and ovens—hooked up to the grid, primed for demand-response efficiency—come online.[36] Cloud and AI research is already a hot topic in the more forward-thinking corners of the utility world; they might genuinely help the electricity system to cope with the real challenges of getting clean power to as many people as possible and to move past bespoke management techniques. Yet it's also not clear what controls, if any, will be placed on what these for-profit firms can do with the new data they collect from our ever-smarter and more energy efficient homes.

At the very least, there needs to be a public option that allows every home in the US to take part in the energy transition and enjoy the lower bills and cleaner, more reliable power that will come with it; if private utilities aren't meeting those goals, there's no reason not to replace them with a power provider run firmly in the public interest. There's all manner of low-hanging policy fruit that can be used to spur more renewable power development, from loan guarantees to extending tax incentives to state-wide 100 percent clean energy mandates, on up through a nationwide renewable portfolio standard. All these send powerful and much-needed market signals and crowd in private investment, guiding it toward wind and solar and away from incumbent fuels. To meet the climate emergency head-on, though, means treating power—from renewable energy to transmission lines to whatever energy data we generate—as public goods subject to public debate, not luxuries controlled by a profiteering few and available only to those who are wealthy enough to afford them. Only democratic oversight—nimble and genuinely independent regulation—can do that, along with taking utilities' seemingly limitless amounts of lobbying cash out of the equation. Leaving a clean energy transition up to the strange array of for-profit actors and ideologies that populate today's power sector means it likely won't happen until it's too late.

CHAPTER 9
A POSTCARBON DEMOCRACY

It's possible that Hydrocarbon Man will be replaced by
Environmental Person.

—Daniel Yergin, author, *The Prize: The Epic Quest
for Oil, Money & Power*

AT FIRST GLANCE, the venues for the wave of strikes that broke out
among public employees in the spring of 2018 seemed odd. The sites of
the first three walkouts—West Virginia, Kentucky, and Oklahoma—were
all deep red right-to-work states that don't recognize the right to collectively bargain, let alone strike.

On closer inspection, something else tied these three events together:
the three states' economies all revolved around fossil fuels. Even the demands of striking public education workers routed back to the fossil fuel
industry. In West Virginia, teachers and school staffers chanted about
raising the state's natural gas severance tax under the capitol dome in
Charleston.[1] The precursor to the strike in Oklahoma was mounting pressure on the state legislature to make the virtually unprecedented move
of going after the oil and gas industry's profits. Finally—under mounting ire from public employees—lawmakers agreed to raise a rock-bottom
production tax on the industry that had helped land the state in a budget crisis despite soaring natural gas profits.[2] The following fall in Baton

Rouge, Louisiana, public schoolteachers voted to hold a one-day strike over billions of dollars in tax giveaways to ExxonMobil. Their district, meanwhile, carried a $22 million deficit, and teachers had gone ten years without either an across-the-board raise or a cost-of-living increase.[3]

There's a basic injustice embedded in these kinds of policies that doesn't require any complicated analysis of why fossil fuels are suffocating the planet to grasp—indeed, climate change was barely mentioned on the picket lines, if it came up at all. As West Charleston, West Virginia, Spanish teacher Emily Comer told me, "People understand that the gas companies can afford it, that there has to be revenue, and that it shouldn't come from poor and working people in West Virginia," referencing a cruel proposal from state politicians to fund public employees' pensions via cuts to Medicare and Medicaid. "We see the effects of poverty face-to-face when we walk into our classrooms every single day. We are the last people who want to get a raise on the backs of poor people."

If there was plenty of oil wealth to go around and make men like Aubrey McClendon, the Chesapeake Energy founder, billionaires, then why weren't the public employees who taught their workers' kids and tended to miners' injuries seeing any of those riches? In some places, public policy meant that the residents of extraction-heavy states saw some benefits. The Alaska Permanent Fund sends an annual dividend out to residents collected from taxes on oil extraction there. North Dakota, where shale drilling first exploded, offers another kind of alternate history for how states can use their resource wealth for the public good. In contrast to Oklahoma, taxes on new wells there were set to 11.5 percent at the start of the fracking boom. Education spending per pupil grew by 27 percent between 2008 and 2016—more than in any other state in the country.[4] The problems plaguing Appalachia's public sphere, though, wouldn't be solved by rejiggering the tax code or sending residents a check every year. The coal industry—which built many Appalachian economies—is collapsing, threatening to further decimate state and local tax bases. In 2011, over 18,000 people were employed by Kentucky's coal industry. As of the second quarter of 2020, that number had dropped to just under 3,760.[5] Coal production overall in Appalachia fell by 45 percent from 2005 to 2015, with eastern Kentucky and West Virginia getting hit worst. By the fall of 2020, the coal industry employed just 43,800 people nationwide.

As the president of the Kentucky Coal Association said of Trump's coal boosterism, "We are stopping the bleeding, but it has not stopped. We're starting to get to that flatlining point." Statewide, excise tax revenues from coal were half of what they had been a decade earlier.[6]

Natural or, rather, fossil gas—which looked like a promising revenue stream in some places, Oklahoma especially—is no long-term fix either.[7] As the last chapter laid out, shale drilling's growth was owed in large part to the low interest rates and easy credit that flowed from the financial crisis of 2008. That's giving way to a faltering economy and shrinking demand, most acutely as the coronavirus and ensuing lockdowns dealt a blow to global energy demand. But the decline started as far back as 2018, when Wall Street investors began losing interest. The Appalachian Basin, where teachers walked out, has seen the effects firsthand. By the end of 2019, eight exploration and production companies with a major presence in the region held $29.5 billion worth of long-term debt and a combined market capitalization of just $12.4 billion.[8] In 2020, Chevron wrote down $10 billion worth of assets rather than continue to invest in Appalachian shale production throughout Pennsylvania, West Virginia, and Ohio, following Shell and other oil majors out of a business where they just couldn't seem to make money.[9]

There's also the fact that addressing the climate crisis requires phasing out fossil gas. It's true that the rise of shale drilling in the United States has kept nearly two hundred coal plants from being built and collapsed the amount of America's electricity that comes from coal. But gas-fired power plants still produce about half the carbon emissions of coal plants, and the potent methane they give off has been responsible for an estimated 25 percent of warming.[10] While methane lingers for less time in the atmosphere than carbon dioxide, it stays roughly thirty times more potent a greenhouse gas than CO_2 a century on, and eighty-four times as potent for twenty years.[11] Should the world's gas-fired power plants follow through on infrastructure build-outs planned in 2019, they would produce 47 percent more gas than is consistent with capping warming at 2 degrees Celsius.[12] Though sold as a cleaner-burning alternative to coal, studies have found that methane leakage, venting, and flaring from gas supply chains—all consistently downplayed by industry and grossly underestimated and underregulated by the EPA—could make it more greenhouse gas-intensive than coal.[13] A 2018 study found methane was leaking at rates 60 percent higher than the EPA reported.[14] A single well blowout at an ExxonMobil drill site in Ohio in that year, another set of researchers found, released methane emissions "equivalent to a substantial fraction of the annual total anthropogenic emission of several European countries."[15] Building up fossil fuel infrastructure in the short term, to get off coal as renewables catch up, creates a kind of path dependence; that is, utilities and gas companies won't be too keen to retire facilities they've invested millions of dollars in just because some dismal climate math says they should.[16] Despite all this, oil and gas companies, Republicans, and prominent Democrats are still attempting to pass gas off as a viable bridge fuel—the only job it might be worse suited for than economic savior.

To figure out how to get from point A to B of the energy transition, Appalachian policy experts are looking to capture fuel wealth before it's snuffed out by economic collapse, decarbonization, or both. Ted Boettner, senior researcher at the Ohio River Valley Institute, has crafted a proposal to bring an Alaska-style permanent fund to West Virginia, financed—as such things are in other states—with severance taxes on extractive industries. Most permanent funds in the US, he explains, were started for similarly practical reasons, as states dependent on oil envisaged a future without it, be that via outside competition from OPEC or supplies simply running out amid peak oil fears that never came to fruition. Creating such a pot in West Virginia would be a way to make sure that funds from its remaining, unsustainable resource wealth can at least help sustain its economy longer term as extraction falters, spreading profits around to more than just the coal, oil, and gas executives who hoard them now. "Shared prosperity and natural resource extraction," he told me, "tend not to go hand in hand." In Belmont County, Ohio, on the West Virginia border, the shale revolution saw GDP grow 150 percent between 2013 and 2018. Compensation for all workers grew by just 12 percent over the same period.[17] It's a tricky line to walk, between letting profits flow to the top and sustaining or expanding public services with revenues that can't flow indefinitely. Alaska's Permanent Fund has encouraged the state's politicians to push for the industry's expansion and oppose most attempts to curb it for climate and environmental reasons; that it's one of the fastest-warming places on the planet has forced its politicians to chase goals that can seem at odds.[18]

There are no easy answers. Boettner's counterpart in Kentucky— Jason Bailey, director of the Kentucky Center for Economic Policy, or KCEP—has worked on a similar proposal, but at this point there isn't much resource wealth to go around in the Bluegrass State, which has very little fossil gas. And though their states' situations differ, Boettner and Bailey were crystal clear on one point: federal intervention is desperately needed.

Bailey argues that Kentucky "can do more" to alleviate economic hardship there but is adamant that revitalizing the economy is beyond the scope of what its constrained state budget would be able to fund. "There's no national policy or commitment to responsibly assisting communities that we don't need anymore for their role in the energy regime. It's really a tragedy what we've done. It's part of the reason that there's so much frustration here and in West Virginia and in other resource-intensive states. It's not going to change. If anything, these trends are going to accelerate as we adopt policies to deal with the climate problem," Bailey said, describing the unjust energy transition happening off coal and, more recently, gas. Adding to the stress of potentially cratered state and local budgets, whole families depend on employer-provided health care. Jobs and state revenue streams from the region's historical economic base drying up stand to devastate communities, further fraying whatever already threadbare safety net might have been there to catch them.

"In the future, you're going to have to look at lowering health care costs through a Medicare for All–type program and making college a lot more affordable," Boettner told me. "It's hard—when you're a poor state with low fiscal capacity—to pay for those things. Especially looking at the future, the federal government's going to have to play a big role once again."

The constrained budgets and ensuing strikes in West Virginia, Kentucky, and Oklahoma were a preview of the years and decades to come. As kids jammed into crowded Appalachian classrooms can attest, it's not just the profits of the fossil fuel industry or the livelihoods of its workers that are at stake in negotiating a just transition to a new energy system. Absent serious changes, declines in coal, oil, and gas production, which have almost nothing to do with climate policy, stand to crater everything from state budgets to pension funds. Regional economies bound up in fossil fuel production—indeed, that know very little else—are a canary in an abandoned coal mine for a broader transition both Democrats and Republicans would prefer to pretend isn't happening. Running for president, Joe Biden repeatedly pledged not to ban fracking. Talk of restricting drilling remains a third rail for even progressive climate proposals, where conventional wisdom holds that threatening fossil fuel jobs in swing states like Pennsylvania will be an electoral death knell. Whether politicians like it or not, though, those jobs are already disappearing. The transition is already happening. Moving forward, that shift can be managed and orderly, on a time line in touch with climate science and supporting communities that have historically depended on extraction to build a sustainable prosperity. Or Democrats can be blamed for the wreckage as vulture funds descend, CEOs raid health care and pension funds, and millions are left behind in states the party should be winning.

The problem isn't confined to red and swing states, either. Capitalism has been built up around fossil fuel extraction and the coal fires of the Industrial Revolution. Beyond just reducing the parts per million of greenhouse gases in the atmosphere, transitioning earnestly off fossil fuels demands fundamental changes in a dizzying number of industries and a rethink of everything from transportation to migration to the very foundations of the welfare state. As extractive sectors continue to shed jobs, they leave behind people and places they have no intention of helping navigate a world without them. By denying the urgency of and need for an energy transition—denying the fact that a slow, painful one is already happening—politicians leave the terms of that societal transition up to executives whose only interest has ever been to turn a profit. The only alternative is big, innovative, democratic government.

RESOURCE WEALTH WASN'T just a vehicle for cash that emergent capitalist economies could use to spur development, as Timothy Mitchell explains in *Carbon Democracy*. The specific nature of the supply chains used for different fuels helped condition radically different types of societies. Coal's dependency on both large concentrations of workers in mines and then teamsters to transport it away created a number of chokepoints that workers could exploit to demand higher wages and more protections, staging strikes that could bring whole countries to a standstill. All this militant energy wasn't just confined to the trains or mines, as unions used their leverage to push Victorian lawmakers to enact everything from male suffrage to the end of child labor to a shorter workweek.

These legacies live on today. In England, the lingering influence of the miners' unions—bludgeoned by Margaret Thatcher's Tory government—has traditionally been a bulwark against the far-right and ascendent xenophobia, even in areas that voted for Brexit as a middle finger to the powers at be; the collapse of Labour's so-called Red Wall in the north of the UK, where many voters flipped to vote Tory in 2019, shows that legacy may be starting to wear.[19] There are still sparks, though. In West Virginia, the first counties to go out on strike—Logan, Mingo, and McDowell—were the site of the brutal Mine Wars of the 1920s, still the largest armed uprising in the United States since the Civil War, sparked by fierce labor disputes. Strikers in the state capital donned red bandannas as an ode to the state's "redneck" union heritage, a callback to those worn by the miners who battled Pinkerton thugs.

There's no linear, deterministic relationship between fuel sources and the sorts of governments they create. But there are certain material realities endemic to different kinds of energy sources or—as author Timothy Mitchell puts it—"arrangements of people, finance, expertise and violence . . . assembled in relationship to the distribution and control of energy." The novel materiality of fossil fuels mixed with its surroundings to build novel political economies in the classical sense, whereby "the economy" isn't some free-floating matrix of rational actors constituting an all-knowing market but deeply embedded in and the product of specific political choices, power arrangements, and power sources.[20] In Appalachia, those forces produced stark stratification and a tiny class of elites, along with a tradition of labor militancy eager to fight back against both. Saudi Arabia, shaped in its earliest days by US oil interests, is a theocratic monarchy that employs armies of slave labor and brutalizes its own residents and geopolitical foes. If we recognize the full sweep of what decarbonization would mean, then we also have to ask the question: What sorts of societies can no-carbon energy build?

That the world will decarbonize, of course, isn't inevitable. If it does, there's no guarantee a greener world will be a fairer or more democratic one. Any decarbonization will also coincide with a crescendo of devastating climate impacts flowing from the inevitably warmer planet that's already locked in. The Internal Displacement Monitoring Service estimates that the "impact and threat of climate-related hazards" displaced around 21.5 million people per year between 2008 and 2015, with tens of millions more to follow from more prolonged droughts, more devastating storms, and ever-rising tides.[21] Resurgent far-right governments and xenophobic movements haven't exactly been welcoming to newcomers. Warming stands to exacerbate existing global inequalities—in no small part because the regions worst hit by climate impacts have also been those on the losing end of Western powers' colonial adventures, military conquests, structural adjustment packages, and foot dragging and obstruction in international negotiations. As Christian Parenti writes in *Tropic of Chaos*, climate change "arrives in a world primed for crisis. The current and impending dislocations of climate change intersect with the already-existing crises of poverty and violence." That "catastrophic convergence," he adds, does "not merely mean that several disasters happen simultaneously, one problem atop another. Rather . . . problems compound and amplify each other, one expressing itself through another."[22]

In such a context, making a low-carbon world a more egalitarian one will be an uphill battle. As discussed in the previous chapter, the renewables landscape in the United States is dominated by the private sector. One of the most recognizable faces of clean energy—Elon Musk—is also an antiunion zealot whose factories of the future have been rife with old-school labor violations, low pay, and debilitating shop floor injuries.[23] It's not hard from there to envision a renewables-based economy mirroring many of the unequal power relations that have defined fossil fuels' reign: hulking monopolies, labor exploitation across continent-crossing supply chains, and a single-minded focus on profits. Far off as it may seem, savvy American fossil fuel firms might even start treating their lip service to renewables as something more than empty symbolism backed up by token investments and transition in earnest to solar, wind, and potentially nuclear, while maintaining their corrosive political influence on governments the world over.

Certain material realities about renewables, though, would seem to make such top-down ownership structures a less easy fit. Solar power especially lends itself to stark decentralization. UV rays are abundant and can be harvested by rooftop energy systems that are delinked from centralized grid networks, albeit bound up in globe-spanning supply chains that furnish inputs like glass and lithium. In Germany, around 40 percent of renewable energy spurred on by the Energiewende, to transition that country to renewables, is publicly owned and operated and has ably competed with the country's Big Six energy providers. Already there are cities and neighborhoods around the US experimenting with community solar arrays and cooperatives and weaning off their monopoly power providers.

There's nothing inherently liberatory about decentralization, appealing as it might sound. Neoliberal policymakers have long seen it as a worthy goal, parcelling essential and traditionally public services like train lines and water utilities out to a bevy of private contractors and middlemen in the name of efficiency and free-market competition.[24] Tea Partiers turned Trump supporters throughout the Southeast have been dogged opponents of antisolar measures pushed by IOUs, citing the need for the "energy freedom" to make their own power and make America's deeply centralized grid less vulnerable to terrorists' cyberattacks. As Debbie Dooley, cofounder of the Atlanta Tea Party and Floridians for Solar Choice, told me not long after defeating an IOU and Koch brothers–backed campaign to kill renewables there, "Capitalism and the free market is a natural fit for solar . . . the government needs to get out of the way."[25] Enticed by the prospect of going off the grid, anarchists and conservative homesteaders could find common ground in mapping out a kind of do-it-yourself energy landscape. Renters and anyone who can't afford to mount solar panels on their homes can easily get left out of a solar separatist vision, particularly if the cost of entry is decided by a market built to suit clean energy shareholders and CEOs.

Justice won't fall out of a low-carbon future, and that future isn't guaranteed. As ever, getting either will mean a fight over not just what kinds of power turn the lights on but who holds it. Much of this book has been spent describing the various ways that fossil fuel executives have undermined democratic politics, bludgeoning hopes for ameliorating climate change and attempting to lay claim to the climate policy debate. The upshot of the story told by the likes of Shell and ExxonMobil now is that the sheer size of their business and the dominance of fossil fuels make their presence in the world's energy landscape—and continued profitability—inevitable. At the very least, we need a public option for energy that challenges that notion. Everyone should get reliable and affordable (if not free) no-carbon electricity, as outlined in the last chapter. But we also need to actively displace the companies that have corroded politics at home and abroad and made a rapid transition seem unimaginable.

THERE IS NO prosperous future for humanity that includes one for the fossil fuel industry. Barring an about-face in its core business model—a change we have no reason to suspect will happen—these companies' mission to dig up and burn as much coal, oil, and gas as possible stands directly at odds with a reasonably habitable planet. Fossil fuel production needs to be wound down at its source, however cheap renewables might get. Renewables—now cheaper than ever—are at this point mostly supplementing fossil fuels in aggregate, not displacing them. Liquified natural gas exports from the US, similarly, have helped spread more fossil fuels around the world without meaningfully displacing coal, Global Carbon Project researchers found.[26] Trillions of dollars of fossil fuel profits will have to go unrealized if the world is going to cap warming at 2 degrees Celsius. As their decades of denial and delay have proved, executives aren't going to want to give those up willingly. Despite all their newfound green rhetoric, fossil fuel companies annually spent $1 trillion building out new supply infrastructure between 2014 and 2018.[27] As late as January 2020, ExxonMobil planned to increase its carbon dioxide emissions by 17 percent through 2025, doubling its earnings by expanding its oil and gas business.[28]

Even Green New Deal proposals, though, have largely neglected the need to constrain fossil fuel supplies directly. Understandably, savvy politicians focus on positive climate policies that poll well: transitioning to 100 percent clean energy (supported by 82 percent of likely US voters) will mean investing to scale up the amount of solar and wind in our energy mix (79 percent support) and electrifying huge parts of the economy to run on them while putting stricter regulatory limits on fossil fuel demand. This will create a lot of jobs, which everyone likes. And those jobs will theoretically make up for the ones lost in coal, oil, and gas. It's noteworthy that a majority of Americans also support more regulation: 80 percent want tougher restrictions on power plant emissions, and 71 percent support enacting tougher fuel efficiency standards on cars.[29] A more ambitious administration could reinstate the crude oil export ban, end existing drilling leases on federal lands and waters, and create setback zones to ban drilling within a certain radius of playgrounds, hospitals, and other sensitive areas.[30]

There's a market logic at work here: by boosting supplies of the good stuff, the bad will wither as it's outcompeted and regulated away by new standards. There are plenty of reasons to be skeptical that this road leads to a zero-carbon economy, as noted above. But let's pretend it will for a moment, and that this suite of demand-side changes alone will drive fossil fuels out of the American economy. How might coal, oil, and gas companies handle their declining prospects?

We don't have to imagine a struggling fossil fuel sector. Worldwide, fuel prices plummeted when energy demand dropped during COVID-19 shutdowns, and the glut of supply on the market saw prices of West Texas Intermediate Crude turn briefly negative that April before limping back to around $40 per barrel the following summer. Mountains of debt and pressure from investors and insurers put the industry on defense, adding to its existing struggles to find financing and profits. By the spring of 2020, investment in the sector was projected to decline by $400 billion through the end of the year—the largest single-year drop-off in history, according to the International Energy Agency. Facing $11.8 billion in debt, Chesapeake Energy—the shale revolution's pioneer—filed for bankruptcy in August 2020. Eighty-four exploration and production firms and oil field services companies, which provide equipment to drill sites, had filed for bankruptcy by November 2020. Their combined debt load was $19 billion higher than that of the 124 companies that went bankrupt after the last fuel price crash in 2016. With hundreds of billions of dollars worth of oil and gas debts coming due in the next four years, many more bankruptcies are coming.[31]

While the pain was deepest for smaller, independent oil producers during the pandemic, oil majors (the big guys) have been hurting too. ExxonMobil's profits fell by 30 percent through 2019. The most valuable company in the world just a few years earlier, it was delisted from the Dow Jones Industrial Average during the pandemic in 2020, after more than ninety years on the index, as its stocks plummeted to eighteen-year lows. In line with their competitors, ExxonMobil executives announced that they would cut capital expenditures by 30 percent and lay off 15 percent of staff worldwide, maintaining generous payouts to shareholders while cutting the company's contribution to employees' health care.[32] How Exxon navigated those cuts offers a preview for how the industry could respond to the kind of dramatic demand collapse an energy transition would bring about. Faced with one of the most profound crises in the history of the sector, it didn't relent or attempt to find new avenues for profit in faster-growing sectors outside fossil fuels. It cut corners, shed workers, and doubled down, writing off its profligate spending in the Permian Basin while pledging anew to produce an additional 750,000 barrels of oil in Guyana by 2026. The top brass at Exxon, it seems, don't know how to do much else. Occidental Petroleum was honest about its limits when outlining a plan to reach net-zero emissions by boosting both extraction and questionably effective carbon capture technology: "We are doing a contrarian approach in that we believe that using our core competence of CO_2-enhanced oil recovery expertise is the best way to go, rather than trying to go learn a new business."[33] BP's Bernard Looney offered greener messaging on his companies' troubles. He pledged a double-digit production decline and more diversified spending, although remained eager to maintain a future for BP's oil and gas business amid peak demand. It laid off 10,000 workers. Shell laid off 9,000.[34]

The think tank Oil Change International has distilled the future of fossil fuels down to three possible outcomes. In two of those, the world manages to restrict emissions below catastrophic levels, in line with the targets set out in the Paris Agreement. In the third, it doesn't. The first is a managed decline of fossil fuel production starting as soon as possible, accomplished through investments that ensure the workers and communities that stand to be most affected by industry closures are made whole and that the world can continue to meet its energy needs. The second—in which warming is also kept below 2 degrees Celsius—sees a "sudden and chaotic shutdown of fossil fuel production" a short time from now that includes "stranding assets, damaging economies, and harming workers and communities reliant on the energy sector." In a final, worst-case scenario, the world fails to restrict fossil fuel production in line with either a 1.5 or 2 degree target, tipping into not just ecological disaster but also, as economists have warned recently, "worldwide economic collapse," where bad weather helps strand assets and torpedoes livelihoods the world over.

Faced with rising temperatures, intransigent, politically powerful CEOs, and an industry already in deep crisis, there's one tool for timely decarbonization and a just transition—for getting to something like that first path—that draws on a proud American tradition: nationalizing the United States' fossil fuel industry. To be specific, the US government should purchase a 51 percent equity stake in fossil fuel corporations to oversee a managed and orderly decline of their core business model within the next ten years.[35]

It's not as crazy as it sounds. Like solar panels or the public ownership of utilities, nationalization is just another technology: it can be wielded toward positive ends or negative ones; to speed up decarbonization or slow it down. It won't solve anything on its own. Twelve of the world's twenty largest industrial greenhouse-gas-emitting companies are state-controlled entities, which account for a much larger share of the world's oil and gas production than private companies. Certainly, no one is looking to Saudi Aramco or Petrobras as models for climate action. Not unlike dynamics in Appalachia, countries that stake most of their revenues on fetching high prices for oil exports stood to be those worst hit from the COVID-19 recession. Nationalization isn't a particularly left-wing idea, either. In a world where Republicans weren't religiously obsessed with drilling up as much fossil fuel as possible, it would be possible to imagine the creation of a US national oil company erected to rationalize production and assert American dominance on the world stage as a true swing producer that keeps prices stable.

After all, higher-order planning in other major oil-producing countries tends to fall to national oil companies (NOCs), which manage production as they interface with global markets to satisfy domestic and international demand. Major investor-owned oil companies have tended to internalize these functions, featuring high degrees of internal planning that are aimed at securing fossil fuel profits for decades to come; instead of protecting state revenue streams like Saudi Aramco or Norway's Equinor, they look out primarily for shareholders, erecting quasi-states like the one run by ExxonMobil.[36] In those pursuits, they enjoy ample governmental support in the form of various subsidies, foreign policy priorities, and complex regulatory regimes designed to encourage so-called energy independence.

Aside from all the foreign meddling detailed earlier in the book, the history of domestic fossil fuel production is one of careful market management and of regulators protecting the sector's longer-term self-interest from individual producers' focus on short-term profits. Historically, that's meant telling companies to stop drilling. As former Obama staffer and industry consultant Robert McNally writes in his book *Crude Volatility*, the breakup of Rockefeller's Standard Oil empire in 1911 was followed by about twenty years of a free-ish market for energy before US regulators enacted what McNally calls "some of the most rigorous economic planning on an industry ever seen to man." New Deal–era interior secretary Harold Ickes took a personal interest in stabilizing the country's then struggling oil industry and sought to establish direct federal control over its operations to conserve the country's natural resources; the National Industrial Recovery Act (NIRA) suspended antitrust laws, allowing oil companies to devise strategies for how to bring up prices that had plummeted as production exploded.[37] So-called Hot Oil laws, of the Connally Hot Oil Act of 1935, restricted the sale of fuel produced in excess of state production limits.[38]

Most of the twentieth century's action, though, came from the Texas Railroad Commission (TRC). In a practice known as prorationing, the statewide regulator placed production caps on every wellhead in the 1930s to prevent overpumping, conserve resources, and stabilize prices. At one point, the TRC even declared martial law, calling in the National Guard to keep oil in the ground and physically prying workers from rigs that were disobeying caps. Working together with the seven sisters, the TRC's heavy-handed planning maintained a cartel in the world's oil markets from the Great Depression through the 1970s before being overtaken in that role by OPEC. That body's cofounder Abdullah al-Tariki had trained at the TRC, and he and its other left-leaning architects saw it as a promising model for both regulating the world's oil markets and ensuring Western capitalists didn't reap all the rewards. "If the tendency towards significant technical and administrative control is justified in the US for its domestic oil industry," Eduardo Acosta Hermoso, the first Venezuelan governor to OPEC commented, referring to the TRC, "would it not be even more justified when the industry is entirely in the hands of foreign investors whose interests are structurally divergent from those of the country that owns petroleum?"[39]

And though outright nationalization hasn't been all that common a tool in the US fossil fuel sector, it has a long history in navigating the country through its most pressing crises. The New Deal displaced all manner of private business, in the power sector and beyond. Several functions of the banking industry were nationalized after Black Monday through the creation of institutions like the Reconstruction Finance Corporation, the Federal National Mortgage Association (Fannie Mae), and the Export-Import Bank, before Roosevelt ultimately nationalized the country's gold reserves. One popular arrangement during World War II was government-owned, contractor-operated factories, where the government had private companies operate facilities used for war production.[40] More drastic actions were taken, too. To meet war production demands as fighting began in Europe, Roosevelt revived World War I's National War Labor Board, allowing him to intervene in labor conflicts that threatened manufacturing on the side of both management and labor. In 1943, Congress passed the War Labor Disputes Act, which let the government nationalize any facilities that might be needed in the war effort. The White House used that authority on coal-mining and oil-drilling operations, railroads, department-store chains, and the US subsidiaries of foreign-owned eyeglass, champagne, and beer companies, among several others. John Ohly, a lawyer in the Office of Assistant Secretary of War, once said "the government was taking over approximately one plant a week" in the three-month lead-up to V-J Day.[41]

Concerted pushback to nationalization as such began during Harry Truman's administration, when the Supreme Court ruled against his nationalization of steel mills to support the Korean War.[42] Importantly, the Court's argument was that the Korean War didn't justify the White House seizing a steel mill without congressional approval—not that future presidents couldn't make similar moves in more pressing emergencies. That ruling, though, did coincide with McCarthyism and the purge of Communists from left-leaning unions more inclined to support various kinds of public ownership. By the 1970s, the Democracy Collaborative's Thomas Hanna told me, "the business sector had gotten itself organized" against the New Deal consensus on the role of the state in the economy. "You have this shift," he said, "where you still have government interventions—significantly more government interventions in the 1970s as economic problems mount—but it's bailouts, not direct ownership. By the mid-1970s, nationalization was not really on the table anymore ideologically." It still happened, just under different names. Passed unanimously through the Senate in November 2001, the Aviation and Transportation Security Act nationalized airport security by creating the Transportation Security Administration, the functions of which prior to 9/11 had been performed by a patchwork of the now-displaced private security companies contracted out by individual airports.[43] The Aviation and Transportation Security Act resulted in the largest single expansion of the federal workforce since World War II.

In 2008, the Obama administration spent $700 billion to bail out troubled banks, insurers, and car companies to prevent a depression. While it demanded some standards and regulations in return, the government chose not to exercise all that much say over how those companies were run. The government similarly took out a 36 percent stake in Citigroup. Instead of conditioning that bailout on a top-level reorganization or change of management, Treasury Secretary Tim Geithner intervened as a relentless advocate for Wall Street, allowing banks to take vast sums of public money with no strings attached, with many pouring it right back into the same kinds of speculative investments that had sparked the financial crisis.[44] As I wrote above, the Obama administration also eagerly rebuffed any suggestions that the government would exercise the shareholder rights that its majority stake in General Motors entitled it to.

The CARES Act, passed amid the coronavirus crisis, left open the option for the government to take an equity stake in companies receiving federal funds. Yet it similarly prohibited the government from exercising its shareholder rights. That bailout included fossil fuels, which received ample support. That spring, too, the Fed specifically expanded the broader emergency Main Street Lending program along the lines that fossil fuel lobbyists had urged. As a result, 83 percent of oil and gas companies' $744 billion debt became eligible for bottom barrel refinancing by the Fed.[45] "The decision to bring oil and gas into the Fed's investment portfolio," former Fed board of governors member Sarah Bloom Raskin wrote, "not only misdirects limited recovery resources but also sends a false price signal to investors about where capital needs to be allocated. It increases the likelihood that investors will be stuck with stranded oil and gas assets that society no longer needs." By the summer of 2020, 62 percent of mining companies, including but not exclusively oil and gas firms, had gotten $4.5 billion bailout funds that were nominally meant to help struggling small businesses.[46] By October, the Fed's bond-buying programs had snapped up $100 billion worth of fossil fuel debt.[47]

Fossil fuel workers, meanwhile, were left out in the cold. Between March and August, 107,000 oil and gas jobs were slashed in the US, 70 percent of which—the management consultancy Deloitte projected— wouldn't come back in an upswing. New jobs projected to be created during a recovery are likely to be in white collar fields like IT and data analysis. Regionally, the picture was even more bleak. Ninety-thousand fossil fuel jobs were shed in Texas alone in March and April of 2020, with some companies laying off thousands of workers in a single day as prices sank. In April, US coal jobs recorded the largest single month-to-month drop-off since the Bureau of Labor Statistics began recording such statistics. For his part, Donald Trump refused calls from mining unions to ensure safety protection for members deemed essential workers, still working underground in close quarters amid the pandemic.[48] The Energy Information Agency projected a 23 percent dip in already declining coal production through the rest of the year. Another analysis found that the number of domestic oil rigs in operation in the fall of 2020 had dropped to just 260. Nearly 600 had been shut in over the course of a year.[49]

Because it bears repeating, tens of thousands of people are already crashing out of the fossil fuel industry with very little policy in place to catch them. Their bosses aren't about to lend a helping hand. Documents made public as part of coal giant Murray Energy's bankruptcy filings revealed the company had funneled hundreds of thousands of dollars to the Competitive Enterprise Institute and Heartland Institute, extending climate denial groups cut off from other funding a lifeline as he laid off workers.[50] When the company's bankruptcy settlement declared it wasn't responsible for claims against Murray under the Black Lung Benefit Act, its $74.4 million worth of liability under that law was shifted to the Department of Labor's Black Lung Disability Trust Fund. In a Stanford Law Review article published just before Murray Energy's and other high-profile coal bankruptcies, researchers Joshua Macey and Jackson Salovaara found that between 2012 and 2017, four of the country's largest coal producers had used bankruptcy to shed $5.2 billion of obligations to employees and environmental remediation.[51] In a representative example from oil and gas, Diamond Offshore Drilling exploited a provision in the March stimulus bill to reap a $9.7 million tax refund and then immediately tried to funnel that same amount to four of the company's top executives through bankruptcy court. Diamond laid off 102 employees the next month.[52] After laying off hundreds of employees through 2018, Whiting Energy CEO, chairman, and founder Bradley Holly secured himself a $6.4 million payout in bankruptcy proceedings. That's nothing new: between 2012 and 2019, Department of Labor investigations recovered $40 million in wage theft for 29,000 oil and gas workers around the country.[53] Potentially worse still is the opportunity mounting fossil fuel bankruptcies present to private equity vulture funds, who were already salivating at the chance to snap up flailing companies and suck them dry before the pandemic.

Nationalization offers an alternative to letting either private equity vultures or CEOs take the money and run, prioritizing communities whose livelihoods have historically depended on fossil fuels. By the time you read this, it's possible ExxonMobil and ConocoPhillips will be riding high again, with oil prices sailing north of $60 or even $100 per barrel. But another bust stalks every boom. At the very least, the next time they come begging to the federal government, it should buy them out.

It might not cost much. As of writing this, energy stocks have begun to rebound. The value of the three largest US-based fossil fuel producers is just over $320 billion—roughly the amount of subsidies the industry could conservatively expect to rake in from states and the federal government before a child born in 2020 graduates from high school, into a world hotter than our own. Whereas a takeover of the world's top twenty-five oil, gas, and coal companies would have once cost some $1.15 trillion, buying them out now would cost somewhere between $550 and $700 billion—or half that with a 51 percent rather than full stake. Nixing market distortions like production-side fossil fuel subsidies could bring that price tag down further still. So could any number of common-sense reforms like those mentioned above, including energy efficiency measures and renewable portfolio standards that would erode fossil fuel demand.

Politically, nationalization is an epic long shot. Logistically, it could be fairly straightforward. While standalone legislation seems unlikely to pass, a broader bailout package could ensure the government can take up equity stakes in whatever firms it bails out. Once funding is authorized, the treasury secretary could set up a stranded assets relief program, buying up stock in US-based fossil fuel companies like it did with major banks in the wake of the financial crisis. If Tim Geithner could spend $700 billion bailing out Wall Street, then surely the planet and hundreds of thousands of livelihoods qualify as too big to fail. On another front, the Federal Reserve's dual mandate for full employment and price stability requires that it take on systemic risks to the economy. Central bankers around the world are starting to recognize that the climate crisis certainly qualifies. That Fed chairman Jerome Powell elected to have the Fed join the Network of Central Banks and Supervisors for Greening the Financial System (NGFS) in the days after Trump's election shows he might agree. And the barrage of novel policies the Fed unveiled after the crisis in 2008 and in the early days of COVID-19 shows there is no shortage of tools in its box, including adding big corporate assets on to the Fed's balance sheet without congressional approval.[54] If the Biden administration takes up demands from the Sunrise Movement and others to create a cabinet-level office of climate mobilization—modeled on wartime planning agencies—its mandate could mirror that of the bodies that seized factories for the war effort. Invoking the National Emergencies Act to declare a climate emergency could unlock additional powers and spending authority. So could wielding the Defense Production Act (DPA), which was used in the pandemic to compel companies to produce PPE. The Pentagon has long stated that global warming is a threat to national security, so utilizing the DPA to spur along the energy transition wouldn't be too far a stretch. The US has nationalized companies to deal with crises before. It can do it again, whether Republicans control the Senate or not.

Some kind of backlash is inevitable. Republicans will call any modest climate measures Democrats try to pass a Stalinist Green New Deal and no doubt use the courts to help stop it; nationalization might see them foaming at the mouth. Whether it enacts climate policy or not, though, a Biden administration will likely be blamed for continued coal, oil, and gas declines. That could spell electoral trouble down the road if the party takes its usual stance: hammer home positives about clean energy and swallow industry talking points that the oil and gas sector is still a steady source of jobs instead of a sinking ship that scrimped on lifeboats. Promising to build a new EV factory in Detroit may not be welcome news to a laid-off roughneck in the Permian Basin.

Nationalization holds some serious political upsides but requires recognizing that a transition isn't some far-off event. A credible plan to keep people on payrolls could head off opposition, potentially peeling off unions and workers that executives have cynically wielded to curry favor for new infrastructure projects and regulatory rollbacks. The idea behind a managed decline is not to shut off all the taps overnight but wind down the fossil fuel industry's core operation along a time line that allows the country to meet energy needs as no-carbon alternatives continue to scale up. And while the top brass won't be setting the strategic direction of their companies anymore, a smart nationalization strategy wouldn't look to boot out either the rank-and-file workers or the higher-level managers, researchers, and IT professionals that keep it running and hold valuable expertise.

There are plenty of ethical and logistical considerations: What's to be done about the inevitable slew of lawsuits and even trade disputes? Which agencies would administer this collection of newly public companies, or should they combine into one entity in the mold of national energy companies elsewhere? Which companies need to be bought up, and which should simply be allowed to go bust? Should fossil fuel executives—who've known about the climate crisis for decades and sponsored denial and delay—really be compensated as their companies come under public ownership? What role should workers and the communities that live on the front lines of fossil fuel extraction have in directing this transition?

THERE'S ONE BIG problem here. Whether it's through deindustrialization or NAFTA, workers have never seen a just transition. That the one away from fossil fuels could be better isn't self-evident. For good reason, fossil fuel workers have seen such calls to keep coal, oil, and gas in the ground and ban fracking as threats to their livelihoods. There are no jobs on a dead planet, of course, but some apocalyptic climate-change-created hellscape seems a lot more abstract than next month's rent check or credit card payment. "You see all these slings and arrows headed your way to your livelihood, climate change being one of them," said Rick Levy, president of the Texas AFL-CIO. The Green New Deal, he added, "is either the panacea or the devil, depending on where you're coming from." In an economy where wages have stagnated as profits have grown, becoming a miner, refinery operator, roughneck, or pipefitter—depending on where you live—can offer a six-figure salary without a college degree. In California, fossil fuel workers on average make almost $38,000 more than the average salary in the state and are more likely to work full-time. It's gritty and often dangerous work but comes with good pay and benefits and predictable hours, if you can get it. That it's possible to make such a decent living in the industry is thanks largely to how densely unionized the extractive sector is and has been historically.

Labor is often, mistakenly, treated as a unified and reactionary bloc on climate. The "jobs versus environment" narrative peddled by the press and policymakers, including many Democrats, tends to assume that the outspoken building trades union leaders—which have bused workers to Washington in support of the Keystone XL pipeline and lashed out at climate campaigners—speak for the 12.5 million members of AFL-CIO-affiliated unions, for large non-AFL-CIO-affiliated unions like the Service Employees International Union (SEIU) and the Teamsters, and for working people as a whole. It's tough to square that picture with the several union internationals and locals, including SEIU, the American Federation of Teachers, the Association of Flight Attendants-CWA, and the New York State Nurses Association, that endorsed the Green New Deal.[55] Or groups such as Trade Unions for Energy Democracy and the Labor Network for Sustainability, which have been rallying labor leaders around climate action worldwide for years.[56] It doesn't fit too well either with the 62 percent of rank-and-file union members that support that framework, per a 2019 poll from Data for Progress.[57]

Fossil-fuel-adjacent unions themselves are hardly a monolith, either. As water protectors at Standing Rock stood in the way of the Dakota Access Pipeline, Terry O'Sullivan, the president of Laborers International Union of North America (LiUNA), penned an irate letter complaining of a "concerted campaign of misinformation" and called the unions supporting protests against the pipeline—including the SEIU, Communications Workers of America, and National Nurses United—"bottom-feeding organizations" that "have sided with THUGS against trade unions."[58] Not long after, LiUNA-City Employees Local 236—a Madison, Wisconsin, local—unanimously passed a resolution in support of water protectors at Standing Rock and against the Dakota Access Pipeline, citing their belief that "there would be more and better sustainable jobs if we invested in other types of energy that were not fraught with so many accidents."[59] As part of the AFL-CIO Energy Committee—a collection of ten building trades and other fossil-fuel-adjacent unions—LiUNA and the International Brotherhood of Electrical Workers (IBEW) executive leadership each came out strongly against the Green New Deal.

"We will not accept proposals," the committee wrote, "that could cause immediate harm to millions of our members and their families." Yet IBEW workers around the country have been enthusiastic about such measures, joining on to supportive resolutions in central labor councils. The headquarters of IBEW Local 103 in Boston is decked out with five hundred solar panels, and it has joined locals across Massachusetts in raising standards for workers in the state's burgeoning solar industry, taking on low-road rooftop installation contracts offering shoddy work and low wages that encourage a race to the bottom in clean energy.[60] Local 103 business manager Lou Antonellis came out in support of a Green New Deal, telling one interviewer that "we've been championing this stuff for a long time" and that he was "glad to see it finally resonate in the

Legislature." IBEW Local 11 in Los Angeles, in similar fashion, has seen a windfall in work thanks to the state's prorenewables legislation in the Golden State.

Ryan Pollock is a member of IBEW Local 520 in Austin, Texas, and through that a delegate to the city's Central Labor Council. While other IBEW members work in coal- and gas-fired power plants, Pollock is an indoor electrician who wires up commercial buildings. Given that decarbonization means electrifying everything, work like his could skyrocket in a concerted push to decarbonize the US economy, as heating and air-conditioning systems switch over to the grid. He worked with members of other unions to get a resolution supporting federal climate policy passed unanimously through the state AFL-CIO—a Green New Deal in everything but the name.[61] In marathon negotiations to build support for that resolution, Pollock said the fossil fuel workers he spoke to weren't in denial either about the climate crisis itself or what's coming. "Even if they don't believe in climate change, they know that green tech is going to be replacing fossil fuels either way," he told me. Pollock recounted talks with fellow IBEW members working in the coal-fired Fayette Power Plant, east of Austin, pushing them to sign on to the resolution. "The ever-increasing cost of fossil fuels means your plant's in danger no matter what we do. We can do this now, or we can do it later, or we can kick this can down the road. This is about labor getting ahead of the ball for once in the last century or so, and it's about us calling the shots and not letting someone else call them for us."[62]

Historically, environmentalists looking to stop fossil fuel infrastructure projects have asked workers to trade their actually existing work for jobs that might exist far off in the future. That's no longer the case. Clean energy jobs surpassed those in fossil fuels in 2017. The occupations projected to grow the fastest in the US are all green, either in clean energy or in the caring economy. Wind turbine service technicians top the list, and those jobs are expected to grow at double the rate of derrick and rotary drill operators over the coming decade.[63] Their median pay is already comparable, if not better. There is no shortage of work available in the move away from fossil fuels. The bigger challenge is ensuring that clean energy jobs are good jobs. Elon Musk and other renewables bosses would be happy to see a race to the bottom in the clean energy workforce that leaves unions out entirely; in the United States today, Tesla is the only nonunion car company. The fact that union contracts for clean energy tend to be weaker than those building fossil fuel infrastructure speaks to the relative weakness of labor when standards for that industry were being set. In rooftop solar, there's little unionized work to be had in clean energy and seemingly scant interest from the relevent unions in organizing those workforces.

As Pollack points out, the labor movement can play a big role in changing this and ensuring a clean energy economy is one where unionized work under strong contracts is the norm rather than an exception. Unions can train workers coming out of fossil fuels for low-carbon work by ramping up their apprenticeship and training programs. Following the lead of IBEW workers in Massachusetts, they can also be a political force demanding the expanded use of Project Labor and Community Benefits Agreements in clean energy development, and that the right to collectively bargain is protected throughout the renewables sector.[64] In Spain, meanwhile, an agreement between their center-left government and trade unions saw the country close down its last coal mines and invest €250 million into the surrounding area. Miners over forty-eight—about 60 percent of the roughly one thousand of them remaining—could retire early on a full pension, with the rest to be retrained in environmental remediation and clean energy.[65] Green New Deal advocates including Bernie Sanders have proposed five years of full wage and benefits parity for workers moving out of fossil fuels, with an option for early retirement to older workers.[66] Free college and universal health care would make that transition easier still. Winning such labor-friendly protections will take ambition and a solidarity with social movements and other unions that the international leadership of the trades has been loathe to embrace in recent decades as they have tried to protect their own narrowly defined turf. It will also mean recognizing that the interests of fossil fuel workers and their bosses are categorically different things.

It wasn't the chemical properties of the bituminous coal British miners unearthed that constructed the British welfare state, after all, but the supply chains they disrupted and the pressure they exerted on the House of Commons. A clean energy economy won't be one that's good for workers without a fight and plenty of hard-won workplace organizing drives.

For moral and pragmatic reasons, climate advocates have a vested interest in making sure green energy is at least as union dense as the one it's replacing, with high standards stretching from lithium mines making battery storage possible to rooftop solar installers to EV charging station technicians. Without a detailed plan to make people whole, drafted alongside workers with the most to lose, any climate plan risks crashing on the rocks that have sunk it before. Today's fossil fuel workers could help to run a new national energy company, along the lines of well-established codetermination models used across parts of Europe. In a survey conducted of 1,383 offshore oil and gas workers in the UK, trade unions and environmental groups there found that 82 percent of those asked said they would consider changing to jobs outside oil and gas, and over half said they would be interested in work in renewables, offshore wind, and decommissioning today's oil rigs.[67]

Liam Cain, a member of LiUNA Local 237 in Cheyenne, Wyoming, joined members of LiUNA-City Employees Local 236, from Wisconsin, to set up Labor for Standing Rock in 2016. It was part of the main Oceti Sakowin encampment in North Dakota. Now a wildland firefighter, Cain spent five years switching off between jobs building oil pipelines and wind turbines and working at refineries. The green jobs he got through LiUNA were often worse than those that "pillage the earth," as he puts it. "When I worked on a natural gas pipeline, that was the best income I ever made," he said, noting that working eight months could earn him as much as $52,000, plus generous benefits and health insurance. The work was long—ten- to twelve-hour days, six days a week—but reliable, and even when work was delayed, a healthy paycheck would come through.

Conversely, Cain said, "If I worked through LiUNA on a green project it was very tenuous whether I could pay my bills or not." He credits that largely to LiUNA leadership, eager to enter the green jobs space but not as willing to negotiate strong contracts. "There's a lot of noise around green energy and green construction, and LiUNA and many other building trade unions want to be on the right side of that." But, he argues, "they've been conciliatory with the windmill companies."

There's plenty of well-paid and unionized work to be done right away that would give today's fossil fuel workers a lucrative role in the energy transition. Much of it just isn't profitable for today's fossil fuel companies to do, either now or in the future. Nationalization offers a pathway to de-emphasize shareholders' skewed priorities in order to get this vital work done. It would deliver a just transition that honors union contracts and what fossil fuel workers have done to build our world and provide tangible ammo against decades of bad-faith jobs versus environment attacks that have undermined climate policy more broadly.

Hundreds of thousands of good jobs are quite literally sitting around the country, waiting to be filled. Plenty of them are sitting right where today's fossil fuel workers already live. The US is home to at least 3.1 million abandoned oil and gas wells, 2.2 million of which are unplugged, potentially leaking prodigious amounts of greenhouse gases into the atmosphere and threatening local air and water quality. Every downturn adds to that count, and as many as hundreds of thousands of abandoned wells have yet to be identified.[68]

Payments for this cleanup are supposed to be furnished by drillers themselves, but the laws governing those amounts haven't changed since the 1960s. A 2018 report by the Center for Western Priorities found that reclamation on federal lands alone will cost $6.1 billion, far outstripping the $162 million oil and gas producers have provided for such cleanup expenses under sixty-year-old federal mandates.[69] Still more dramatically, the Government Accountability Office has estimated that just 0.01 percent of wells under federal jurisdiction have the necessary cleanup funds available.[70] Companies have already socialized the risk of mounting price tags and public health hazards created by abandoned wells. Especially given that the price tag is now effectively a sunk cost, drillers could be kept on the payroll of their now publicly owned employers with full wages and benefits to fill them; after all, who knows those wells better? A 2020 study from Resources for the Future and Columbia's Center on Global Energy Policy found that a federally coordinated effort to identify and plug 500,000 abandoned wells could create 120,000 jobs—more than the total number of jobs lost in oil and gas between March and October of 2020.[71]

Since these wells are clustered in places that have active drilling, state and local tax bases wouldn't need to face an apocalyptic loss of tax revenue as residents pick up and leave for work elsewhere. Under new, public management, drillers could stay on payroll to clean up their own states instead of generating profits for out-of-state shareholders. Beyond former sites of extraction, about half of the country's 450,000 contaminated lands, known as brownfield sites, are believed to owe that designation to petroleum infrastructure, including from the leakage of storage tanks buried under gas stations.[72] Modest reforms to the EPA's Brownfields Program could create a more direct hiring path for workers, who can now partake in federally supported training programs but whose job prospects still depend on often elusive private sector interest in their area.

There are other opportunities, too. While geothermal energy currently provides just 1 percent of power to the US, a 2019 Department of Energy study found that it could generate between 16 and 20 percent of the country's electricity by 2050. Despite lingering technical barriers and geographic limitations, the still-fledgling US geothermal sector could be a major complement to renewables, since it's available around the clock. Some firms have also explored the possibility of recovering lithium from the superheated brine unearthed in geothermal power production before it's reinjected underground, potentially producing a critical ingredient in the batteries that store solar power and increase reliability. Known hot spots for geothermal—which requires superheated underground pools—don't map cleanly onto the sites of former fossil fuel infrastructure, but the equipment and expertise used to get it out of the ground can be virtually identical. Every member of the start-up geothermal company Fervo Energy got its start in the oil and gas business, including founder Tim Latimer. "The skill I had developed to that point was poking deep holes in the ground," he told me. "What I learned in researching geothermal was that I could do that for no-carbon energy."

As mentioned above, retrofitting buildings and housing stock for energy efficiency could create millions of jobs, many of them unionized. In 2019, housing and climate justice activists joined with organized labor in pushing New York City to pass its own Green New Deal prototype, requiring buildings in the city—among the state's biggest contribution to global warming—to slash emissions by 40 percent through 2030. Upgrades to just 55,000 buildings there are projected to create 161,000 new jobs.[73] Scaled up nationwide, such work could be a virtually unlimited source of well-paid union work. Unions themselves could train ex–fossil fuel employees to do it and swell their ranks and political power in the process.

The bottom line is that after more than a century of corporations treating the environment as a dumping ground, there is a vast amount of work to be done not just plugging wells but repairing all manner of landscapes scarred by fossil fueled development.

There's been some interest in these ideas already. Canada announced in 2020 that it would pour $1.7 billion into an abandoned well cleanup fund spanning three provinces, expected to create 5,200 jobs overall.[74] In the US, a consortium of thirty-one oil-producing states requested that the Trump administration attach terms to fossil fuel bailouts as part of the CARES act that would require oil and gas companies to keep workers on staff amid the coronavirus downturn to plug wells.[75] Unsurprisingly, the White House didn't oblige. Future administrations needn't make the same mistake.

To oversee a managed decline and re-employment for today's fossil fuel workers, researchers at the Democracy Collaborative have suggested the creation of a federal Just Transition Task Force and ensuing agency, modeled on the cross-agency planning bodies that stewarded the domestic mobilization around World War II. Another, complementary possibility would be forming a national energy company, inspired only loosely by the NOCs elsewhere. It could house and set rules for companies newly brought under public ownership, in coordination with the Departments of Energy and Labor. Loosely following the structure of state-owned energy companies elsewhere—and along the lines of Tennessee Valley Authority's founding mission—a national energy company could be organized at the regional level by subsidiaries adhering to national standards, drawing on geographically rooted knowledge and expertise to meet the unique energy and employment needs of specific areas; with codetermination, workers could sit on the board and have a sizable role to play in management and planning. Rather than jockeying against OPEC and Russia (OPEC+) on the world stage, a US equipped with both a national oil company and stronger regulations on its energy sector could be an active participant. Seeing OPEC as a venue for climate diplomacy—more in line with the conservationist vision of that body's idealistic founders—the US could collaborate with other major fuel producers to develop multilateral solutions for both confronting the climate crisis and ensuring that countries dependent on exporting fossil fuels don't crash violently out of the oil age.[76]

Political scientist Paasha Mahdavi undertook a comprehensive study of national oil companies, tracking the perverse political incentives that accrue to leaders whose countries depend heavily on fossil fuel revenue.[77] Talking about the potential of a national oil company in the United States, he wasn't worried about it falling victim to the kinds of authoritarian power grabs often associated with petrostates. For one, such countries have been on the losing end of centuries of Western powers' predatory adventurism, with fossil fueled, export-led growth in many cases being one of the few paths to development on offer. The US also wouldn't face the same threat that other countries that have nationalized private energy companies tend to: of a US-backed coup or sanctions. Despite industry talking points, the economic well-being of the US is not intimately tied to its oil exports. Nationalization wouldn't change that, particularly when undertaken at a moment when the industry is in serious decline and with the baked-in mission to wind down extraction. "If the government wants to pursue a certain objective, it would seem that a national oil company is the best way to do it," Mahdavi told me, noting the positive aspects, in particular, of NOCs being beholden to the state rather than shareholders. In some ways, the question of whether or not to nationalize fossil fuel assets is even simpler: Do we trust the companies that have spent decades delaying action on climate and spreading misinformation about its existence to steward a transition off fossil fuels, as they claim they will? To value the urgency of the climate crisis and the needs of their workers over the interests of their shareholders? If the answer is no, nationalization is our best option to decarbonize as quickly as is needed to avert catastrophes both economic and ecological.

Improbably, though, transitioning the United States' fossil fuel workers into new jobs might be the easy part. Twenty-four percent of the global equity market is bound up in fossil fuels, totaling $18 trillion. On top of that is the $8 trillion worth of corporate bonds, with unlisted debt that could be far larger. Should countries follow through on their existing commitments to the Paris Agreement, Carbon Tracker has predicted that a combination of falling oil demand and investment risk could shave the value of coal, oil, and gas reserves down to a third of their former value, from $39 billion to $14 billion. While the US is well equipped to put the narrow sliver of the American workforce bound up in fossil fuels back to work, the challenge of transitioning off them entirely is orders of magnitude larger.

That conversations about a just transition have traditionally been concerned only with today's energy workforce ignores both the profound challenges and opportunities of decarbonizing a society built by and around fossil fuels. Nationalization is no silver bullet for building a better world without them, or for absorbing the shockwaves that switch will send through the global economy; it also shouldn't be understood only as a narrow pragmatic tool. It's also a statement of intent, making a claim to a postcarbon democracy every bit as abundant as the sun and wind on which it will be built.

CHAPTER 10

TOWARD A NONVIOLENT ECONOMY

THE CLIMATE CRISIS is the terrain on which all politics will play out in a twenty-first century rocked by more fearsome floods, droughts, and epidemics. Few bat an eye when tropical depressions brewed in unusually warm waters spin out to sea or over unpopulated islands or when diseases spread among animals in the wild. These events turn into disasters when they crash into people who are organized in particular ways. Any just transition off fossil fuels—of which today's fossil fuel workers are just a small part—will also necessarily be a transition into a hotter and wetter world, too. The climate crisis is now crashing into a United States that has been organized to protect minority rule against majoritarian threats since well before James Buchanan and Milton Friedman started teaching economics. Even if an energy transition were only about subbing out the stuff that turns lights on and makes cars run, it would still mean going up against some of the most powerful and politically entrenched companies on earth. In league with the rest of the 1 percent, their best strategy for warding off such conflicts has been to keep majorities of people who could challenge them divided against one another. Especially in the US, racism has been their best means of doing so. The biggest barrier to a just transition is less about particular power sources than undermining that wealthy minority's power to determine what fuels the world and how it responds to crises. Only the kind of multiracial, democratic majorities they've spent centuries trying to crush can reorganize society off fossil fuels to make way for a more egalitarian and inevitably climate-changed future.

It's worth understanding a bit about why that seems so difficult. The US was built on the foundations of two original sins: the genocide of Indigenous people and chattel slavery, atrocities integral to the development of the American and even global economy. Cheap land and labor—extracted through violence—were start-up capital for Europeans looking to stretch beyond the limits of their tiny continent. The depiction of the South's plantation economy as a feudal backwater, far removed from more dynamic northern industries, belies just how bound up the two were in one another. As C. L. R. James put it succinctly, "Negro slavery seemed the very basis of American capitalism."[1]

After the slave uprising of the Haitian Revolution shook off French rule, the nerve center of slavery in the Americas shifted from the Sugar Islands of the Caribbean north to the frontiers of a fledgling United States. Instead of sugarcane, the most important commodity crop in this corner of the New World—the one that would help expand the wealth and territory of the United States—was cotton. As historian Edward Baptist writes, cotton production in the US "linked technological revolutions in distant textile factories to technological revolutions in cotton fields, and it did so by combining new opportunities with the financial tools needed to make economic growth happen more quickly than ever before."[2] While in 1802 cotton accounted for 14 percent of the value of all US exports, the expansion of slave labor camps in these new territories had grown that figure to 42 percent in 1820, bolstered by innovations in finance and productivity that linked the coal-fired mills of Manchester, England, and Rhode Island to Wall Street to the whip. The key technology of the planter class was brutality, enforcing ever-stricter production quotas through routinized torture that made US-produced textiles cheaper and more competitive in international markets. All this, as Baptist writes, helped cotton for a time to become "the dominant driver of US economic growth," and by the 1840s, it had enabled the young country to grow "into both an empire and a world economic power . . . all built on the back of cotton." By 1836, he estimates, "$600 million, almost half of the economic activity in the United States . . . derived directly or indirectly from cotton produced by the million-odd slaves—6 percent of the total US population."[3]

Well-meaning northern liberals found slavery distasteful but ultimately essential to the cultural and economic fabric of the United States, too entrenched and thorny to be unwound on any sort of rapid time line. Thomas Jefferson—a slave owner whose political maneuvering had been key to slavery's development in the US—opined that it might simply fade away on its own, eventually. "The revolution in public opinion which this cause requires is not to be expected in a day or perhaps in an age," he lamented, "but time, which outlives all things, will outlive this evil also."[4] Yet by 1850, enslaved people in the United States would account for a full one-fifth of the nation's wealth, much of that bound up in sophisticated financial products on offer from banks on both sides of the Atlantic.[5] Plantations were more efficient and profitable than ever, to the point that there were relatively few jobs on offer for whites in the cotton South. Abolitionists held limited sway over national politics. With hope for legislative abolition scant, Abraham Lincoln encouraged slave-owning states to voluntarily abolish the practice, with hopes that it might start a trend. He disagreed with radicals in believing for a time that slave owners were owed "just compensation" for any assets left stranded as a result of emancipation.[6] England, after all, had paid out $26.2 million to some 46,000 slave owners when it ended slavery in its Caribbean colonies following years of rebellions; France had demanded reparations from Haiti with a punishing debt load. There was still a thriving domestic slave trade when the Civil War kicked off and likely would have been for decades to come. For more moderate Republicans, dissolving the value of all those assets—declaring people who had to that point been property human— seemed unimaginable. Doing so without compensation seemed fiscally irresponsible.

That's exactly what happened. Ending slavery required abolishing a system of production that had been central not just to the development of the United States but to capitalism itself, stranding the modern equivalent of tens of trillions of dollars of assets.[7] It would take more than a century of organizing and rebellions among enslaved people, tireless work by Radical Republicans in Congress, and, finally, a brutal war to bring about the largest single evaporation of wealth in modern history.

Reconstruction—generally considered the period that came directly after the Civil War, from roughly 1865 to 1877—was in its most ambitious phase an attempt led by people who had just won their freedom to build a new and more egalitarian society in the wake of that expropriation. It involved an enormous expansion of citizenry and in the scope and size of the federal government, along with experiments by the state and ordinary people alike to transform the United States into what W. E. B. Du Bois, in his 1935 epic *Black Reconstruction,* dubbed an "abolition-democracy." Perhaps that era's most remarkable institution—the Freedmen's Bureau—attempted to build a South in which four million freed slaves were full participants in society, which required suppressing a planter class (to use Du Bois's term) that had just lost its battle to keep them enslaved and that was eager to return to that status quo by any means necessary. "The very name of the Bureau," Du Bois wrote in the *Atlantic,* "stood for a thing in the South which for two centuries and better men had refused even to argue—that life amid free Negroes was simply unthinkable, the maddest of experiments."[8]

For a brief moment, freed people, joined by allied whites, started down a new path. "In the Freedmen's Bureau," Du Bois wrote, "the United States started upon a dictatorship by which the landowner and the capitalist were to be openly and deliberately curbed and which directed its efforts in the interest of a black and white labor class. If and when universal suffrage came to reinforce this point of view, an entirely different development of American industry and American civilization must ensue."[9] The South's first state-funded public schools emerged out of this period, as did a more egalitarian tax code, protections for Black and white workers, and a range of economic development initiatives. After the hard-won Fifteenth Amendment legally enfranchised Black men, an estimated two thousand held office throughout the South, where in many places freed people vastly outnumbered their former masters; two of the ten Black Senators to have ever served in the US Senate were elected during Reconstruction, from Mississippi. Black political representation in Congress was higher in the 1870s than at any point in American history.[10] Called a Second Founding of the United States, Reconstruction constituted a thoroughly radical effort to expand democracy.

It was brutally attacked for its duration and eventually defeated by, among other factors, organized campaigns of terror by white southerners organized into paramilitary formations that included the Ku Klux Klan. Multiracial "fusion" governments were smeared as hopelessly corrupt and incompetent, as they would be for much of the next century in depictions like 1915's *The Birth of a Nation*.[11] Politicians furthered the same ends through law in measures like the Black Codes, modeled on fugitive slave laws to restrict the movement of free people, whose labor planters still depended on. The goal and success of the backlash, as Du Bois put it, was to "establish a new dictatorship of property in the South through the color line." The Democratic "Redeemer" regimes that followed ousted not just fusion governments but looked to dismantle the Reconstruction state itself as a complement to their campaign of organized violence. Historian Eric Foner describes a governing approach that might sound familiar today: "Redeemer constitutions reduced the salaries of state officials, limited the length of legislative sessions, slashed state and local property taxes, curtailed the government's authority to incur financial obligations . . . and repudiated, wholly or in part, Reconstruction state debts. Public aid to railroads and other corporations was prohibited, and several states abolished their central boards of education."[12] When they took control in 1875, Mississippi Democrats slashed the state budget in half over the next decade.[13] While Redeemer governments made a new class of planters, merchants, and industrialists rich, "the majority of Southerners of both races," Foner writes, "sank deeper and deeper into poverty. For the South's yeomanry, the restoration of white supremacy brought few economic rewards."[14]

The descendants of efforts to kill Reconstruction and entrench white supremacy have been legion, from Jim Crow laws that segregated the South and undermined Black suffrage to a brutal War on Drugs and the bloated system of policing and mass incarceration it helped create. Various systems of domination that have sprung up in the decades since have served to discipline and manage Black life and labor and, in the process, sew political coalitions between poor whites and capitalists whose interests couldn't seem more at odds. This strange alliance modeled by yeoman farmers and the planter class in the South assigned value to whiteness, what Du Bois called a "psychological wage": as poorly paid and treated as white workers might have been, the color of their skin afforded them limited rights (i.e., to vote) and protections (i.e., against vigilante violence) they would fight like hell to defend. During Reconstruction, that defensive fight was waged against a fledgling multiracial democracy that could have greatly expanded those limited rights. Du Bois—writing in the early days of the New Deal—described this phenomena as a "color caste founded and retained by capitalism," noting that "the plight of the white working class throughout the world today is directly traceable to Negro slavery in America, on which modern commerce and industry was founded. . . . Thus the majority of the world's laborers, by the insistence of white labor, became the basis of a system of industry which ruined democracy and showed its perfect fruit in World War and Depression."[15]

There is no comparing the centuries-long horror of chattel slavery and the climate crisis. But how the former was abolished—and the long backlash against that—provides a critical context for abolishing another mode of production central to capitalism and building a better world on the other side. Fossil fuels aren't only a fuel source but have structured modern society in deep ways. Given the centrality of fossil fuels to the global and US economy, a just transition in this country toward both a warmer world and one without fossil fuels means ensuring such basic rights as health care aren't tied to people's employment, making it possible for them to move to avoid rising seas and pursue work in lower-carbon fields. Adequate planning will be needed to ensure those on the move have housing, jobs, and good schools in their new homes. As state and local tax revenues from fossil fuels wane, social services—from schools to hospitals—will need to be more amply funded by the federal government. Markets will have to be more tightly regulated than they are now, with a special eye toward the carbon content and political meddling of certain industries. As just discussed, some of those will need to be brought under public ownership. Any remotely equitable transition off fossil fuels and into a climate-changed world will require building something like a social democracy. Slavery, the promise and defeat of Reconstruction, and the systems of racial domination that followed are key to understanding why the United States has never had one.

As a moral and strategic matter, making the twenty-first century a livable one will mean picking up the unfinished work of Reconstruction. For one thing, racism distributes negative climate and environmental impacts disproportionately onto communities of color. Race is both a stronger predictor of a person's wealth than their income class and continues to be the main predictor of whether someone lives in the vicinity of toxic waste. An astounding 68 percent of Black Americans, for instance, live near coal-fired power plants. In most major cities those same communities endure the country's highest energy burdens; a full half of Black households in the mid-Atlantic are energy insecure.[16] Historically segregated communities of color in more than one hundred US cities experience temperatures up to 7 degrees higher than in other neighborhoods and are more vulnerable not just to extreme heat but also to storms and floods, and are less able to recoup losses when extreme weather hits.[17] The basic facts of climate justice globally hold true within our borders: that those who have contributed to and profited least off the drivers of the climate crisis—having had land, labor, and resources stolen in the process—are suffering its worst impacts. That's in large part because capitalism has always required what environmental justice organizers have long referred to as "sacrifice zones," places and people whose value can be discounted or written off entirely in the name of profits, be that by performing cheap or free labor or being made to swallow pollutants wealthier and whiter communities don't.[18] This has been less a matter of neglect than of careful, often violent management of nonwhite people to maximize extraction.

But there's another reason climate justice requires fulfilling the promise of Reconstruction. The 1 percent has long used racism to continue to increase its profits and prevent a more equitable society from taking root, keeping eyes off the bosses by dividing the working class among itself. As Heather McGhee and Ian Haney López have written, "In our diverse society, racism has been the plutocrats' scythe, cutting down social solidarity to harvest obscene wealth and power."

These tactics have deep roots in the fossil fuel economy itself. As historian Robin D. G. Kelley describes in *Hammer and Hoe*, cheap labor was the means coal operators and other industrialists used to build fortunes in the cotton South through the decades that followed Reconstruction. By 1910, some 90 percent of the unskilled labor force in Birmingham, Alabama, was Black.[19] Coal mines in particular were largely integrated and in some cases predominately Black workplaces, and in Birmingham 53 percent of miners were Black. The mines became hotbeds of organizing for multiracial unions like the Knights of Labor and United Mineworkers of America. Alabama workers carried out 603 strikes between 1881 and 1936. More than half of those strikes happened between 1881 and 1905, around the same time when white officials disenfranchised nearly one hundred thousand Black Alabamans. Employers fought back. "Taking advantage of the large black presence in the UMWA [United Mine Workers of America]," Kelley writes, "employers adeptly used racist propaganda, violence and black convict labor [as strikebreakers] to weaken unionism in Alabama's coal fields."[20] Following consecutive UMWA strikes in 1919 and 1920, the Tennessee Coal and Iron Company (TCI) crushed both the strikes and union with the backing of state troops called in by Governor Thomas Kilby. The Communist Party would pick up the mantle a decade later and again face regular repression, with organizers at one point arrested on charges of "advocating social equality between whites and negros."[21] A majority Black organization, the CPUSA (Communist Party USA) in Alabama was the first party to endorse a Black candidate for office since Reconstruction and at one point had organized as many as twenty thousand people through rural and urban areas, who all the while faced coal operators' attempts to divide workers on the basis of race. Such tactics paid dividends for mine owners. By the end of the twentieth century, just 15 percent of Birmingham's miners were Black and the unions shells of what they'd been; today the UMWA has fewer than eight thousand members.[22] That coal mining is now considered a breeding ground for Trumpism's brand of white identity politics is owed in part to mine owners' work to pit Black and white workers against each other.

The Black freedom movement's 1960s-era struggle for inclusion into American democracy and into the largely segregated New Deal economic order prompted right-wing politicians to pick up the same strategies, which fit like a glove onto the neoliberal project. They used dog whistles about criminality and eroding family values to cast people of color as the primary and undeserved beneficiaries of activist government and cleave off working-class whites from the New Deal coalition. For politicians like Barry Goldwater and Ronald Reagan, it was "welfare queens" who were siphoning off hard-earned (white) tax dollars through welfare— nevermind that the GI Bill had fueled white veterans' college educations and upward mobility, or that Federal Housing Administration loans had subsidized white wealth building through homeownership. Eventually, Bill Clinton and fellow New Democrats took up that call in the 1990s, ending "welfare as we know it" and passing a disastrous crime bill supposedly aimed at the "superpredators" said to be stalking America's inner cities. One representative War on Drugs provision saw crack carry a sentence one hundred times greater than cocaine, which flowed through Wall Street and elite private schools. The result of stoking those divisions has been a disaster for just about everyone but the ultrawealthy. Policing and mass incarceration have exploded as wages have stagnated, with Clinton himself presiding over the largest increase in state and federal prison populations of any president in US history. Black Americans are 13 percent of the US population and 40 percent of those incarcerated. All the while, poor whites are dying "deaths of despair" linked to poverty and a widespread lack of social services: from drug overdoses to suicide. Corporate profits and productivity have soared as the type of government needed to take on the climate crisis has gotten harder to imagine.

Backlash to the movements of the 1960s and '70s—to nonwhite people staking a claim to stolen wealth and wages, in the US and internationally—took another form, too. New York City's fiscal crisis in the 1970s is among the most stark examples of the federal government striking back against attempts to construct an actually existing social democracy. A city of 7.6 million people was placed under the management of an austerian and unelected fiscal control board. But similar antidemocratic punishments were doled out to cities around the country as a check on Black political power and the rising power of workers.[23] This wave of Black mayors faced a combination of impossible, interrelated binds. Some saw major industries move South in search of cheaper labor or shed workers thanks to automation. Cities were left to the mercy of bond markets, forced to take on their own debts and make painful cuts while the federal government peeled back War on Poverty programs and what was left of the New Deal state. The Volcker Shock's sudden spike in interest rates plummeted an increasingly aspirational Third World into debilitating sovereign debt crises. Stateside, sky high unemployment and shuttered factories compounded existing troubles. So did white backlash.

In Gary, Indiana, for instance, the election of the thirty-four-year-old civil rights activist Richard Hatcher in 1967 triggered a wave of white flight from residents and major businesses alike, including U.S. Steel. The city's dwindling tax base severely constrained the kinds of ambitious changes Hatcher had in mind.[24] Harold Washington's election as the mayor of Chicago in 1983 kicked off what would become known as the "Council Wars, in which twenty-nine reactionary white aldermen siphoned power out of the mayor's office and into the city council. He would later be investigated by the FBI on trumped-up bribery charges that he would battle until his death in 1987.[25] While hardly without their flaws, Washington and other Black mayors faced charges that echoed those levied against the fusion governments birthed under Reconstruction: too generous, too corrupt, and, more than anything, simply too Black. "Many black mayors are denied substantial control over the policy-making process of city government by city charters," the legal scholar William E. Nelson observed, adding that the "crucial powers of budget control and appointment were assigned either to city councils or to city managers."[26]

Rev. Dena Holland-Neal, a lifelong Gary resident—and Richard Hatcher's goddaughter—put it in starker terms: "I often compare Gary to Haiti. In both instances those in power wanted to send a strong message that said they did not want the people to think we were in a position to determine our own destiny."[27]

IF THE LEGACIES of the backlash to Reconstruction live on, so do the ideas that animated that project. An enduring strain of social movements in the United States has questioned what it is that government is for and which lives it should protect. Radical Republican fusion governments, the Alabama Communist Party, the postwar Black Freedom struggles, and today's abolitionists have all offered visions for a fundamentally different, freer, and more democratic society than even some of the most ambitious New Dealers could have imagined. For the most part, though, the Black radical tradition has been a less common reference point for Green New Deal campaigners than its namesake.

There are plenty of lessons to draw from the New Deal, as previous chapters have illustrated, as well as from the domestic mobilization for World War II. Few other episodes in American history offer a better showcase for the sheer scale of what industrial policy can accomplish against seemingly impossible odds: wartime planning bodies created entire supply chains virtually from scratch, nationalized defiant factories, and put millions to work as the federal government dragged the country out of a recession with record low levels of unemployment. The supposedly rational self-interest of everyone from corporate executives to middle-class families was subsumed through public policy by a singular goal: winning World War II.

Even ardent supporters of the war mobilization, though, recognized that it was no blueprint for a sustainable society. By the time Axis forces surrendered, the federal government had spent more than a decade expanding its role in the economy. Labor leaders saw peace as a chance to democratize economic life, giving the country's brimming unions a role in both industrial production and economic planning oriented not toward war but to building a more equitable country. United Auto Workers (UAW) president Walter Reuther, arguably the most powerful labor leader of the postwar era, drew up his own ambitious plans to mobilize Detroit for the war effort. In 1945, less than a month after atomic bombs were dropped on Hiroshima and Nagasaki, he called for an "unswerving will to plan and work together for peace and abundance, just as we joined forces to work for death and violence."

"If we fail," he concluded, "our epitaph will be simply stated: we had the ingenuity to unlock the secrets of the universe for the purposes of destruction, but we lacked the courage and imagination to work together in the creative pursuits of peace."[28]

On Reuther's terms, we did indeed fail. The transition out of the war mobilization was a difficult and uneven one, even while popularly remembered as a Golden Age of American capitalism. Left-leaning unionists were purged as the US lurched toward a Cold War and McCarthyism, and Reuther himself abandoned his most ambitious visions as the UAW sought peace with automakers and politicians. Rather than looking for less belligerent ways to maintain the full employment economy that had lifted the country out of the Depression, politicians deferred to industry to shape the economy. Generous social spending like the GI Bill enabled soldiers returning from World War II and, later, the Korean War to attain college degrees. Federal Housing Administration loans opened up paths to homeownership and wealth building to millions of families, easing the transition for millions as wartime levels of production rapidly wound down. These programs largely excluded Black workers and veterans, many of whom returned from fighting for their country abroad to either Jim Crow in the South, or to segregation and violent white backlash in the North.[29] They would also be among the first to spot signs of decay in the postwar prosperity they'd been largely excluded from.

In Detroit, Black autoworkers were on the front lines of the transition to a new era of production. Having been slotted into the lowest paying factory jobs—those most likely to be replaced by more streamlined industrial processes—they experienced the throttle of automation before their white counterparts in higher-ranking and better-paid positions. This predated later trends of the industry moving to the South and then, through trade agreements like NAFTA, out of the US entirely in search of cheaper labor and higher profits. Building on long traditions of Black radicalism and militant labor organizing in the Motor City, workers on the front lines of these innovations created the League of Revolutionary Black Workers, which militated both against their bosses at the Big Three automakers and white union leadership—including Walter "Pig" Reuther, as they called him—who ignored threats league members saw as imminent. The theorist, organizer, and socialist autoworker James Boggs wrote in a 1963 pamphlet, *The American Revolution: Pages from a Negro Worker's Notebook*:

> Automation displaces people, and you don't just stop growing people even when they have been made expendable by the system . . . in the United States, with automation coming in when industry has already reached the point that it can supply consumer demand, the question of what to do with the surplus people who are the expendables of automation becomes more and more critical every day. There is only a limited number of these old workers whom capitalism can continue to employ in production at a pace killing enough to be profitable. The rest are like the refugees or displaced persons so familiar in recent world history. There is no way for capitalism to employ them profitably, yet it can't just kill them off. It must feed them rather than be fed by them.[30]

There were too many people, and too few jobs profitable enough for the private sector to create. War on Poverty programs carried out under both the Johnson and Carter administrations sought to handle deindustrialization through a range of self-help, job training, and education programs intended to uplift communities left behind by corporate and white flight. As historian Elizabeth Hinton has detailed, this era would also lay the groundwork for the War on Crime pursued by the Reagan and Clinton administrations, as Johnson poured federal funds into police departments in response to the urban uprisings of the 1960s.[31] Insofar as they were bound up in one another, the War on Poverty and War on Crime were each answers to the problem of surplus and mostly nonwhite populations that capitalism had deemed nonessential. Reagan and the neoliberals who surrounded him killed off the former and expanded the latter. Geographer Ruth Wilson Gilmore explains in her study of California's prison explosion that a combination of recessions and monetary policy manipulation in the 1970s saw the state seek to warehouse its increasingly nonwhite surplus populations by expanding the carceral system of both prisons and the policing that sent people to them.[32] "The federal retreat required subnational politics and institutions to take responsibility for social problems whether they wanted to or not, forcing them to deal with the newly dispossessed, who ranged from unemployed youth to financially needy students and homeless families," she writes. As social spending was pulled back, she notes, rapid expansions of the carceral state in cash-strapped rural communities were pitched on the false promise of providing an increasingly rare source of jobs and economic development. Cages, Gilmore has written, became "catch-all solutions to social and political problems."[33]

There were alternatives. In the spirit of Reuther's hopes for a postwar economy, one of the animating demands of postwar liberals and the left was for a full employment economy, where the federal government was willing to serve as an employer of last resort. Today, the same idea can help inform visions for a holistic just transition—one that extends well beyond fossil fuel workers and encompasses the many kinds of work that stand to be disrupted and expanded as the world moves off fossil fuels. For its most visionary proponents, full employment wasn't just a jobs program. It was an effort to redefine what sorts of economic activity were valuable and who the government should serve and plan accordingly.

It also wasn't a strange idea. The architects of the WPA had intended it to serve as a permanent source of public sector jobs, not a temporary relief program forced to battle for funding each year. The very first tenet of Franklin D. Roosevelt's 1944 Economic Bill of Rights is the "right to a useful and remunerative job in the industries or shops or farms or mines of the nation."[34] In a dissenting Supreme Court opinion in 1972—in a case on alleged discrimination in the firing of a Wisconsin State University-Oshkosh professor—Justice Thurgood Marshall contended that the right to a job was guaranteed by the Fourteenth Amendment.[35] "Every citizen who applies for a government job," he wrote, "is entitled to it unless the government can establish some reason for denying the employment. This is the 'property' right that I believe is protected by the Fourteenth Amendment and that cannot be denied 'without due process of law.' And it is also liberty—liberty to work—which is the 'very essence of the personal freedom and opportunity' secured by the Fourteenth Amendment." The 1945 Full Employment Bill would have officially established the government as the country's employer of last resort, only to be killed by an alliance of Dixiecrats, the US Chamber of Commerce, and the National Association of Manufacturers.[36]

Twenty-plus years later, the report of the Kerner Commission—convened in response to the riots of the 1960s—recommended job creation as "the single most vital question to the Negro community."[37] Certain voices within the Johnson administration had advocated for direct job creation to be a central feature of the War on Poverty, and the Democratic Party Platform Convention in 1968 recommended that the federal government act as "the employer of last resort."[38] Even Richard Nixon reluctantly signed the Comprehensive Employment and Training Act (CETA) in 1973, a direct hiring program that employed hundreds of thousands of people per year through 1981. Ultimately, more conservative Keynesians won out and left the focus on priming struggling communities for entry into a job market still controlled entirely by the private sector.

Heavily influenced by Du Bois and his writing on Reconstruction, labor and civil rights leader A. Philip Randolph would become one of the century's most dogged advocates for a job guarantee. As head of the Brotherhood of Sleepingcar Porters, he had led the fight to desegregate the defense sector in the lead-up to World War II. He was also one of the key organizers of the 1963 March on Washington for Jobs and

Freedom, at one point slated to be called the Emancipation March for Jobs. As historian David Stein has detailed, that iconic demonstration was a response, in part, to the little-remembered recession of 1958, which like most economic downturns hit Black workers the hardest.[39] Among the march's demands was a "massive federal program to train and place all unemployed workers—Negro and white—on meaningful and dignified jobs at decent wages."[40] Yet despite the size and historic nature of the March on Washington, Randolph and fellow organizer Bayard Rustin didn't see the momentum they generated translate into influence over lawmakers. Added to this was a frustration that the gains of the civil rights movement on issues like desegregation and voting rights had not, by and large, translated into improved economic conditions for Black Americans more broadly.[41]

Reflecting on the state of the civil rights movement in 1965, Rustin wrote, "I fail to see how the movement can be victorious in the absence of radical programs for full employment, abolition of slums, the reconstruction of our educational system, new definitions of work and leisure. Adding up the cost of such programs, we can only conclude that we are talking about a refashioning of our political economy."[42]

From there, civil rights leaders, economists, and unions worked to create the Freedom Budget, released in 1967 by the A. Philip Randolph Institute, where Randolph served as president and Rustin as executive director.[43] The budget sketches out an ambitious economic program for everything from education to monetary policymaking. "For the first time," its authors stated, "everyone in America who is fit and able to work will have a job. For the first time, everyone who can't work, or shouldn't be working, will have an income adequate to live in comfort and dignity. And that is freedom. For freedom from want is the basic freedom from which all others flow."

In his foreword to A "Freedom Budget" for All Americans, a policy document published in 1967, Rev. Martin Luther King Jr. urged that policies be universal, rather than either means tested or directed solely at vulnerable populations. As he wrote, "We shall eliminate unemployment for Negroes when we demand full and fair employment for all," and he committed his Southern Christian Leadership Conference to pushing for it. "We must dedicate ourselves to the legislative task to see that it is immediately and fully achieved. I pledge myself to this task and will urge all others to do likewise."

The need for a nonviolent economy became more urgent during the Vietnam War, as US guns were trained both on the Vietcong and participants in the urban uprisings of the 1960s through an uptick in bipartisan efforts to criminalize Black and brown communities. Unemployment rose sharply during a recession that began in 1969; as has been the case since World War II, unemployment for Black Americans was double that of whites. Nonetheless, Stein writes, Randolph and other labor leaders worried presciently that a general economic downturn would fuel white backlash and undermine multiracial solidarity, fearing that "competition for decreasing jobs" was "certain . . . to eventuate racial tensions."[44] Looking abroad and around at her own country, Coretta Scott King argued, "This nation has never honestly dealt with the question of a peacetime economy and what it means in terms of the development within the country," seeking to fight what she called the "triplets of racism, extreme materialism, and militarism."

In 1974, Scott King founded the National Committee for Full Employment/Full Employment Action Council (NCFE/FEAC) to fight for legislation that guaranteed jobs to all Americans along those lines. Demands included expanding public service jobs, fighting the Ford administration's budget cuts, and pushing the Fed to lower interest rates. Noting the role of the oil crisis in fueling inflation and sapping workers' purchasing power, Scott King and NCFE/FCAC pressed Congress to "encourage the development of new energy sources and prevent the abuse of monopoly power by energy corporations."[45] For Scott King and others in NCFE/FEAC, war represented a violent misallocation of budgets and priorities: "If we stop the war 2 months sooner, 230 housing units could be built; one hour could create a new school or housing center. I must remind you that starving a child is violence . . . suppressing a culture is violence. Neglecting school children is violence. Punishing a mother and her family is violence. Contempt for poverty is violence. Even the lack of will power to help humanity is a sick and sinister form of violence."[46]

It's no exaggeration to say that the state support continually provided to the fossil fuel industry constitutes its own sick and sinister form of violence, heaped on to all the others Scott King identified that still exist today and that, in many cases, have gotten worse. Reallocating to-day's budgets and priorities, per Rustin, will entail a "refashioning of our political economy." His and Coretta Scott King's demands for full em-ployment—a means toward a nonviolent economy—can offer some les-sons in how that might happen.

FULL EMPLOYMENT HAS historically meant different things to different people and institutions. Throughout its history, the Federal Reserve has tended to define its target on that front as the nonaccelerating infla-tion rate of unemployment, or NAIRU, usually set at somewhere around 5.5 percent. The theory posits that unemployment below a certain level— a so-called natural rate of unemployment—will trigger accelerating inflation, potentially overheating the economy. When jobs aren't hard to come by, the thinking goes, workers enjoy more bargaining power, freeing them to push bosses for better wages and quit if they don't get them. In turn, bosses are more vulnerable to pressure and more likely to comply with their employees' demands. They then pass increased pro-duction costs down to consumers, who—as employees themselves—will start to demand even higher wages.

The closer you get to full employment, according to the NAIRU the-ory, the higher the risk of inflation. It also means that the fix to existing or impending inflation is for central banks to engineer higher levels of unemployment by raising interest rates. Rohan Grey, founder and presi-dent of the Modern Money Network, calls NAIRU "an intellectual edifice designed to reinforce and justify keeping labor weak and undermine their ability to bargain," arguing that it "puts the burden of macroeconomic policy management on the most vulnerable people in society." For the millions of people embedded in that permanent class of unemployed workers that NAIRU demands, not ensuring full employment feels like violence through neglect, as Scott King posited. The Fed had dealt with economic downturns of the 1950s and 1970s through regressive "tight money" policies to curb inflation at any costs, its top or even sole priority throughout the postwar era.

To correct that, NCFE/FEAC would take its landmark fight to plan a more humane economy to Congress. Augustus Hawkins—a founding member of the Congressional Black Caucus and author of Title VII of the Civil Rights Act—had initially introduced his Equal Opportunity and Full Employment Act in 1974 as a means of wholly eliminating involuntary unemployment and building out a network of local planning councils to determine which jobs were needed around the country. Economist and Hubert Humphrey aide Leon Keyserling convinced him to temper some of the bill's more ambitious visions, including defining full employment as 3 percent unemployment, but Hawkins insisted on it establishing "the bedrock responsibility of the Federal government to finance directly supplemental public service jobs for all." Introduced in 1976, the Humphrey-Hawkins Act (HHA) remained a wide-ranging measure intended to put teeth behind the Federal Reserve's nominal but often ignored mandate to promote full employment. Humphrey-Hawkins—for NCFE/FCAC—was to be a response to the crisis of rising unemployment, both changing the mandate of the Federal Reserve and establishing truly full employment as an official priority for the White House. There were reasons to be optimistic. Direct employment was central to the antipoverty program Jimmy Carter ran on in 1976, and he supported an early version of the HHA. For a brief time, the Full Employment Coalition backing Humphrey-Hawkins brought together the Congressional Black Caucus and civil rights leaders like Scott King with labor leaders including UAW president Leonard Woodcock, progressive economists, New Left activists, and feminists including Gloria Steinem. A little-remembered group called Environmentalists for Full Employment saw it as a means of opening up new economic opportunities for workers in the extractive industry, especially those living in places where refining and mining jobs were some of the only ones around. With the momentum building, Democratic majority leader Tip O'Neill called it "the centerpiece" of the party's platform that year, helping Democrats control the House, the White House, and a filibuster-proof majority in the Senate.

In public and within the White House, though, Keynesians battled one another and a rising crop of neoclassical economists about whether HHA's combination of economic planning and guaranteed jobs would trigger ruinous inflation. Ronald Reagan—who ran for the GOP presidential nomination in 1976—likened the HHA to both fascism and slavery, stating that it was "telling free Americans where they would work and what kind of work they'd do."[47] Still, the GOP had lacked the numbers in Congress to kill the bill outright. The HHA was the Carter administration's to lose, and so it did. "Rather than a case of the conservative winning congressional votes or infiltrating a Democratic administration," Steve Attewell explains in his history of full employment in US politics, what felled HHA was "a loss of faith within liberal circles, as Keynesian economists lost confidence in their nostrums in the face of conservative counterarguments, and Democratic policymakers reacted . . . by instinctively shifting to the middle."[48] Carter, for his part, had supported the idea of direct hiring in his campaign—but also enthusiastically backed the conservative hallmarks of fiscal discipline, balanced budgets, and cutting red tape. "To the extent that Carter had any personal convictions on these issues," Attewell writes, "it was the attitude of a fiscally conservative technocrat who was willing to undertake humanitarian reform as long as it did not cost any money. . . . But if Carter was a technocrat, he was one who had little confidence in the power of government or economic planning."[49]

A man ahead of his time, Carter was eager to kill the HHA in pursuit of his own favored approach, which centered on targeted cash assistance rather than government-wide planning and direct employment. His administration feared doing so too loudly would alienate critical voters, since, per his advisers, it was "one of the few bills in which we are clearly aligned with our major constituencies—labor and the minority community."[50] Administration officials worked to water down key provisions, pushing successfully to degrade enforcement power and add in anti-inflationary measures. Conservative Republicans and Democrats would add in language that went even further, including a 0 percent inflation target and a balanced budget amendment. The bill passed by large margins in the House and the Senate as a shell of itself, stripped of enforcement authority. Carter would use a spike in oil prices in 1979 as an excuse to deal the HHA a final blow.[51] Carter's labor secretary Ray Marshall informed Hawkins that "achievement of four percent unemployment by 1983 is simply untenable given the recent OPEC oil price increases." As unemployment soared, a Democratic governing majority had just killed its last, best shot at a holistic response to the painful economic stress settling in, spurning some of its most important political allies out

of fealty to an arcane and empirically suspect set of economic orthodoxies. In 1979, Carter appointed Paul Volcker as Fed chairman, who administered his painful round of shock therapy to the US economy in the name
of cutting inflation, leaving millions out of work and with little recourse.
Ronald Reagan won the 1980 election in a landslide. Besides climbing
greenhouse gas emissions, the forty years that followed saw a 500 percent
increase in the country's prison population.[52]

After that, full employment was excised from top-line Democratic
Party politics as the party shifted further right and toward balanced-
budget orthodoxies and "race-neutral" policy under the influence of
wonks like Robert Rubin, purging itself of any professed commitments
to planning or direct hiring. Jesse Jackson's 1984 and 1988 presidential
campaigns would continue demanding full employment through the
Rainbow Coalition but had lost sway in the halls of power. "Direct job
creation," Attewell writes, "was abandoned by liberals who placed their
faith in a technocratic vision of Keynesianism as a science beyond partisan conflict, who believed that human capital improvements and anti-
discrimination would defeat poverty." It wasn't even debated as a response
to the Great Recession, despite pleas from progressive economists. "Government," Obama said in 2010, "can't create jobs to replace the millions
that we lost in the recession, but it can create the conditions for small
businesses to hire more people, through steps like tax breaks."[53]

A decade later, the idea reemerged among progressive elements in
the Democratic Party as a federal job guarantee. Backed by virtually every candidate in the 2020 Democratic primary besides Joe Biden, and a
core element of the Green New Deal, the basic idea is the same as it was
half a century ago: creating a public option in the job market. Economist
Darrick Hamilton has done some of the pioneering work on how to make
the job guarantee a twenty-first-century reality.[54] "The private sector," he
explained, "does not absorb stigmatized workers—those that are formerly
incarcerated, black, disabled—at the same rate that it does nonstigma-
tized workers. A job guarantee would enable workers, particularly at the
lower end of the labor market, but throughout the labor market too. It
would remove the threat of unemployment and of being destitute."[55] By
making living-wage work available to anyone who wants it, the program
would also establish a de facto wage floor, forcing private sector employers

to match the kinds of wages, working conditions, and benefits available to workers through the public sector. "It gets rid of involuntary unemployment altogether," Hamilton notes. As to what those guaranteed jobs might be, Scott King herself laid out an evergreen road map. "They must be meaningful jobs," she said when asked what government-provisioned work might entail, "in areas where there are human needs and in areas of education, medical care, housing. Those areas where there is a great shortage in terms of meeting people's needs."[56]

There are potentially millions of shovel-ready jobs around the country that can help mitigate greenhouse gases and prepare for the levels of warming already locked in, be they remediating wetlands to make coastal cities more flood resilient, creating green spaces to alleviate the urban heat island effect, or reclaiming vulnerable shorelines for outdoor recreation instead of condo development—a modern Civilian Conservation Corps—in cities and rural areas alike. Federally hired workers in the South could plant mangrove trees along the water, protecting against erosion as they suck up carbon dioxide.[57] There are any number of ways to make agriculture more resilient to increasing temperature fluctuations and turn it into a carbon sink. The sorts of projects derided during the New Deal as boondoggles should be firmly on the table, too. Talented, out-of-work food industry workers could be paid to write recipes to popularize low-carbon diets or host cooking classes in community centers run by job guarantee recipients that host weekly pickups of produce from local farms, happy hours, and free electric vehicle charging stations. From oral history projects to avant-garde theater, there's plenty of valuable and low-carbon work to be done that simply isn't valued by the private sector. It's hard to imagine any company, for instance, being able to make a profit off of building playgrounds or keeping elderly people company to help ward off loneliness, which has been linked in several studies to premature death.

What feeds a profit margin and what makes for a good society tend not to overlap. Newsrooms are hemorrhaging staff positions and being gobbled up by hedge funds, if not shuttered altogether, as online outlets compete for clicks in a sea of disinformation. Corporations are churning out cheap, carbon-intensive junk and selling it through poorly paid service sector jobs at places like McDonald's and Walmart, the largest

employer in twenty-two states. Through paying a living wage and offering robust benefits, a job guarantee—properly formulated—could give
workers another option and help redefine what valuable, productive work
looks like.[58] It could revitalize the public sphere in the process, providing
funds and people power to help build a country where people are not only
better paid but happier.

By no means does a job guarantee resolve the myriad supply and
demand–side policy challenges posed by climate change, but it could go
a long way toward rethinking what kinds of outcomes an economy is
supposed to work toward. Decarbonizing the economy along the time
line physics demands means making low-carbon work the norm for all
workers. And ideally, a low-carbon world isn't one where people spend
all their waking hours picking up trash or manufacturing useless zero-
carbon widgets in the name of economic growth.

To administer the job guarantee, researchers at Bard College's Levy
Institute suggest a Public Service Employment program that utilizes
the expansive, already existing American Job Center Network to match
job seekers with meaningful, living-wage work in their communities.[59]
"Municipalities, in cooperation with community groups, conduct assessment surveys, cataloging community needs and available resources,"
economist Pavlina Tcherneva wrote in a 2018 report, while the Labor Department itself would make "'requests for proposals' indicating that it will
fund employment initiatives by community groups, nonprofits, social
entrepreneurial ventures, and the unemployed themselves for projects
that serve the public purpose," with an eye toward not displacing existing
employment. Standards for such jobs would likely be set by the Department of Labor "to serve the public purpose and not compete with private employment. The focus should be on delivering public goods, aimed
broadly at three areas: caring for people, caring for the planet, and caring
for communities," Stephanie Kelton, former chief economist for the Senate Budget Committee and a high-profile job guarantee advocate, said.[60]

A JOB GUARANTEE challenges several tenets of neoliberal economic orthodoxy head-on, opening up fiscal space for less easily understood programs under the Green New Deal framework, from mass electrification

to energy efficiency upgrades to mass transit. In the wake of a painful recession brought on by COVID-19, it can improve livelihoods for millions in the here and now and help build the movements and electoral coalitions for deep and equitable decarbonization to continue for years to come. Democrats should learn the lessons from 1979 and 2009 and not let arcane balanced budget orthodoxy get in the way of popular policy.

Since the passage of the Congressional Budget and Impoundment Control Act in 1974 and the ensuing creation of the Congressional Budget Office (CBO), any given piece of legislation's financial viability has been assessed mainly by one question: How much money will it add to the federal deficit? Nominally, the CBO provides an independent analysis of proposed legislation's impact on federal spending and revenues. In effect, CBO analyses—or "scores"—can damn legislation to purgatory before it ever comes to a vote. Yet the sticker price of a policy is just one of many factors that contribute to its overall effect on the economy and society more broadly: Does it, for instance, meet a pressing public need? Will it increase or decrease inequality? What about its carbon emissions?

"If we had had CBO in 1935, we wouldn't have Social Security. If we had it in 1964, we wouldn't have Medicare," Kelton told me. "It has become in many ways the key impediment to the progressive agenda and just good economic policy generally." To get that, she argues, progressives will need to be willing to disregard their proposals' CBO scores or work to reform the way that they're calculated. A common response among progressives has been to present their policies as a two-step process: tax the bad—corporate profits, financial sector speculation, offshore earnings, and so on—to pay for the good, be it health care, free education, or renewable energy. As Kelton asks, "Why not decouple those fights? Go after the rich—fight to return the estate tax to some reasonable level, for instance. But don't hold the nation's infrastructure, the poor, and the sick hostage to increasing tax revenue."

The push for full employment—and the scale of state-led planning and spending that implies—also forces a broader conversation about the purpose and scope of the public sphere. It encourages a contest over budgets at each level of government less as issues of scarce resources than as statements of principles about where cities, states, and the federal government should be invested.

In the US, the police departments and prisons bolstered by Wars on Crime and Drugs are remarkably resilient to austerity compared to our already threadbare safety nets. A 2017 report by the Center for Popular Democracy, Black Youth Project 100, and Law for Black Lives found that spending in major US cities on policing far outpaces that on vital services, with huge chunks of city budgets devoted to outfitting officers, as vital services starve: in the years surveyed, Minneapolis devoted 35.8 percent of its general fund expenditures toward policing.[61] As they nourish systemic racism, massive police and prison budgets stymie attempts to build a more sustainable society. This parallels federal politics, where there are seemingly unlimited funds available for new fighter jets, border enforcement, and military adventurism but talk of tight belts when it comes to improving people's lives—the lives of Black Americans, especially. The kinds of Humvees that roamed the streets of American cities in the summer of 2020 to intimidate protesters have long been used to secure US access to oil abroad. Inflated budgets for policing, borders, and military are increasingly being wielded to beat back climate migrants, recognized as such or not.[62]

Water protectors at Standing Rock learned firsthand that police are perfectly willing to use force to protect fossil fuel interests; as protests against both police brutality and fossil fuel infrastructure picked up in 2020, state legislators pushed nearly two dozen laws seeking to further criminalize protesters.[63] Journalist Amy Westervelt has pointed out that Chevron—which made a big show on social media about its commitment to Black lives in the weeks after George Floyd's killing—helps fund a police force 50 percent larger than those in neighboring cities in Richmond, California, the predominately Black and brown city housing its polluting refinery. Carceral infrastructure is toxic in more ways than one. Prisons leach chemicals into nearby water supplies and sustain themselves by polluting Black and brown communities with police forces. That's been cause for solidarity among environmental justice and antiprison activists challenging carceral expansion in California, New York, and elsewhere, fighting the "three p's," as youth in the San Joaquin Valley told scholars and organizers Craig Gilmore and Rose Baraz: police, pollution, and prisons.[64] These kinds of fights for a nonviolent economy are openings toward a bigger contest over what the core competencies of government should be. They can build solidarity across and break down issue silos, forcing debates about a just transition to a postcarbon world to stretch beyond their typically narrow focus on workers in the fossil fuel industry.

At the local level, especially, the types of climate policy that mayors and city councils have leverage over are also the investments campaigners in the movement for Black lives have urged are needed to make communities safer and stronger: affordable housing, rapid bus transit, and jobs programs. Demands to defund the police that proliferated after the police killings of George Floyd and Breonna Taylor show what demands for a more broadly nonviolent economy can look like. The Movement for Black Lives had long pushed an invest-divest policy platform, demanding "investments in Black communities, determined by Black communities, and divestment from exploitative forces including prisons, fossil fuels, police, surveillance and exploitative corporations."[65] In many places, this involved explicitly demanding investments in sectors that are already low carbon, often unionized, and—in some places—vulnerable to cuts as fossil fuel revenues dry up, including in education and health care. A coalition led by Black Lives Matter-LA, after months of consultation with thousands of Angelenos, proposed a People's Budget for the city, focusing on a framework they call #CareNotCops, allocating just 5.72 percent of unrestricted funds to law enforcement and policing, as opposed to Mayor Eric Garcetti's proposed 53.8 percent.[66] After public outcry over those figures in 2020, Garcetti announced days later that he wouldn't authorize an increase in the LAPD's budget after all and would reallocate $250 million to address health and education in Black communities, though he offered few specifics.[67] In 2019, the coalition Durham Beyond Policing had won its campaign for the North Carolina city to invest in "life-affirming services, not an unjustified expansion of the police force."[68] Budgets that revolve around criminalizing Black communities are ill-suited for taking on the climate crisis. Those that reject those functions can make a more sustainable and democratic world a reality.

The benefits of a rapid energy transition should be clear to workers now employed in the fossil fuel industry, along the lines discussed in the previous chapter. But that transition also shouldn't be held hostage by the leadership of a handful of unions that have consistently acted in bad faith on climate and in the bosses' interest. In their decades of backlash to militant, multiracial organizing, fossil fuel executives are by now old hands at wielding the white, male identity politics of coal, oil, and gas work in defense of business as usual. Splitting off fossil fuel workers from their bosses is a worthwhile goal. It also means going up against about a century of concerted ideological priming. Focusing debates about a just transition solely around them excludes a much broader and more diverse working class who—thanks to the ubiquity of fossil fuels—are all bound up in their own ways in the extractive economy. Teachers, nurses, sanitation workers, electricians, cocktail waitresses, mailmen, bartenders, and train engineers (among many others) deserve as much of a say over what the twenty-first century looks like as today's fossil fuel workers.

Transitions have invariably been ugly things for working people in the United States. Pushing the transition off fossil fuels down the road—or underestimating its vast scope—can only make for a hotter and uglier world. Taking on one of the most powerful industries ever will require public pressure at least as concerted as that put on city governments in the summer of 2020 to defund the police. The fight to get off fossil fuels will fail if switching energy sources is its only goal, or if it in turn fails to build the sorts of broad-based, multiracial solidarity the 1 percent has gone to such great lengths to prevent.

Excising fossil fuels from the global economy necessarily requires rethinking its very foundations, not just in the burning of coal and oil but in more deep-seated extractive ideologies. Their dominance relies on managing people of color and excluding them from democracy—political, economic, and otherwise. To borrow a phrase from Rev. William J. Barber II, the fight for a low-carbon future is a fight for a Third Reconstruction.

"We cannot separate the question of whether we can survive together on a warming planet from the question of whether we can redeem the promise of liberty and justice for all in this nation. We are, in the powerful image from the biblical story of Noah's ark, all in the same boat." he writes. "White supremacy had a beginning and it will have an end, whether in the eternal silence of an uninhabitable earth or on a planet far more just, loving, free, and peaceful than the one we are trying to save."[69]

CHAPTER 11
MANAGING ECO-APARTHEID

THE QUARTER, A wing of Atlantic City's Tropicana casino, is designed to look like the bustling streets of Havana. Its cobblestone halls are lined with candy stores and nightclubs and theme restaurants under a ceiling painted blue to resemble a Caribbean sky. A massive statue of Vladimir Lenin in front of Red Square—a high-end Soviet-themed chain restaurant—lent an authentic touch before it was removed when the company went under. As a kid, walking around the casino was an occasional treat, a forty-five-minute drive in the back of my parents' car that would end in a fancy meal and maybe a new toy. In my twenties, with several casinos having gone bankrupt in recent years, it's all a bit eerie. It's also, to be fair, nine in the morning, and I'm weaving through blaring empty slot machines and busty computerized blackjack dealers to find my way to the New Jersey Emergency Preparedness Association's (NJEPA) annual conference.

A gathering for emergency management professionals and vendors from around the state, about half the programming through the week is devoted to informational lectures from contractors and state agencies and half to professional development sessions for practitioners. That fact didn't sink in when I registered for a training session called Emergency Management Basic Workshop weeks before. Having gotten lost under the faux Cuban sky, I arrive red-faced and late and take the first seat I find. Like all the others, it's at a table between two burly men sporting polo shirts and crew cuts. A few other women are in the room, but nobody else, so far as I can tell, under the age of thirty-five; nearly everyone is white. For the emergency managers in the room, coming to these sorts of confabs is professional development that comes with drink tickets and discounts to the steak restaurant downstairs. We're seated at tables in small groups facing presenters and their PowerPoints at the front, who periodically lead us through interactive exercises to break up the lectures on disaster response protocols.

The Frankenstein monster known as the Department of Homeland Security (DHS)—the body that governs most of these protocols—was cobbled together out of twenty-two previously separate federal agencies in the anxious, uncertain weeks after 9/11. It has too many responsibilities to do any of them well. That origin story has also frozen the way the US responds to emergencies at one of the most paranoid moments in American history, as the federal government turned its guns more obviously than ever on enemies both foreign and domestic.[1] While to that point most emergency response had been administered by local and state governments, George W. Bush in 2003 ordered the newly formed DHS to create the National Incident Management System (NIMS), a unified framework for disaster planning that could be deployed at the federal, state, and national level, creating language and protocols held in common with the NGOs and private sector actors involved in emergency response as well. NIMS lays out an "all hazards" approach, allowing for a coordinated response to "prevent, protect against, respond to, recover from, and mitigate the effects of incidents, regardless of cause, size, location, or complexity."[2] Such threats are now broken down into three categories: natural hazards, including hurricanes, floods, and other so-called acts of God; technological hazards, including oil spills or dam breaks; and human-caused incidents, like chemical weapons attacks, school shootings, and any "demonstration, riot, or strike that disrupts a community and requires intervention to maintain public safety." Our handbook for the day states that "all disasters begin and end locally. Municipal government is recognized as the first line of public responsibility for emergency management activity."

That responsibility, then, falls to county and municipal Offices of Emergency Management (OEMs), either independent agencies or housed in other departments. Often, they're lumped into police departments; many emergency managers are former military personnel or police officers themselves. Chronically underfunded and understaffed afterthoughts of local governments, OEMs depend inordinately on resources from DHS. Thanks to its founding ethos, much of that funding and training has been weighted toward counterterrorism, not the sorts of everyday disasters OEMs deal with regularly. Every two years, OEMs are required to create Emergency Operations Plans (EOPs), handbooks for how they'll navigate all manner of disasters in accordance with FEMA (Federal Emergency Management Agency) preparedness guidelines. These are approved by the state; though in New Jersey, they aren't subject to the state's open records request law. In 2002, then governor Jim McGreevey issued an executive order exempting EOPs from "public inspection, copying or examination" on the grounds that transparency could "substantially interfere with the State's ability to protect and defend the State and its citizens against acts of sabotage or terrorism." Emergency managers are also tasked with declaring states of emergency following the protocols laid out by their EOPs, which, in their respective jurisdictions, can include areas as small as a city block.

County OEMs can further request the governor to declare a state of emergency across New Jersey and—if need be—send in the National Guard. As we learn in the workshop, it's not necessary to consult elected officials in this process. New Jersey is one of two states—the other is Michigan—in which emergency management falls under the jurisdiction of the state police. If an evacuation is necessary, emergency personnel are required to overlook any lawbreaking, whether due to noncompliance with the emergency declaration or other crimes. As the presenter says in this portion of his talk: "We'll arrest them later." He's echoed by a cheeky voice from the crowd that's followed by a few laughs: "We'll come back for 'em later!"

The next exercise walks us through how to assess whether a given situation should be declared a state of emergency. Trainers present three sample situations. Table groups then discuss for a few minutes each. The first is a fire at an apartment building, followed by potential flooding behind a school during a heavy rainstorm. Like the rest of the room, everyone at my table's mixed on whether either qualify; as it happens, the building fire doesn't, but the flood does. The third scenario is different: "Civil unrest in your community due to a recent police-involved shooting. There have been protests in the area for the last 2 nights." There are reports, the PowerPoint says, of a liquor store having been looted. Thanks to national attention, officials expect "a high number of people from outside the community to join the protest." This time the room is unanimous: this is a state of emergency. The presenters reassure us that "as emergency management coordinators, you have access to police resources." It may be worth putting in a request to Trenton to send in the National Guard, too. Just to be safe.

How emergencies are designated and get managed are inherently political questions, Those rules—enforced at the most mundane levels of government and taught in dreary casino ballrooms—form part of the expansive amount of climate policy already in place in the US. It's not governing carbon so much as picking the winners and losers of a warming world. The various management techniques developed in the last few centuries have stacked the deck in favor of those looking to get rich while the world burns.

ATLANTIC CITY HAS been home to its share of history for labor and capital alike: the International Chamber of Commerce, a parent organization for national chambers of commerce still working to make the world safe for profit, was founded there in 1919. Sixteen years later, at an American Federation of Labor conference, a rankled John L. Lewis—the United Mine Workers of America (UMWA) president and an influential voice in Roosevelt's ear during the New Deal—landed a punch on the face of one Big Bill Hutcheson, the president of the conservative United Brotherhood of Carpenters. Former poker buddies, the two had parted ways in the early days of the Depression around the need Lewis saw to organize the unemployed and underemployed industrial workers who lay outside the traditional bounds of the AFL's largely white, male craft unions. The punch, commemorated on the boardwalk, all but secured that Lewis would spin off with the UMWA and several radical garment workers' unions to found the Congress of Industrial Organizations (CIO).[3] On the same ground in 1964, Fannie Lou Hamer—an organizer with the Student Nonviolent Coordinating Committee and congressional candidate of the Mississippi Freedom Democratic Party (MFDP)—gave a blistering testimony to the Democratic National Convention in a bid to upstage her state's all-white delegation there. The daughter of a sharecropping family, Hamer told of being imprisoned, evicted, brutally beaten, and sexually assaulted for trying to register herself and others to vote. "All of this," she concluded, "is on account of we want to register, to become first-class citizens. And if the Freedom Democratic Party is not seated now, I question America. Is this America, the land of the free and the home of the brave, where we have to sleep with our telephones off the hooks because our lives be threatened daily, because we want to live as decent human beings, in America?" The MFDP was not seated.[4]

The city has witnessed more recent history, too, and had a front-row seat to most of what's ailing the twenty-first century. Despite bullish national news reports claiming total destruction, Atlantic City and its iconic Boardwalk were left mostly intact by Hurricane Sandy; an older section of the Boardwalk was battered but had been slated for demolition before the storm. Most of the damage was more mundane: flooded homes and basement apartments, with victims going through an onerous and, in some cases, years-long process for reimbursements from FEMA. That kind of quieter damage is inflicted regularly, with fewer news crews around to observe it. Like climate change's other impacts, it's spread out unequally. Atlantic City is majority Black and particularly vulnerable to climate change; high tide there now is nearly a foot higher than it was a century ago and disproportionately affects some of the poorest parts of town. Nuisance flooding increased eightfold between 1963 and 2013.[5] By 2050, one in three of the city's homes is expected to be inundated by floodwaters every year. Although FEMA is tasked with regularly updating flood maps, the responsibility for ordinary planning and budgeting around rising tides falls to local governments and budgets, many of them increasingly cash-strapped and dependent on predatory bond markets to fund basic services.

In recent years, Atlantic City's finances have taken a harder beating than the one delivered by Sandy. That's thanks in part to a crumbling black glass edifice a short walk from the Tropicana, chunks of which have been crumbling off in the years since its closure. Its namesake was diligently scrubbed off, though a few letters were still visible before its demolition on the side facing the beach: Trump Plaza.

Like Donald Trump's other properties in the city, the Plaza struggled not long after its 1984 grand opening to turn a profit, even as other Atlantic City casinos thrived. Like the fracking boom, Trump's gaudy Jersey shore ventures were financed by selling off junk bonds to the unseemly Wall Street buyers willing to take them. By the early nineties, default on that debt seemed like a distinct possibility as Trump missed nearly $50 million worth of interest payments. His saving grace would be Carl Icahn, the quintessential corporate raider and largest holder of Trump's Atlantic City bonds, who snapped them up as things started going south. Through a prepackaged bankruptcy settlement, Trump was allowed to maintain control of the board in 1990 while bondholders got a 50 percent equity stake; future commerce secretary Wilbur Ross represented one of the largest bondholder groups through Rothschild Inc., and Icahn—who twisted arms to make the deal happen—would go on to advise a future president Trump on regulatory reform; as the New York Times pointed out, the value of an oil refinery company he owns, CVR, doubled shortly after his advisory role was announced, netting Icahn some half a billion dollars.[6]

Trump, the self-proclaimed "king of debt," bragged early and often about the killing he made off his three casinos in Atlantic City and the genius he showed in exploiting bankruptcy for personal gain. "The money I took out of there was incredible," he told the New York Times in 2016. The extraction that ended as a win-win for Trump and his bondholders-cum-political allies came at Atlantic City's expense. In 1998, he convinced the Casino Reinvestment Development Authority to condemn three city lots for a parking lot and stretch limo staging area on the grounds that it was a public good; the move was later deemed illegal, but the residents and business owners on the condemned lots had already been moved out. When Trump restructured his debt, contractors were shorted millions of dollars' worth of work and supplies that built his properties.

He wasn't the only one on the take, of course. New Jersey politicians had long heaped generous tax breaks onto the casino industry. After a debt-financed building frenzy by the industry in the lead-up to the Great Recession—and an ensuing round of layoffs—a new Morgan Stanley–backed casino called Revel became the darling of state and local lawmakers. Its developers promised to revive the city's stagnating gambling business with an ambitious design and profitability projections premised on attracting the gamblers who'd stopped coming to Atlantic City as Pennsylvania and New York got gambling licenses. By 2007, lawmakers had already changed zoning laws to allow Revel to build what would become the state's tallest building, part of a ballsy plan to build the most expensive casino the city had ever seen. Notably, it'd be Atlantic City's only nonunion casino—and so attracted the attention of UNITE HERE Local 54, the muscular union representing workers on every floor in every one of the city's casinos, from housekeepers to cocktail waitresses.

"The whole plan was completely fucked up," said Bob McDevitt, the union's president, who started working in the industry as a barback in the early eighties. With Revel's antiunion stance having piqued their attention, Local 54 researchers started taking a closer look; the company's rosy projections didn't hold up to union researchers' scrutiny. "We were the only people saying this is not gonna fucking work. And they all thought we were just pissing on it because we were having a dispute about whether it was going to be union or not. And yeah, some of the fact that they were nonunion drove this, but we were gonna organize it anyway. The whole goddamn thing was a Potemkin village," said McDevitt.

Local 54 launched a ballot petition to stop the city from issuing a $50 million bond for Revel to raze and streetscape seventeen acres in advance of the grand opening and collect some $300 million in tax breaks over a decade. Response to the petition drive was enthusiastic, and it looked poised to block the bond issuance, tax breaks, and streetscaping. When the state moved to guarantee the money instead, Local 54 launched another successful petition initiative aimed at Trenton. Led by Chris Christie, the state legislature responded by passing a law to invalidate the ballot process entirely and plowed ahead with the tax breaks and demolition. Meanwhile, investors in Revel, including Morgan Stanley, began pulling out while the building was still under construction.[7]

Sure enough, Revel debuted as one of the state's lowest-performing casinos.[8] Rather than bringing in new customers, it siphoned off business from existing casinos. And less than a year after opening, it filed for bankruptcy in February 2013, closing down its casino and hotel the next September after being unable to find a buyer. It was the start of an ugly period for Atlantic City's casinos. Four closed in 2014, followed by the

Trump Taj Majal in 2016. Between 2011 and 2016, city tax revenues fell by 70 percent.[9] "It was like a catastrophic weather event," McDevitt recalled. "In the blink of an eye we had six thousand people out of work. It was like a fucking tsunami hit." The union set up an emergency response center at the Atlantic City Convention Center, bringing in 120 computers and other resources for newly unemployed workers within and without Local 54 to apply for unemployment.

Private equity involvement made things worse. An umbrella term, private equity is defined as a negative—essentially, ownership of companies that exist outside the stock exchange. For years, private equity firms were known as leveraged buyout shops. They'd take over companies, strip them for parts, and then sell at a profit. After the financial crisis, private equity has transitioned from unseemly sideshow to ubiquitous fixture of the global economy, snapping up everything from beloved retail chains to major tech firms to newspapers to fossil fuel companies. Icahn himself made more money as a private equity investor in energy than any other sector. Early on in his career, he orchestrated takeovers of Texaco and Phillips Petroleum, crafting his trademark: to buy enough of a company to drive up its share price in the short term as an outspoken "activist" investor, pumping up profits at the expense of long-term value, and then selling the inflated shares.[10]

In a similar vein, private equity firms now load up their targets with debt to finance takeovers. The collapse of interest rates after the financial crisis—the cheapening of new debt—made this a gold mine. In theory, the idea behind such deals is to render companies more functional on the other side with McKinsey-style tactics that include breaking union contracts and finding any unexplored opportunities for profits, usually at the expense of employees and for the benefit of investors—and all with less oversight than is found in public companies. In reality, it means already struggling companies are saddled with copious debt, which adds to their troubles.[11] It's a bit like flipping a house: buying cheap and selling high after some improvements. Another comparison is strip mining: finding more elaborate, capital-intensive, and often damaging ways to extract value without much care for consequences down the line.

For already troubled industries, private equity firms' tactics can be a kiss of death. Ten of the fourteen largest retail sector bankruptcies since

2012 happened after private equity takeovers, killing 1.3 million jobs in the process. Seventy percent of the shale drillers that filed for bankruptcy in 2020 were backed by private equity firms.[12] Casinos have fared poorly, too. As neighboring states collected gambling licenses, private equity investors flooded in and acquired Caesars Entertainment and MGM Resorts International, shuttering even well-performing properties that didn't meet their high standards for profitability. To put a cherry on top, Icahn led a charge the next year to have Atlantic City pay him tens of millions of dollars worth of back taxes, arguing that he'd overpaid on properties that were now worth much less; in an early such lawsuit, in 1994, Donald Trump himself had won a $5.9 million tax refund.[13] If public ownership seems like an unsavory way to deal with fossil fuel interests circling the drain, consider the fleets of private equity vultures salivating to take over those companies and strip them for parts.

The approach was contagious. In May 2016, the New Jersey legislature passed the Municipal Stabilization and Recovery Act requiring city officials to create a plan for reducing Atlantic City's $500 million in municipal debt, incurred after years of casino closures. Early on in negotiations, then governor Chris Christie, having rejected the city's restructuring plans, threatened to divert $30 million in casino tax revenue away from the city until it followed the "true path to economic revitalization and fiscal stability."[14] Republican mayor Don Guardian likened Christie's takeover plan to a "fascist dictatorship." At one point, a bar off the boardwalk behind the Plaza was selling one-dollar shots of Fireball to anyone who signed a petition to stop it.[15] Christie ultimately installed his own appointee to oversee the state takeover, empowering an unelected law firm with the authority to break union contracts and sell off public assets, superseding the authority of the mayor and city council. To lead the takeover, he appointed his longtime ally and former state attorney general Jeffrey Chiesa, in a position blurring the line between political appointee and private contractor. Chiesa's law firm would bill the state $1.2 million for less than half a year of work.

New Jersey's state takeover reportedly happened on the recommendation of a real estate investor close to Christie, but it meshed with a strain of legal scholarship that was just hitting its stride. In 2014, law professors David Skeel and Clayton Gillette began working on an argument that would be published two years later in an influential 2016 *Yale Law Journal* paper.[16] Taking inspiration from New York state appointing a fiscal-control board to oversee the city's fiscal crisis in the 1970s, they contended that "deep financial distress is emblematic of the failure of a city's democratic processes. Displacement of those processes in an effort to restore the financial stability that a well-functioning democracy would pursue," they add, "arguably is far less problematic than it might be with a city that is already providing the local public goods that localities are created to deliver."[17] Skeel and Gillette propose a provocative solution: dictatorships for democracy.[18]

In an earlier paper, Skeel reasoned, "States are like people. When they find themselves in an insoluble financial predicament, it is often because of systemic distortions in their decision making. Policymakers' 'bias' toward constituents—including to public sector unions—could make them 'unsustainably generous,'" echoing James Buchanan. He dismisses the possibility of the Federal Reserve guaranteeing the debt of public institutions facing financial distress, as it has in recent rounds of stimulus for corporate bond markets. "A direct bailout would, among other things, externalize the costs of the state's profligacy to other states," he writes. Instead, appointed managers can use bankruptcy to overcome distortions like union contracts and so-called nonessential public goods. State oversight in turn could function much like the IMF in imposing "structural reform in the state's financing." In one interview, he mused about taking the approach federal; empowering the president with a line item veto over congressional budgets could, he argued, be a "pretty good constraint on the capacity of legislators to engage in over-expenditures."

Passed over the objections of voters, Michigan's Public Act 436 enabled then governor Rick Snyder to appoint emergency managers to oversee cities and school districts with troubled finances, empowering his appointees to override elected officials in the name of constraining spending in what just so happened to be Michigan's blackest cities.[19] While roughly 10.1 percent of Michigan's population is Black, African Americans accounted for 51 percent of the population under emergency management at its peak in 2012.[20] On the advice of handsomely paid management consultants, Detroit cut funding for emergency services, turned off its streetlights, and even considered selling off art in the city's museums.

Michigan seemed to be a blueprint of sorts for Chris Christie's appointee in Atlantic City. After Chiesa recommended laying off one hundred of the city's firefighters, his collaborator, Ronald Israel, told the *Press of Atlantic City*, "If we don't have everyone sacrificing, we're going to be Detroit." They made quick work of slashing other public services. Following a tense debate, the city council tabled a vote on whether to privatize trash collection services. Before they could revisit the issue, though, Christie's appointees made the decision for them, unilaterally handing a three-year, $7.2 million contract over to a private company.[21] The state's next target was the city's municipal water utility. Residents wouldn't let it be sold off without a fight.

As a city with a proud union history and present, organizing has been at least as central to Atlantic City as gambling. The city's Black residents had fought off persistent segregation that stretched back to the city's founding, lending it the nickname the "plantation by the sea" for the stark division of labor, wages, living standards, and even beaches between visitors and the white and Black residents of this working-class resort town.[22] The civil rights movement brought sit-ins to desegregate the city's businesses in the 1960s, and the Atlantic City Youth Council removed discriminatory W (white) and C (colored) designations on voting registration forms. The group Direct Action Youth stopped Sun City Casinos—headquartered in apartheid South Africa—from setting up shop in 1983.[23] Veterans of those battles filled out AC Citizens Against the State Takeover, a coalition of some forty local, statewide, and national groups. In fighting the privatization of the Atlantic City Municipal Utilities Authority (ACMUA), residents were keen to avoid the fate of another majority Black city placed under state control as its flagship industry sputtered: Flint, Michigan. There, an appointed emergency manager made the cost-cutting move to switch the city's water source from Lake Huron to the Flint River, despite knowing its pipeline infrastructure and water treatment facility were dangerously outdated. Levels of lead and bacteria shot up almost immediately as hundreds of residents got sick, and there were dozens of cases of Legionnaires' disease, a severe bacterial pneumonia. Residents

spoke up almost immediately, but Veolia—the French water company—
assured city and state officials they had nothing to worry about, advising
them to treat taste and discoloration issues instead of underlying toxins
or infrastructure problems.[24] "If a private company comes in, that which
happened in Flint could happen here," Charles Goodman, a longtime organizer with Atlantic City's NAACP chapter, who helped lead the fight
against water privatization, told me. After more than a year of protests,
delegations to Trenton, a City Council Resolution, weekly door-to-door
canvassing drives, and phone banking to stop the sell-off, Chiesa finally
agreed in December 2017 that ACMUA would remain in public hands.

ATTEMPTS TO PRIVATIZE water utilities and trash collection were small
bore compared to the dictatorship for democracy imposed on Puerto
Rico. The island has been treated as a laboratory for capitalist management techniques since it first became a colonial possession of the US in
1898. As the effects of climate change have hit the island, that extractive,
entrepreneurial spirit has only sped up, serving to cement both Washington's and Wall Street's control against calls for self-determination. Today,
Puerto Rico's colonial status is distilled in the Fiscal Oversight and Management Board, the unelected, Washington-appointed body known on
the island as La Junta. It was created in 2016 through Congress's passage
of the bipartisan Puerto Rico Oversight, Management, and Economic
Stability Act (PROMESA), which set up a bankruptcy-like procedure to
deal with the island's estimated $74 billion in debt. A test case for Gillette
and Skeel's thesis, PROMESA handed broad authority over the island's
budget and finances to La Junta; as Skeel told journalist Simon Davis-
Cohen of PROMESA, "what they did looks quite a bit like what we were
proposing."[25] Only one of the body's seven members is required to be from
Puerto Rico. The Puerto Rican governor is technically a member but can't
vote on any of its final decisions; the body answers to the House Natural Resources committee. Skeel was named to the board by the Obama
administration. So was Andrew Biggs—an American Enterprise Institute resident scholar, former George W. Bush administration official, and
longtime proponent of privatizing social security—and Carlos García,
who in running Puerto Rico's Government Development Bank helped to

engineer the debt he was tasked to rein in. Those deals were underwritten by Santander Bank, where García previously served as a top executive.[26]

"They say that the US has to come to manage the structure and the budget because you Puerto Ricans have not been efficient," Lourdes Torres Santos told me in the San Juan classroom where she teaches middle school, explaining that La Junta fits into a long history of forces outside the island arguing its residents can't be trusted with their own government. "That's been true of the Spanish colonizers and of the US invasion. It's been the same message over and over again."

In the 1940s, Operation Bootstrap sought to transition Puerto Rico's agricultural economy into an industrial one through a managed shrinkage of the farming sector, offering generous tax exemptions to corporations for things like capital investments, exporting, and industrial licenses.[27] The move effectively fashioned Puerto Rico into the world's first special economic zone, a term that—since then—has come to refer to any number of policies applied to a certain area to attract outside investors.[28] For a time it seemed to work: consumer goods industries flooded in, wooed by labor costs lower than on the US mainland. But the results were rosier for CEOs than for the people they were ostensibly employing. In the textile industry, for example—previously a cottage industry in Puerto Rico—overall employment was nearly halved while corporate profits more than tripled, jumping from $18.5 million to $60.3 million between 1950 and 1960. Hundreds of thousands of people fled the island as unemployment rose across several industries.

As the policy drove Puerto Ricans into poverty, radical social experiments were put on the table. A collaborative program between the Puerto Rican government and the International Planned Parenthood Federation (IPPF) sterilized at least one-third of women on the island by 1968, and pharmaceutical companies tested out new and often dangerous forms of birth control.[29] Although a number of Puerto Rican women did seek out both new forms of birth control and sterilization during this period, many did not, and those behind population control experiments had their own aims. Speaking to an IPPF conference in San Juan in 1955, Margaret Sanger asked, "How long can the American people be expected to pay confiscatory taxes to support these over-populated lands forever and ever?" She complained about foreign aid to Asia, "a territory and a society, in the words of the *Wall Street Journal*, possessing neither the economic nor the political sagacity to handle any such sum sensibly. . . . Must the *haves* be called upon to support the *not-haves* indefinitely?"[30] As a Pentagon researcher explained bluntly in 1971: "Capital investment from local resources is a function of individual business, and government saving rates and the profitableness of investment in Puerto Rico is oriented toward exports to the mainland and hence would not be influenced by the size of the Puerto Rican market. Individual saving is more likely to decrease than increase with a higher birth rate."[31]

A more recent round of business-friendly federal tax reforms in Puerto Rico came in 1976 through the addition of Section 936 to the federal tax code. It exempted US companies from paying federal taxes on income earned in Puerto Rico. Painting it as corporate welfare, President Bill Clinton scheduled the Section 936 exemptions for a decade-long phaseout starting in 1996. They finally ran out in 2006, leading many manufacturers to leave the island, triggering a recession the island has yet to recover from. After its bond rating collapsed as a result—followed soon after by the global financial crisis—thousands of Puerto Ricans moved elsewhere in search of more economic opportunities, putting further strain on an already stretched economy. Wall Street saw an opportunity to swoop in, seeing its own opportunity in the tax code. Since 1917, bonds issued by the Puerto Rican government have been "triple tax exempt," meaning their buyers don't have to pay federal, state, or local taxes on the returns on them. Once its bonds had been downgraded to junk level after the recession, the vultures descended.

Desperate for quick cash, the Puerto Rican government and public corporations kept issuing junk bonds with the promise of tax-free returns. In turn, major investment banks like Goldman Sachs and Wells Fargo engineered risky, complex new financial products that allowed Puerto Rico to exceed its borrowing limit and continue taking on new debt.[32] Some of these operated like payday loans, complete with astronomical interest rates. Now, nearly half of Puerto Rico's at least $74 billion debt—$33.5 billion—is owed on interest; while many have alleged that tens of billions of dollars of this debt could be illegitimate, there's never been a full audit performed to find out whether that is the case despite frequent calls and protests from island residents.

In exerting control over the island's budget, the fiscal oversight board has operated under the premise that cutting local taxes, selling off public assets, and scaling back labor protections and social programs would create a welcoming business climate for investors, increasing islanders' incentives to work for cheap and thus, they say, fuel economic growth and restore its bond rating. While US corporations operating on the island pay few local taxes, Puerto Ricans pay some of the highest sales taxes in the US and its territories, set at 10.5 percent. Antidemocratic austerity has been better at delivering lucrative fees and contracts to banks and consultants than recovery to Puerto Ricans, who are living in a painful recession that has lasted well over a decade.

That was the situation when Hurricane Maria slammed into Puerto Rico in 2017. Some parts of the island were without power for nearly a year. The official death toll of the storm was 2,975, but real numbers could be far higher, as people died in the aftermath of the storm cut off from essential services, aid, and life-saving services like dialysis.

For La Junta and Puerto Rico's right-wing government, the storm was an opening. After Maria leveled the island's grid, then governor Ricardo Rosselló joined the board in support of privatizing its beleaguered public electric utility, known in English as the Puerto Rico Electric Power Authority (PREPA). Citibank was tapped by La Junta to oversee the process. Besides having underwritten some $9 billion worth of PREPA bonds, the bank was also the second-largest underwriter of so-called scoop and toss financing deals for Puerto Rico, which—while allowing bond issuers to push payments off into the future—also mean heaping additional fees and interest onto preexisting principal and interest payments. From 2000 to 2016, the bank collected $302 million in fees off underwriting $11.3 billion in scoop and toss arrangements, researchers at the Action Center on Race and the Economy found.[33] Like other big banks, Citibank carefully insulated itself from the risks shouldered by the island's biggest creditors in hedge and mutual funds. Investment banks don't generally own many Puerto Rican municipal bonds directly, instead reaping profits by underwriting debt and collecting huge fees during the restructuring agreements that tend to follow. It's thanks to Wall Street's involvement that Puerto Rico's debt grew both bigger and impossibly complex, pitching risky products like auction rate securities and interest rate swaps, which imploded after the recession and left Puerto Rico's government on the hook for $40 million.

As in Detroit and Atlantic City, law and accounting firms have descended on Puerto Rico, too, charging premium rates to manage the debt restructuring process. "If there's a winner out of this whole debacle, it's the consultants," said Lara Merling, economic research officer at the International Trade Union Congress, who has tracked the situation in Puerto Rico closely. "They've all been paid a lot of money, and they've continued to be paid."[34] All of the money paid to consultants—for either the board or the island's government—has come out of Puerto Rico's operating budget. McKinsey & Co. made $3.3 million furnishing advice on how the island could sell off its public sphere for quick cash, cut wages and services, and restructure its debt. Conveniently, they suggested the island pay $1.5 billion over five years in fees to the consultancies, banks, and law firms tasked with getting its fiscal house in order. In 2020, PREPA—having run through a string of directors ousted over corruption—hired none other than Chris Christie as a consultant, paying him nearly $30,000 a month.[35]

McKinsey consultants were boosters of a poorly received plan to close hundreds of schools in the year after the storm hit, aimed at saving $63 million; by the following January, McKinsey had already billed the island $72 million. Closures were carried forward with enthusiasm by Julia Keleher, secretary of the Puerto Rico Department of Education, who cited New Orleans's wholesale, Heritage Foundation–devised school system privatization post-Katrina as a "point of reference."[36] While 17 percent of Puerto Rico's population fled after the storm, she aimed to close 40 percent of its public schools and use those buildings to house charters.

Keleher hit more than a few snags. Torres Santos's Montessori middle school was one of several targeted for closure. As was the case for schools throughout the island, teachers and parents joined forces in the days after the storm to get schools up and running, which in some cases were serving as community resource hubs. When privatization calls came, they came together again with students to stage rolling protests and strikes across the island as part of the Federación de Maestros de Puerto Rico, or FMPR. Ire at Keleher was so pitched in the years after Maria that it inspired an island-wide hashtag: #JuliaGoHome. Aside from the details of the plan and spate of recent closures, much of the anger directed at her came from her earning $250,000 per year, which means Keleher actually made more than Betsy DeVos, and more than twelve times Puerto Rican teachers' base salary of around $20,000. In 2019, Keleher was arrested by the FBI and indicted on corruption, conspiracy, and fraud charges.[37] Keleher's legal counsel had motioned for her trial to happen outside Puerto Rico, citing that she "became the face of corruption" in the Puerto Rican government and couldn't receive a fair trial there.[38] The motion was rejected. As of writing this, her trial is set to begin on the island in February 2021.

In battling to keep schools open, teachers in Puerto Rico have fought against not just closures and privatization but also the suite of austerity measures being pushed on the island more broadly. "Puerto Rico is facing the biggest attack on all public services at once," Mercedes Martinez, president of FMPR, told me. "Children lost their homes and their friends. Children have lost family members that have flown away as well. The government of Puerto Rico has stopped our children from truly recovering by trying to take away their teachers midsemester." Martinez and other teachers were inspired by the wave of "Red for Ed" strikes that occurred in West Virginia, Oklahoma, Kentucky, Arizona, and other states. FMPR's approach has plenty in common, as well, with United Teachers Los Angeles, who struck in 2019 under the banner of "Bargaining for the Common Good," demanding not just better wages and working conditions but broader improvements in the areas where teachers, students, staff, and parents live, too.[39] Like those fights, FMPR's fight was as much about raising teachers' salaries and protecting benefits as it was about defending a vision of what it is the public sphere is for and what it should provide. Both go to show why teachers—an organized, low-carbon workforce—can be such an invaluable part of the coalition fighting for a Green New Deal.

After over a decade of a painful recession, the tragedy of Maria, and austerity measures imposed by La Junta, Telegram chats published by the Center for Investigative Journalism were the final straw that broke the back of former governor Ricardo Rosselló's right-wing administration.[40] Hundreds of pages of a group chat on the messaging app Telegram between Rosselló and members of his staff were leaked. In addition to the governor joking with his top staff about shooting San Juan mayor and then gubernatorial candidate Carmen Yulín Cruz and calling former New York City Council speaker Melissa Mark-Viverito a *puta* (whore), the chats showed that confidential information on government contracts and operations was being shared with ex-officials now working for corporate interests.

Roughly a third of the island's population took to the streets calling for more than just Rosselló's resignation. As one popular chant put it: "*Ricky renuncia y llévate la junta*," *Ricky resign and take the board with you*. Amid the island-wide protests in 2019 that ousted Rosselló, the *Washington Post* editorial board took the opportunity to argue that "Congress should take steps to strengthen the board," pointing to governmental corruption.[41] It was a tricky line for organizers on the island to walk—much less to communicate beyond its bounds. "We've had corrupt governments and that is true. But you cannot evaluate the Puerto Rican government without looking at the constant and continued presence and intervention of the US in Puerto Rico," Julio López Varona, a lawyer and San Juan–based organizer with Construyamos Otro Acuerdo and the Center for Popular Democracy, told me. "The continued intervention of the US hasn't made life better for everyday Puerto Ricans. It's made life better for people who are rich who can come to Puerto Rico and pay no taxes."

"We are not asking for a better control board," he told me. "We are asking for investment in Puerto Rico. What we need is a Marshall Plan that invests money in the island and community-initiated programs that are overseen in Puerto Rico—a real process of self-determination that can help Puerto Rico pave its way to whatever it chooses."

As with shale drillers, cheap debt helped Trump build his fortune and reputation. When the house of cards he'd built in Atlantic City started to fall, there was an understanding billionaire ready to lend a helping hand, generously letting him retain a sizeable stake in the empire he'd helped drive into the ground. Those left holding the bag were the people who'd built Trump's fortune: contractors left unpaid and employees out of a job. Creditors haven't been nearly as forgiving with governments that incur debts to provide essential services to the public rather than boost profits.

The narratives of personal responsibility that have plagued thinking about the climate crisis—that humans are just too wasteful, greedy, and short-sighted—closely mirror those on public debts and deficits. When corporations fall on hard times, the laws of supply and demand are to be blamed. When individuals and governments do the same, it supposedly demands a quasi-moral reckoning with their reckless ways and evidence of the failure of self-government.

Financial distress in Atlantic City, Michigan, and Puerto Rico has been used as an excuse by Wall Street and the right to dismantle democratic rule and enforce a thinly veiled ideological agenda, empowering unelected managers to sever union contracts and privatize public goods in the name of instilling fiscal discipline, all the while siphoning obscene amounts of money off to law firms, Wall Street banks, and management consultants. That these arrangements have been imposed on majority nonwhite populations is less a matter of coincidence than continuity, from France's demands for compensation from the Haitian Revolution to the long backlash against Reconstruction. When the 1 percent's best sources of cheap land and labor are choked off—by bottom-up claims to things like public planning and resources or even just basic democratic rights—it will go to great lengths to restore them. In effect, emergency managers do for struggling governments what leveraged buyouts do for struggling corporations and what capitalism has done for capitalists: smooth paths to extracting as much value as possible, whatever the cost to people or the planet.

We can consider Republicans' more recent drives to commandeer the courts, attack voting rights, and gerrymander themselves electoral majorities with smaller and smaller vote shares in a similar light. The Electoral College and Senate have been major allies in that minoritarian project, crafted as they were to keep the masses out of politics. The latter significantly underrepresents nonwhite voters while overrepresenting white voters and empty land. When Democrats lost the Senate in 2014, Republicans won 52 percent of the vote and nine seats in that body. When Democrats got 52 percent of the vote two years later, they picked up just two seats; when they won the same vote share in 2018, they lost two seats.[42]

The coronavirus has laid bare a cruel reality that will only become more obvious if business as usual keeps up through a warming twenty-first century. The people most likely to be killed look a lot like the people who keep society running, most of them Black and brown, and who've had the fewest avenues to participate in democratic processes. What makes obscene wealth possible is an enormous amount of cheap life, land, and labor. That societally necessary work—growing food, raising children, caring for the sick—tends to be done by people of color and women who are either underpaid or simply not paid at all.

Climate dystopias tend to point to some underhanded scheme elites will devise as an escape hatch to save themselves from the ravages of climate change: floating cities removed from it all, space colonization, underground bunkers. Some or all of these experiments might well be tried. None other than Milton Friedman's grandson founded the Seasteading Institute with Peter Thiel.[43] But the belief that these projects will be *the* answer from the 1 percent to rising temperatures ignores that all that demographic's hoarded wealth is contingent on keeping everyone else in check. The management techniques they use to enforce business as usual span from the ordinary tyranny of the workplace to the punishing laws that allow debt to be used as a tool for disciplining not just people but entire cities, states, and countries. A tiny crust of haves relies on a well-managed army of have-nots, which, in the US, is disproportionately and by design nonwhite. Racism has been capitalism's most profitable management technique.

And despite corporate odes to free markets, that management has always required enlisting the state. As prison abolitionist thinkers and organizers have long pointed out, decades of neglected social services in communities of color can't be understood cleanly as a matter of disinvestment or even of small government ideology. Its flipside has been a bloating of budgets for policing and prisons, leaving the carceral state to take on a number of functions for which it is manifestly unqualified, from mental health to education. As the world warms, the list of functions that bloated carceral workforces take on will increasingly include the work of disaster response. California's and Arizona's reliance on inmates to fight climate-fueled wildfires makes that clear.[44] So does FEMA, which functions as an agency within the Department of Homeland Security (DHS) and has acquired its counterterrorism lens: in the wake of Hurricane Katrina, National Guard troops navigated washed-out New Orleans streets like they would a battlefield in the War on Terror. The priority of the US government now in place to respond to climate-fueled disasters is to protect property from foreign and domestic threats—as reflected in laws and budgets at every level of government. There's no need, in other words, to look to science fiction to imagine climate dystopias. Racial capitalism is already delivering them through guns and budgets as grifters skim off whatever wealth they can along the way. Almost always, though, the people on the losing end of those antidemocratic schemes are fighting back—sometimes, facing long odds, they win.

Within the US, as journalist Todd Miller has noted, internal migration resulting from unbearable heat and catastrophic storms and fires could soon be treated as an excuse for immigration officials to erect new barriers to basic resources and state protections. In his book *Storming the Wall*, Miller imagines a scenario in which the majority Latino population of Phoenix is forced to flee through interstate checkpoints controlled by US Border Patrol agents. "Borders can be enacted quickly through road blockades and interrogating agents," he contends, noting that during the Depression farmers fleeing the Dust Bowl were blocked from entering California.[45] Indeed, new borders are already proliferating in the twenty-first century.[46] Since its creation in 2003 as part of the War on Terror, the Immigration and Customs Enforcement (ICE) has effectively extended the militarized southern border into towns and cities throughout the United States, often to terrifying ends for those targeted. ICE's inception made it far more likely that people could be picked up—in a routine traffic stop, at their workplace or school, or from their own home—and deported, forced at a Kafkaesque moment's notice to leave behind families and communities. As one-time ICE acting director Thomas Homan warned, "You should look over your shoulder, and you need to be worried."[47]

As temperatures continue to rise, the military and the combined capacities of DHS are increasingly becoming America's front-line responders to direct and indirect climate impacts. Allowed to continue, this situation is poised to create a kind of climate apartheid, endowing these agencies with the power to decide who gets to recover from devastating storms and floods and which families are allowed to stay together. Rebalancing the US government for a climate-changed future should entail dismantling the Department of Homeland Security, abolishing ICE altogether, defunding the broader carceral system, and scaling down the bloated US military—itself a major contributor to climate change. At the state and local level, this can free up resources for everything from education to health care to climate resilience. The federal government doesn't need to raid the military's coffers to fund a full-throated energy transition. Like all budgetary choices, it's a statement of principles.

Optimistic talk of the GOP having a come-to-Jesus moment on climate all too often ignores the party's other policy commitments to criminalizing Black and brown communities, fighting forever wars abroad, and closing off the country's borders or even expanding them to include towns and cities hundreds of miles from them. The substance of a Republican response to the climate crisis will be defined less by an embrace of market mechanisms or conservation rhetoric than its overwhelming desire to uphold white supremacy by any means necessary and ensconce minority rule of the few over the many. Daunting as it is, decarbonization in that context is much too narrow a goal. Clean energy isn't an alternative to the vultures and violence. The choice of the twenty-first century is between a postcarbon abolition-democracy or an extractive eco-apartheid, whether the latter happens to be called green or not.

CHAPTER 12

EMERGENCY INTERNATIONALISM

The tradition of all dead generations weighs like a nightmare on the brains of the living. And just as they seem to be occupied with revolutionizing themselves and things, creating something that did not exist before, precisely in such epochs of revolutionary crisis they anxiously conjure up the spirits of the past to their service, borrowing from them names, battle slogans, and costumes in order to present this new scene in world history in time-honored disguise and borrowed language.

—Karl Marx (1852)[1]

We might build the wall, literally.

—Garrett Hardin (1977)[2]

GARRETT HARDIN, THE late University of California ecologist, is listed by the Southern Poverty Law Center as a white nationalist. It's a designation he worked hard to earn. Throughout his long career, Hardin worried obsessively about nonwhite birth rates, serving as an active board member for the nativist Federation for American Immigration Reform (FAIR). "Those who breed faster will replace the rest," he once wrote, previewing the chants of a later generation of white nationalists who rallied in Charlottesville, Virginia, in the summer of 2017. "In a less than perfect world, the allocation of rights based on territory must be defended if a ruinous breeding race is to be avoided."[3]

Hardin is better remembered for his still widely cited 1968 article in *Science*, "The Tragedy of the Commons." Using the metaphor of a common pasture, Hardin argues that it's in the best interest of each individual herdsman to let his or her own cattle graze as much as possible so as to grow the herd, and that each herdsman pursuing his own rational self-interest would eventually overwhelm the pasture: "Freedom in a commons brings ruin to all." Hardin saw private property as an answer to this dilemma but also believed that states should carefully manage their populations: welfare, he said, unduly supported the "freedom to breed," a right inscribed in the UN Universal Declaration of Human Rights that he saw as an "intolerable" strain on the planet.[4] As an alternative, Hardin advocated a "lifeboat ethics," imagining countries as vessels on the open seas, locked in a brutal war over the earth's scarce resources.[5] "Continuously, so to speak, the poor fall out of their lifeboats and swim for a while in the water outside, hoping to be admitted to a rich lifeboat, or in some other way to benefit from the 'goodies' on board." The only reasonable course of action, he contended, previewing scenes of European border police beating back migrants in the Mediterranean, is to let them drown—and if need be, using the oars to bat away anyone who dares to climb up.

Against the background of calls from postcolonial states for a more equitable world order, Hardin would write that countries were each wholly responsible for their own fate, despite the enormous amount of wealth rich countries had taken as land, labor, and resources from their colonial holdings. A "wise and competent government," he wrote, "saves out of the production of the good years in anticipation of bad years that are sure to come," and immigration and foreign aid both rewarded bad behavior. Hardin cast leaders making these demands as shifty rent seekers. Like children or terrorists, giving in would only encourage them. "What gifts do they demand," he asked. "And what will they do with the gifts if we accede to their demands?"[6]

"The Tragedy of the Commons" remains an influential text in environmental studies, but most everyone pushing for some kind of climate policy in the US today would reject Hardin's hard-line racism and xenophobia. Population control became a hot topic for environmentalists following the oil crisis and the publication of Paul M. Ehrlich's *The Population Bomb* in 1968. Sierra Club activist John Tanton would go on to lead some of the country's most influential nativist groups—including FAIR and NumbersUSA—with generous funding from an heiress to the Mellon family fortune.[7] To his and Hardin's dismay, the Sierra Club disassociated itself from such efforts decades ago, and revanchist immigration politics have mostly fallen out of favor among US greens. Thankfully, the same people who back climate action now are by and large those who have been appalled by Donald Trump caging migrant children at the border. Even still, many progressives cling to a more genteel lifeboat ethic.

As mainstream discussions about climate policy in the US have thankfully progressed beyond narrow talk of carbon pricing—embracing core tenets of a Green New Deal—their commitments to a more equal world have tended to stop at the border. Like carbon emissions themselves, though, any talk of a muscular industrial policy and climate justice can't bind itself to lines on a map. There's some comfort in imagining that the US could navigate through the twenty-first century in atmospheric conditions of its own design, experiencing only the climate impacts created by the emissions generated within its own borders, taking care to defend against increasingly sweaty enemies. That's unfortunately not how any of this works.

The atmosphere doesn't much care whether greenhouse gases rise above Houston or Shenzhen. Oddly, that the US is responsible for around 15 percent of global emissions has been deployed as a talking point by centrist and right-wing pundits, who point to rising emissions in China (28 percent) and India (7 percent) as proof that anything the US does is ultimately futile. That masks the fact that the US continues to send a great deal of the fossil fuels produced here to be burned abroad, having in the last few years become the world's largest exporter of oil and gas. It also ignores the outsized power the US still wields globally. As home to the world's largest economy and its most powerful central bank, with de facto veto power in the Bretton Woods Institutions, the US could potentially remake a global order that today is weighted heavily toward polluters' interests. Thanks to how this country has historically used that tremendous power, though—on wars, extrajudicial killings, coups, and more—it's running up a dangerous trust deficit with much of the world.

At home and abroad, the US government is currently designed to do some things much better than others. COVID-19 has shown that it's not great at handling public health crises or providing its residents with basic economic stability. Where it does excel is in the often interrelated goals of protecting private investment and ginning up conflicts abroad, each of which have tended to involve managing nonwhite populations through varied, often violent means. So long as these remain among the few core competencies of the US government, anything called climate action will be filtered through them. Unchanged, then, zero-sum foreign policy as usual will leave the United States in a race toward a dangerous future. The alternative is to direct its enormous power and resources toward building a global economy that's better at keeping people alive than killing them.

CONSIDER BORDERS THEMSELVES. Climate change is poised to cause the largest mass migration in human history, as millions are forced to leave homes rendered uninhabitable by rising sea levels, unbearable heat, and declining crop yields.[8] The best estimates hold that warming could displace anywhere between twenty-five million and one billion people. Border and immigration policies, in other words, are climate policies, and efforts to restrict access to temperate parts of the world will be a defining political issue of the next century. Still, even some progressive messaging on that front imagines climate-fueled migration primarily as a security threat, with borders as one of many national security assets to be defended against the effects of rising temperatures. John Kerry's World War Zero initiative, founded in 2020, brought together military officials, retired politicians, and celebrities from both sides of the aisle in a bid to win over conservative skeptics through a series of talks, in part around the idea of the climate crisis being a national security threat.[9] As Kerry has put it, "We have to heighten our national security readiness to deal with the possible destruction of vital infrastructure and the mass movement of refugees, particularly in parts of the world that already provide fertile ground for violent extremism and terror."[10]

He's not alone in that thinking. The Pentagon has been authoring reports on global warming's security implications since 2003 and sees it as an "urgent and growing threat to our national security."[11] Their *2014 Climate Change Adaptation Roadmap* envisioned that, as climate change accelerates, "the Department's unique capacity to provide logistical, material and security assistance on a massive scale or in rapid fashion may be called upon with increasing frequency."[12] Former Trump White House defense secretary James Mattis has been a longtime proponent of the idea that the military wean itself off fossil fuels.[13] Military climate risk assessments generally fall into one of two buckets: threats to Defense Department infrastructure and operations (rising sea levels putting coastal bases at risk, wildfires disrupting training activities) and analyses of what sorts of future conflicts the military is likely to get involved in as a result of rising temperatures, described in the defense community as a "threat multiplier." For instance, will drought-induced food shortages require more boots on the ground to handle the armed insurgencies and mass migrations that follow them? That conversation expanded amid the Syrian civil war, which several academics have attributed in part to devastating, climate-fueled droughts that pushed farmers into cities starved of resources and investment, sparking some of the initial protests during the Arab Spring in 2011.

The idea of a climate-smart national security state emerged during the Trump administration as a potential source of bipartisan agreement. Elizabeth Warren introduced legislation in the Senate in 2019 requiring military contractors to account for their carbon footprints, citing concern about "climate threats to operations, and the potential downstream impacts on military readiness."[14] The House Select Committee on the Climate Crisis's 538-page climate plan states that climate change "amplifies geopolitical threats as resource scarcity and catastrophic events fuel conflict, mass migration, and social and political strife."[15] Though its authors seek to ease wildlife migration across borders, human migration is treated mainly as a menace. They urge directing the Department of Homeland Security—including ICE and the Border Patrol—to prepare for "climate-driven internal and cross-border migration." Laudably, the report's authors do recommend admitting and offering formal protections for 50,000 "climate displaced persons per year." Given that

16.1 million people were displaced in 2018 by weather-related disasters, arriving at that definition could be fraught and shaped in untold ways by the United States' ugly and violent history of immigration quotas. As in the past, that modest quota would be quickly overwhelmed. And the designation of climate refugees as singularly deserving of entry to the US in a warming world should be challenged. The front row seat that people across the Global South have had to centuries of northern policy, legal scholar E. Tendayi Achiume argues, should trouble such narrow conceptions of who's allowed in. "Citizens in Third World countries in which First World countries intervene," she writes, "should not be considered political strangers; they are political siblings or cousins, with rights that should include entry and inclusion through migration."[16]

To be clear, the basic diagnosis put forward by those calling for a green national security state isn't wrong: climate change *is* a threat multiplier, exacerbating all manner of religious and political conflicts the world over. But the proposed solutions—often, a greener military better prepared to shoot down climate-fueled threats—ignore *why* it is that rising temperatures so often fuel violence and waves of migration, which don't just happen to fall over cities like a summer rain. Central American farmers have for years been forced northward thanks to a combination of prolonged drought and increasingly neoliberal agricultural policies in the region, in some cases brought about by cuts pushed by the IMF.[17] The North American Free Trade Agreement (NAFTA) flooded Mexico with US-subsidized corn almost as soon as it passed, devastating that country's agricultural economy and forcing many to pursue work elsewhere, often in factories along the borderlands or in the US.[18] Following NAFTA's passage and the ensuing surge in immigration, then president Bill Clinton doubled the budget of the Immigration and Naturalization Service and swelled security forces and technology along the border. *Tropic of Chaos* author Christian Parenti refers to the blinkered US security state response as the "politics of the armed lifeboat: the preparations for open-ended counterinsurgency, militarized borders, aggressive anti-immigrant policing, and a mainstream proliferation of rightwing xenophobia."[19]

When faced with a first wave of climate refugees from Latin America in the 1990s, Democrats followed the same strategy they'd tried in other policy fights that decade. As elsewhere, it mainly succeeded in putting the national debate on the right's terms. Decades before she scolded youth climate activists who called on her to embrace a Green New Deal, California senator Dianne Feinstein was one of the leading voices in the Democrats' ill-fated accommodation to right-wing nativism. As Clinton attempted to appear tougher on crime than Republicans, Feinstein sought around the same time to outflank them by getting tough on immigration herself. In an ad for her Senate race, she boasted about having "led the fight to stop illegal immigration," making sure to single out the bad, criminal immigrants from the good, law-abiding ones. The Democratic establishment in California ceded the premise of a virulent anti-immigrant policy even as they opposed it, in this case a draconian ballot measure known as Proposition 187. The *Los Angeles Times* noted that Feinstein's moderate hard-line stance during this time gave "'cover' for politicians of both parties, lending respectability to a sensitive area where it is easy to be branded a demagogue and a bigot." As author Daniel Denvir put it, "the liberal and mainstream establishment . . . simply amplified the nativists' politics and advanced their policies."[20]

Attempts to woo Republicans over to the climate fight with promises of stronger bases and borders play dubiously into a longstanding Republican frame—one held in common with Hardin—about encroaching non-white masses. There's also just not much evidence the appeal will win over anyone but a handful of balance-of-power realists in the national security establishment. For little gain, then, this strategy mainly promises to drag climate politics into a different corner of the culture wars than where it has lived in so far.

Ecology, according to Hardin, was a means to the end of suppressing nonwhite birth rates and erecting a state that could curtail population growth through contraception and harsh immigration restrictions.[21] There's no reason to think a new generation of xenophobes won't take a similar route, exploiting increasingly widespread popular concerns about the climate crisis to drive through their nearer and dearer priority of restricting immigration. Seventy-seven percent of young, right-leaning voters now say the climate crisis is an important issue to them. More than half said it would impact how they voted in 2020. For any number of reasons laid out so far in this book, Republicans will change their party line on climate sometime soon.[22] White supremacy is far more fundamental to the GOP than climate denial. Any pivot on the latter will likely leave the former fully intact and birth a horrifying new brand of climate politics of the sort already sprouting abroad. Any accommodation to nativists in a climate-changed twenty-first century is playing with fire.

Imagined border crises, though, have relatively little to do with what's actually happening at the border. They usually serve to reflect real crises elsewhere. Migrants, in that sense, have always been useful scapegoats for politicians eager to deflect from their own failures. As unemployment skyrocketed, Herbert Hoover's labor secretary William Doak ordered raids on Mexican communities in California and elsewhere, the justification being to preserve jobs for native-born citizens. As many as two million Americans of Mexican descent—an estimated 60 percent of them US citizens—were "repatriated" (deported) to Mexico as the Great Depression set in. The main criteria for being rounded up—whether at a park, hospital, or social club—was having brown skin. Complementing the raids were Department of Labor orders against local governments employing people of Mexican descent. Ford, U.S. Steel, and the Southern Pacific Railroad complied with Doak's push and laid off thousands in order to create "jobs for needy citizens."[23]

In 2020, as recession set in and the death count from the novel coronavirus soared, there were disturbing parallels. The Department of Homeland Security deployed federal agents to suppress Black Lives Matter protests in Portland, Oregon, rounding up anyone who looked suspicious off the streets and into unmarked vans. Trump threatened to kick students on education visas out of the country, and Border Patrol conducted a military-style raid on No More Deaths, a Tucson-based humanitarian aid group providing relief to migrants crossing the US-Mexico border.[24] At one point, Trump floated postponing the election.

As the administration failed to contain the pandemic and its impacts, people in the US looked for leaders that seemed to rise above the fray of politics. Anthony Fauci, director of the National Institute of Allergy and Infectious Diseases, took on a kind of cult status. New York governor Andrew Cuomo became another unlikely hero, despite overseeing one of the country's most devastating outbreaks. One institution, though, seemed to hold the global economy on its shoulders. While

Federal Reserve chairman Jerome Powell never became a technocratic sex symbol like Fauci or Cuomo, the Fed earned some deserved plaudits for undertaking one of the most remarkable interventions in the history of central banking.[25] In March 2020, it injected $1.5 trillion worth of liquidity to keep credit markets from seizing up, cut interest rates to zero, and bought $500 billion worth of Treasury bonds and $200 billion in mortgage-backed securities. When that all fell short, Powell announced the unprecedented step of buying $1 trillion worth of "commercial paper," including high-yield corporate debt, on top of $450 billion worth of "swap lines," exchanging lines of credit to other central banks for foreign currency. The goal was to make dollars—the global reserve currency—less scarce.

On the one hand, it was evidence of the dazzling powers of skilled technocrats against a backdrop of gross incompetence from elected officials. For some, it seemed to bear out the double government thesis: that competent experts can accomplish what democracy can't. That the Fed has so much power over the fate of people half a world away, though, should be cause for alarm about the health of the international order.

The Fed's sweeping actions in 2020 did indeed keep the world from descending into a total depression. They also reflected and redoubled on longstanding priorities. Fossil fuel companies, for example—flailing amid the crash in fuel prices—made out well. As of the summer of 2020, energy company purchases made up nearly a fifth of the bonds the Fed had purchased; by contrast, energy companies make up around 2 percent of the S&P 500. The UK-based think tank InfluenceMap found that the Fed could be saddled with $19 billion worth of high-risk fossil fuel bonds if the program continued down the same path.[26]

State and local governments weren't nearly so lucky. By the end of July, just one of the 255 cities, states, or counties that qualified for the Fed's $500 billion Municipal Lending Facility—intended to prop up struggling governments—had even applied: Illinois. A report by the Center for Popular Democracy found that a stunning 97 percent of eligible municipalities for the program were functionally excluded thanks to loan terms less favorable than the ones they could get in markets. Many states contain no eligible cities or counties at all, and the criteria excluded all but one of the five cities saddled with the country's highest unemployment rates, including Atlantic City.[27] States, which can't deficit spend, faced a $315 billion deficit by 2021, a crisis that would leave many at the mercy of private bond markets and extractive creditors. Between February and July, state and local governments fired or furloughed nearly 1.5 million people, nearly twice as many as during the entirety of the Great Recession. So as fossil fuel executives squirreled away relief funds, public schoolteachers and sanitation workers who had kept society going in the pandemic faced punishing cuts and the prospect of handling new waves of outbreaks with even less support than they had during the first.

The inequalities of existing Fed policy in the US extend beyond its borders. The Bretton Woods Institutions, including the IMF and World Bank, were initially designed as a means to start moving away from a gold standard, the built-in scarcity ethic of which was credited with fueling fascism in Europe. In 1944, John Maynard Keynes had proposed an International Currency Union to be run on a new currency called the *bancor*, meant to facilitate easy trading and price stability without handing inordinate power to a single country's central bank. Instead, the US steamrolled over the idea, and its dollar became the world's reserve currency.[28] Today, some 60 percent of all central bank reserves are held in US dollars, and the Federal Reserve can turn the spigot on and off for them. Given the superabundance of dollar-dominated debt, dollar diplomacy has been a powerful tool for the US to get its way. In times of crisis, the Fed gets to choose when to extend or withhold an unlimited number of swap lines to other central banks, allowing access to dollars that can stave off collapse and keep basic services afloat.[29] So as the US bumbled through its own COVID-19 response, its central bank got to determine how many resources other countries would have on hand to devote to theirs.[30]

More capital fled out of poorer nations in the first three months of the pandemic than in any year on record, and the value of some southern currencies slid as much as 30 percent against an increasingly strong US dollar.[31] Countries that were already struggling to make sovereign debt payments faced an abyss. For export-intensive economies the situation was more dire, still. Nigeria—where oil accounts for over 60 percent of government income—scrambled to make $7 billion in debt payments. Iraq had planned to cover 95 percent of its budget with oil revenues, assuming a $56 price per barrel; after a modest rally, Brent Crude by the following fall was hovering around $40 per barrel. Ecuador, a dollarized economy that became home to one of South America's deadliest COVID-19 outbreaks, had just committed to paying off a $4.2 billion loan from the International Monetary Fund when people started getting sick. Toward that end, President Lenín Moreno had slashed investments in the country's public health system by 36 percent in 2019, part of a wave of cuts that sparked nationwide protests.[32] In 2020 alone, 3,680 people had been laid off from the country's health ministry.[33] Buckling under a total $17 billion debt load, much of it held by Wall Street, the government struck a deal with bondholders that may only postpone what many fear is coming: default.[34]

The OECD has found that meeting the goals set out by the Paris Agreement and UN Sustainable Development Goals will require $7 trillion worth of sustainable infrastructure financing. Though leveraging private capital has been a popular buzzword in climate finance, development finance institutions—bodies like the Development Bank of South Africa and the KfW of Germany—have only been able to leverage $60 billion a year, at most.[35] Amid calls from the Global South for both debt relief and adequate funding for climate adaptation, mitigation, and recovery, today's multilateral institutions, including the World Bank, have consistently pushed for loans to be a substantial piece of the climate finance picture, bound to add new principals, interest, and fees onto the balance sheets of already indebted countries.

"It's their knee-jerk response to all of these problems," Lidy Nacpil, director of the Asian Peoples' Movement on Debt and Development, told me. "You can't solve a problem by creating another problem, and that's what's being created by lending for climate action." It's also, she said, "a subversion of the principle of climate finance," which should be "a form of compensation for the fact that all these rich countries are primarily the ones responsible for climate change."

In desperate situations, fossil fuel reserves offer quick cash. Without a transformation of a global financial infrastructure that prioritizes the free flow of capital above all else, these reserves will continue to be exploited toward disastrous ends. Indeed, faced with paying back the $57 billion debt to the IMF incurred by his right-wing predecessor, left-leaning Argentine president Alberto Fernandez had hoped to double down on trying to develop shale oil and gas reserves, exploring how to exempt foreign oil and gas investors from capital controls and furnish generous tax benefits with the goal of boosting exports.[36] When investors balked after the oil price collapse, the government showered them with still more subsidies to try to prime the pump. Fracking—overleveraged and unprofitable in Texas oil country—was a dangerous boondoggle for Argentina. So long as there are debts to be paid, though, creditors come first.

Managing carbon is neither distinct from nor any less thorny a subject than managing the world's economy; as has been suggested by former Fed officials, the Fed could exercise enormous power by treating the climate crisis as a systemic risk. Dollar diplomacy could theoretically be wielded against rogue actors on climate, too, greening the so-called rules-based international order, which, while Trump was in office, establishment types have longed for the return of. Yet the fact that such unilateral disciplinary action is possible should make anyone interested in a more collaborative and democratic world order uneasy. Climate and energy policy similarly demands skilled bureaucrats that can reconfigure grids and design well-targeted industrial policy, along with a host of ethical questions. How accountable will these experts be to the full range of people their decisions affect? How insulated should their work be from successive governments that might be hostile to aggressive climate action? And so long as it exists, are there any responsible ways to deploy US hegemony in the climate fight without further entrenching it?

THROUGH THE TRUMP era, Democrats' central ambition for international climate action has consisted of a single call: getting back into the Paris Agreement, maybe to be complemented by some token funds for climate investments through USAID. Or resuming Obama's $3 billion commitment to the Green Climate Fund, the climate finance mechanism set up by the UNFCCC process. Democrats have both over- and underestimated the UN process: there's plenty more that can be done within it to drive decarbonization and adaptation forward, but most of the financing and enforcement mechanisms needed to make a global transition a reality lie outside its nonbinding structures, in trade agreements and multilateral institutions with actual teeth.

Any genuine assertion of US leadership, moreover, entails a broader reckoning with and accounting for the most painful parts of this country's history—a recognition that centuries of government policy continue to hobble the ability of mostly nonwhite communities within and outside the United States to participate fully in an energy transition and adapt to the levels of warming already locked in. Real resources are needed to make the cooperation required for any really global decarbonization possible and to build trust shattered—if it ever existed—by decades of engineered debts, extraction, wars, lynchings, coups, cuts, torture, and assassinations that have defined the Great Acceleration at least as much as exploding greenhouse gas emissions.[37] In that bigger picture, US participation in the UNFCCC process is one small but important part of making it to the end of this century with our humanity intact. Let's start there.

As the world's largest economy, the US can exercise tremendous power within the UNFCCC and frequently does—whether to downplay reports from the IPCC, elevate the concerns of polluting corporations, or bury conversations about climate finance. If it hasn't already by the time this book is released, the US should reenter the Paris Agreement with a science-based plan for reducing its emissions—a nationally determined contribution, or NDC, in UNFCCC speak—and use that as the basis to encourage compliance among wealthy nations including the G20, whose member countries account for 78 percent of global emissions. Rapid decarbonization, carried out equitably, would mean aiming for net zero emissions as close to 2030 as possible, leaving space in the global carbon budget for developing countries that still lack the technical and financial capacity for such a stark transition to catch up. With all its vast resources, the United States should also commit to equitable climate financing of at least $200 billion USD through the Green Climate Fund, to be used in speeding along decarbonization around the world. Accompanying that should be support for institutional reform within that body to ensure it can handle a rapid influx of funds that are distributed promptly and equitably. As with NDCs, this pledge should be a basis for the US to encourage other wealthy countries that have benefitted from centuries of

South to North resource transfers to provide their fair share of financing to developing countries. Any talk of account for historical responsibility for the climate crisis should look within US borders, too, and earnestly consider demands for climate debts internationally alongside those for reparations and giving land back to Native tribes domestically. As philosophers Olúfẹ́mi O. Táíwò and Beba Cibralic have written, "climate reparations are better understood as a systemic approach to redistributing resources and changing policies and institutions that have perpetuated harm—rather than a discrete exchange of money or of apologies for past wrongdoing."[38]

Ultimately, any reparative global climate action will be premised on a new multilateralism that can move resources and political will at the macro level on the basis of cooperation rather than domination. Decarbonizing the global economy, after all, will require deploying vast resources from many countries across borders and into places that currently lack access to fair financing. The US can join with like-minded countries that have committed to rapid decarbonization to make progress where existing multilateral institutions are unable or unwilling. As the world's largest economy, the US can lead the effort to press those institutions into catalyzing a green transition not built on extractive debt.

Rather than pitting the US against the world, its leaders could convene a "New Bretton Woods" summit aimed at reforming the international financial system to reflect the challenges of the twenty-first century, of which the climate crisis is the most pressing. Such a gathering could endeavor to set up new mechanisms for multilateral financing, coordinated research and development, and crackdowns on global tax evasion, with the aim of making people at least as free as capital. These conversations should explore widespread debt relief, opening up valuable fiscal space for governments to recover from already existing climate impacts. As recommended by the Civil Society Equity Review, adequate funds should be made available for not just mitigation but for adaptation and loss and damage, as well—areas where private sector interest is far harder to come by.

Exciting work is being done on this front already. Through extensive consultations in 2018 and 2019, the UN Conference on Trade and Development (UNCTAD) developed what it calls Geneva Principles for a Global Green New Deal, for a new multilateralism that sets social and economic stability, shared prosperity, and environmental sustainability as its top goals; recognizes the common but differentiated responsibilities of different countries in meeting twenty-first-century challenges; and seeks to democratize existing multilateral institutions to make them "accountable to their full membership," rather than handing veto power to the wealthiest participants. "Global regulations should be designed both to strengthen a dynamic international division of labor and to prevent destructive unilateral economic actions that prevent other nations from realizing common goals," the plan's authors write.[39]

For now, the existing world order remains more of a hindrance to decarbonization than a helper. Powerful international law threatens to unravel the most ambitious climate plans at the national level, having been written to give corporations and governments acting on their behalf a mighty line of defense against democratic reforms. Modern interpretations of the 1947 General Agreement on Tariffs and Trade (GATT) have already subjected even modest, innocuous rules like buy local provisions in state-level clean energy measures to Investment State Dispute Settlement (ISDS) arbitration.[40] This onerous system allows companies to sue sovereign governments, ensnaring them in costly, open-ended legal battles for the crime of infringing on corporate profits.[41] While the US-Mexico-Canada Agreement (NAFTA 2.0) was lauded for mostly nixing its punishing ISDS system, a sizeable loophole preserves the ability of oil and gas companies that have contracts with the Mexican government to challenge any potential threats to their profits posed by state policy. The single most invoked treaty in ISDS disputes is the obscure Energy Charter Treaty (ECT), created in 1991 to assure nervous Western fossil fuel executives that their investments in former Soviet bloc countries wouldn't be expropriated.[42] Today, it threatens to undermine the Paris Agreement and any progress the fifty-three European, Asian, and Middle Eastern countries that are full signatories to it might want to make, opening states up to multimillion-dollar suits from companies and even individual shareholders. After Italy moved to ban offshore drilling, for instance, the UK-based oil company Rockhopper claimed it was owed $350 million in compensation. Even the threat of a challenge under the

ECT can undermine sound energy policy. In 2017, former French environment minister Nicolas Hulot backed a measure that would have phased out fossil fuel extraction in the country by 2040 and banned new drilling permits and renewals.[43] After the Canadian company Vermillon threatened to bring an ECT complaint, the law was watered down, allowing drilling companies to continue renewing their permits through 2040. Without a major overhaul, the total number of fossil fuel assets protected by the ECT could grow to $2.4 trillion in the coming decades. The ECT may be an extreme example, and the US is thankfully not a party to it, but there's simply no place for today's ISDS system or anything like it in a low-carbon tomorrow.

A new trading order can instead create mechanisms for accountability from US-domiciled companies operating abroad. While abolishing its current form, US policymakers should advocate for an inversion of the ISDS system, allowing communities to mount legal challenges against US-based companies in US courts on the basis of their human and ecological impact.[44] More broadly, any trade deal the US negotiates should center climate as a top priority, pursuing collaboration across a range of fronts—on intellectual property, manufacturing, and more—rather than race-to-the-bottom competition with allies and adversaries alike. This could all be pretty popular, too. As of 2020, 83 percent of Democratic voters believe trade agreements should address climate change.[45]

Concretely, economist Todd Tucker has proposed that extant international trade obligations be suspended for the duration of a decade of a Green New Deal, instead using the WTO to *advance* decarbonization and adaptation.[46] "During the mobilization," he writes, "Global Green New Deal countries would perform and make public a full audit of their progress towards these commitments every six months. If they fall short, they would be subject to legal challenge at the WTO's Dispute Settlement Body, which would be converted to an enforcement body for the Global Green New Deal for the ten-year period." At the end of that period, he suggests holding negotiations to decide whether the Global Green New Deal will remain in place or be subbed out for "an entirely new treaty framework."[47] If any country can rewrite the rules of globalization, it's the United States.

AMONG THE BEST things the US can do to speed along decarbonization is to bring down its own emissions as quickly as possible. For that, returning to the Obama-era energy status quo simply isn't an option. Thorough decarbonization means a truly economy-wide transformation and a government that factors a changing climate into every decision it makes. To reiterate, making that politically palatable means pairing it with a credible promise to improve the lives of ordinary people: in other words, a Green New Deal. Enacting such a plan within the US at the speed and scale science demands can be an electoral boon to Democrats looking to maintain majorities in the White House and Congress—and recast climate as an issue of public investment rather than collective sacrifice. It's also hard to imagine US residents supporting green spending internationally if they haven't seen it benefit them at home. Accomplishing that entails a full-scale mobilization of the US government, thoroughly integrating decarbonization into every branch.

As during World War II, a state orbiting around a central goal would have few policy fields that *aren't* involved at some level in achieving it. Today, midlevel Treasury appointees liaise with the State Department when crafting trade policies that govern how the United States imports and exports emissions and can set the agenda for international institutions like the IMF and World Bank. Pursuing decarbonization, diplomats would be as persistent in prioritizing it in those spaces as they are now about protecting US pharmaceuticals' intellectual property. The Department of Agriculture would plan how to maintain food production through droughts and higher temperatures and tackle the methane-spewing, pathogen-spawning world of industrial meat production. The Department of Labor would administer a climate-minded federal job guarantee, or even sectoral bargaining in low-carbon fields. Bodies that seem mostly irrelevant to climate exercise a deciding vote over decisions made in others, foreclosing on changes elsewhere. The head of the National Credit Union Administration, for instance, has a voice on the Financial Stability and Oversight Committee, which plays an instrumental role in assessing the risks posed by fossil fuel companies and how to handle the coal, oil, and gas assets those companies abandon as they become too costly or risky to dig up out of the ground.

. Federal procurement policies could be a workhorse for decarbonization. The General Services Administration (GSA)—which handles government procurement—can steward America's massive federal vehicle fleet and buildings to run on clean energy. As of 2019, the US government had a fleet of 645,047 cars that traveled 4.5 million miles that year, consuming 386 million gallons of gasoline.[48] The GSA owns and leases over 376.9 million square feet of space in 9,600 buildings, including office buildings, post offices, courthouses, national laboratories, data processing centers, and ports of entry. Green retrofits for all those federal facilities would create millions of unionized construction jobs in most counties and open up additional opportunities for creative uses of federal infrastructure. Per a proposal from the Canadian Union of Postal Workers, retrofitted post offices could be hubs for more than just mail. In addition to electric vehicle charging stations—already there thanks to newly electric fleets—post offices could host community supported agriculture pickups and offer postal banking services, serving the whopping 25 percent of Americans unserved or underserved by banks and without access to basic services like savings accounts and check cashing.[49]

Rather than blindly pumping up new demand for clean energy, a Global Green New Deal should transform domestic consumption in ways that reduce ballooning demand for materials and precious minerals extracted in the US and from the rest of the world, as industrial policy encourages the recycling and reuse of metals that currently go to waste. But as demand for EVs and battery storage expands, some amount of additional extraction for technology metals such as lithium and cobalt is virtually inevitable. While working to develop whatever of those resources it can domestically, the enormous purchasing power of the US can offer leverage to establish industry-wide labor and environmental standards across clean energy supply chains, making sure that any new projects are fully compatible with the UN Declaration on the Rights of Indigenous People and that they bar development in protected areas such as the Amazon.

As noted above, the Federal Reserve—the steward of the world's reserve currency—might be the United States' most important institution for foreign and domestic policy alike. University of Chicago historian Destin Jenkins has suggested the Fed eliminate its population thresholds, whereby cities of less than 200,000 are ineligible for relief, and instead prioritize "cities that are majority black, territories with preexisting debts and municipalities already suffering from hospital closures, overcrowded living facilities, homeless shelters and county jails."[50] All of these changes, he argues, "would lessen the racial and spatial inequalities that have reigned for decades" and keep already cash-strapped states and cities from dropping off a fiscal cliff. Besides addressing historic inequalities, this makes it more possible for those governments to be able to respond to the climate crisis, fully funding independent Offices of Emergency Management to take on the kind of long-term climate risk planning now out of reach for so many of these departments.

Jasson Perez, an organizer with the movement for Black lives and analyst with the Action Center on Race and the Economy, similarly argued that the Fed should buy state and local government bonds through the duration of the COVID-19 crisis, lend at zero percent interest rates, and commit to refinancing existing local and state debt—measures at least as generous as those offered to corporations.[51] It could also live up to its mandate for truly full employment, targeting the egregiously high Black unemployment rate and abandoning NAIRU full stop—and not try to sabotage a job guarantee or other green stimulus measures. Taken together, all this would represent a dramatic shift in Fed policy: intervening on behalf of the public instead of the private financial system. Given the global importance of the Fed, such changes could also be an entrée to more democratized central bank planning on the world stage—a monetary multilateralism, as Daniel Bessner and David Adler have called it, to provide a check on the US and "ensure that all nations have the ability to exert their right to self-determination."[52]

The crux of a wartime style mobilization lies in the transformed relationship between public and private production. During World War II, it meant a rapid build-out of the federal government to enable a more rapid build-out of weapons, clothes, and other supplies needed to battle the Axis powers. As new planning agencies proliferated, so did arms: before entering the war, the US annually produced 12,500 aircraft per year, fewer than 100 tanks, and just 0.3 million tons of merchant ships. By the time the war ended, it had 550,000 aircraft, 88,000 tanks, and at least 18 million tons of merchant ships and had created whole supply chains from virtual scratch. As economic historian Andrew Bossie and economist J. W. Mason found in their study of the era, "The more rapidly the economy must be reorganized, the greater the direct role for government must be." Where price signals may be enough to shift corporate behavior over years or decades, doing so rapidly demands more sweeping interventions, with a lot of carrots for the private sector and a few formidable sticks. The federal government was largely responsible for administering prices, wages, and sourcing in sectors deemed vital to the Allies. And the wartime economy was also a more egalitarian one by virtue of creating such a tight, hot labor market, as well as having come at the end of a decade of militant labor organizing that welcomed millions of Americans into unions.[53]

A folktale of the war mobilization is one of heroic American companies revving up to take on the war effort. Their main loyalty, though, was to profits. A few years earlier, during the Spanish Civil War, oil company Texaco had funneled supplies to General Francisco Franco's insurgent fascists. As Adam Hochschild points out in *Spain in Our Hearts*, Texaco head and Nazi sympathizer Torkild Rieber wrote blank checks to Franco's far-right military coup.[54] The company, since subsumed under Chevron, gifted Spanish Nationalists—who were already backed by Hitler and Mussolini—with lavish lines of credit, strategic intelligence, and all the oil money could buy, plus some that it couldn't. Fueling trucks, boats, tanks, and planes, that oil helped propel Franco's forces to victory and over thirty-five years of autocratic rule. One Franco-era official Hochschild quotes laid out just how good Texaco was to the aspiring regime: "Without American petroleum and American trucks and American credits, we could never have won the civil war." Rieber wasn't alone. Hochschild notes that Dow Chemical gave forty thousand bombs to Hitler just before the war began, and the heads of Ford, General Motors, Eastern Kodak, and other companies met at New York's Waldorf Astoria to talk "about the prospects for American cooperation with the Nazi regime."

Though plenty of companies made out well in the war, even those with fat contracts to gain from the war effort had to be dragged into compliance. Nationalization, as discussed in Chapter 9, was one of many tools used to deal with rebellious CEOs. Among the most iconic images of the period's changed power dynamics was a widely circulated image of Sewell Avery, the president of Montgomery Ward. During World War II, Montgomery Ward, a mail-order corporation, produced everything from uniforms to bullets for soldiers abroad. In 1944, the National War Labor Board ordered Avery to let his employees unionize to ward off a strike and potential disruption in war production. When he refused, Roosevelt ordered the National Guard to haul him off, chair and all, and seize the company's main plant in Chicago. The government took over operations at the company's factories in several other cities by year's end. Although instances of full nationalization weren't uncommon, they were essentially a conflict resolutions strategy. Most production happened through more amiable partnerships, especially government-owned, contractor-operated (GOCO) arrangements wherein the government built factories and leased them for cheap out to companies that would run them.

But it's worth being realistic about how much industrial production can happen in the US and how much it will need to rely on international partnerships. After decades of US leaders passing the buck on climate policy and clean energy development, the world's most cutting-edge expertise on green manufacturing simply exists elsewhere, in factories halfway around the world that in many cases employ relatively few people.[55] As Jonas Nahm, a political scientist at the School of Advanced International Studies at Johns Hopkins, told me, "China controls so much of the manufacturing capacity for so many of the technologies we need to combat climate change that—in the time frame that we have to make a difference—the world cannot replicate this ecosystem elsewhere." They are, he explains, far ahead of us: "It took China thirty years to build the capacity to do this, and so now it's making 60 percent of the world's solar panels, nearly 70 percent of the batteries, and a third of the wind turbines. We don't have any of the basic ingredients that you would need to do the same thing China is doing. To try and replicate that elsewhere will just delay us infinitely." The US has comparatively little of the technical expertise, trained workers, or institutional support in place to mirror China's booming clean manufacturing sector anytime soon. If a central goal of the next decade is to deploy as much clean energy as quickly as possible, the most rapid and equitable route forward means that many components of America's new green economy will be imported.

An American-made, export-led green manufacturing renaissance may not be in the cards for the US, even if there are still many things that can be made here and supply chain vulnerabilities that can be addressed. Thankfully, millions of new union jobs and a robust social safety net—the things people remember fondly about the decades after World War II—aren't mutually exclusive with clean energy imports. Today, there are any number of ways to run a healthy and hot economy that doesn't involve the mass production of bombers and artillery shells, let alone appealing to nostalgic imagery of midcentury production lines or the "family wage" flowing to male breadwinners. A twenty-first-century industrial policy can take a wider scope than what factories are pumping out. Producing solar panels and wind turbines is a piece of that. Just as if not more important are the millions of jobs in building retrofits and weatherization, grid transformation, and regenerative agriculture, not to mention sorely undervalued and essential, already low-carbon sectors like teaching and care work. A green manufacturing plan doesn't need to be synonymous with a green jobs plan.

As I write this, lawmakers on both sides of the aisle in the US are fumbling toward a new era of great power conflict with China, with Democrats hammering Trump for being too soft on the country. "US leaders increasingly argue that Chinese businesses must not be allowed to challenge the dominance of US corporations. Facing a disintegrating domestic political consensus, they hope to use the fear of China to unify the population around their favored agenda," organizers and researchers Tobita Chow and Jake Werner argue.[56] Under the guise of concerns about Chinese premier Xi Jinping's plainly egregious turn toward autocracy and the plight of the American workers, US politicians are following in the proud national tradition of projecting domestic anxieties onto a foreign enemy. Demands in the trade war, they explain, orbit mainly around pressuring China to give up the tenets of the industrial policy that's fueled its rapid economic development. "The reason that a large part of America's elite is deeply concerned about China has nothing to do with trade imbalances or blue-collar jobs," Adam Tooze has written. "What matters is the sheer weight of state power conferred on Beijing by China's spectacular economic growth. As far as America's hawks are concerned, every dollar added to China's GDP, every piece of technology that China acquires, shifts the geopolitical balance in the wrong direction."[57]

The only alternative to a new Cold War is a progressive internationalism of the sort the first Cold War spent decades trying to violently extinguish. Working people in the US and China have more in common with one another than they do with the elites now in charge of their own countries. The problems confronting each—the ones faced by most people who don't happen to be millionaires under capitalism—is an economic system that funnels rewards to the top at the expense of both people and planet.

It's possible, of course, that the US will cede leadership on climate: that the country's administrative rot, partisan politics, and commitment to trade supremacy are simply out of synch with a timely response to the climate crisis. After its disastrous response to the coronavirus, it's not impossible to imagine that other, better organized countries will realize the United States will not be the driving force behind decarbonization and move on to deal with the climate crisis without it. I hope it doesn't come to that, just as I hope—against the odds, admittedly—for a new, more collaborative style of US leadership. There is some strange comfort in the fact that the United States' failure to curb the climate crisis doesn't have to be the world's.

"MEN MAKE THEIR own history," Marx once wrote, "but they do not make it as they please; they do not make it under self-selected circumstances, but under circumstances existing already, given and transmitted from the past." US hegemony is not a circumstance many progressives would self-select, but it certainly does exist. Building a more democratic world—one capable of carrying out decarbonization—will likely entail the US wielding the undemocratic, arguably intolerable power it's amassed, at least for some time. A warming world is also one filled with similarly uncomfortable contradictions.

Despite its populist origins, the Fed has historically acted in the best interest of corporations, not ordinary people.[58] For as long as the climate crisis has made headlines, the Bretton Woods Institutions have chased the same goals. As that crisis worsens, what precisely corporate interests are vis-a-vis carbon has become a more complicated question. Increasingly muscular climate movements and those climate changes already wreaking havoc—including in monied enclaves like Miami and Los Angeles—are starting to craft an odd, maybe even untenable alignment with arms of capital that stand to profit from an energy transition. There's hardly been a month in writing this book that a new insurance company, bank, or monopolistic tech firm hasn't come out with a new pledge to reach net-zero emissions by some too far-off date. That these pledges are at best too slow—and at worst meant to preserve toxic business models indefinitely—doesn't mean there's no hope to be found in them. In a country where corporate money in politics is ubiquitous, having every capitalist in the country in lockstep against decarbonization is an obvious dead end. Division among the 1 percent on climate is a welcome sight. And we can choose our enemies wisely. After their decades of denial, delay, and obstruction, fossil fuel executives should have no place in polite society, including among elites now recognizing the material and reputational benefits of wanting to do something about the climate crisis. It's a sum positive for the world, that is, if oil executives stop getting invited to the hottest parties in Davos. The trick is to not let the energy transition happen on terms devised in the Alps.

Neither corporations nor skilled technocrats are going to provide the world with a comprehensive, internationalist, and equitable plan for decarbonization. Democratic majorities winning and wielding state power can force them to abide by one. The way to get there, I argue—and undercut rising authoritarianism in the process—is to use climate policy as a means to deliver tangible improvements in people's lives.

Economic planners during and in the direct aftermath of World War II made deliberate choices about what types of economic activity were worth spurring through state support. In the 2020s, those choices will be different, albeit no less creative in crafting a global order fit for a climate-changed century. There's something admittedly tempting about transposing the war mobilization onto decarbonization: of factories humming along, filled with unionized assembly-line workers. For nonwhite Americans and those on the losing end of a global Cold War, among others, that era offers a lot less to be nostalgic about. Even if it were desirable to revive the midcentury United States, that vision doesn't have much in common with what production or supply chains actually look like today, least of all for clean energy. Manufacturing jobs weren't good because of something endemic to factory production but because workers fought for unions that could make them good. For the same reasons, the many jobs that will be critical to a Green New Deal—in childcare, solar installation, teaching, energy efficiency, and more—can be just as good. Workers here, in turn, can forge ties to their counterparts along global green supply chains to ensure the gains of a Green New Deal don't stop at the border.

The driving mission of the next ten years needs to be decarbonizing the world as fast as possible and creating a world better able to adapt to those changes already heading toward us. If that's a success, that world will eventually be built. Regular updates and improvements to it won't take the kind of all-out, frenetic push demanded by a decade of the Green New Deal. So what should life look like on the other side?

CONCLUSION

We Can Have Nice Things

Some folk think that freedom just ain't right,
Those are the very people I want to fight.

—Langston Hughes, *Freedom's Road*[1]

BRONNIE WARE HAS watched a lot of people die. As a palliative caregiver in Australia, she has counseled hundreds of people through the last weeks of their life. Some asked her to share their parting wisdom with the world, a request Ware obliged through a blog that she eventually turned into a book, *The Top Five Regrets of the Dying*, which compiled the themes that emerged from conversations with those she cared for. "Most people had not honored even half of their dreams," she writes. The top regret people expressed was wishing they'd lived truer to themselves rather than expectations placed on them. Unsurprisingly, the list didn't include having not bought a bigger, more expensive house or a faster car. The people she cared for wished they'd kept up with friends instead of losing touch, and that they'd let themselves be happier. Nearly all of Ware's male patients wished they had worked less; she had mainly looked after an older generation for whom it was rarer that women worked outside the home. "They missed their children's youth and their partner's companionship," Ware writes. "All of the men I nursed deeply regretted spending so much of their lives on the treadmill of a work existence."[2]

For all the terrible things humans can do, the things most of us want in the end are pretty simple and low carbon: food, water, shelter, sex, good health, joy; to love and be loved. Capitalism is not particularly good at providing them. Beyond the basic number of calories needed to fuel our contributions to the GDP, the things that make us well and truly happy have to be sneaked in around the edges of an economic system that treats them as unprofitable waste. Long dinners with friends and family; late, wine-soaked nights spent dancing and playing music or arguing about politics; getting lost in a forest or finding a quiet spot in the park to read or a corner of a bar to make out with a new fling. We want these things so much that companies have to sell their products through them, from that shiny H&M shirt needed for a successful night out to the Ikea end table that will bring you domestic bliss, too. If happiness is valuable to capitalism, it's en route to productivity. A budding industry known as Wellness has packaged human contentment into corporate retreats and self-help guides, wherein a new app offers ten minutes of meditative bliss to help power you through another ten hours of meetings and emails.

Despite all the evidence to the contrary, capitalism's most stalwart defenders on the right have insisted that it's the surest path toward freedom and have spent centuries trying to build its monopoly on that term. For America's Founding Fathers, freedom meant boundless expansion into and propagation on the Western frontier—and the violent displacement and murder of the people who already lived there. Following in that tradition, Andrew Jackson would define freedom primarily as the freedom from restraints on slavery and dispossession. Historian Greg Grandin has called this "the country's founding paradox: the promise of political freedom and the reality of racial subjugation."[3] And as Ruth Wilson Gilmore writes, "The practice of putting people in cages for part or all of their lives is a central feature in the development of secular states, participatory democracy, individual rights and contemporary notions of freedom."[4]

The coronavirus showed us the limits of a society where capital is freer than people. With roughly one thousand people a week succumbing to COVID-19, the pharmaceutical company Gilead Sciences announced it would cost $3,120 for hospitals to sell its proprietary treatment regimen to patients with private insurance.[5] Horrifically, at least 28 million people in the US didn't have insurance before the pandemic. The ensuing crisis left many to fall through this country's Swiss cheese safety net. As 45.6 million people filed for unemployment, the wealth of 643 billionaires in the US swelled from $2.9 trillion to $3.5 trillion between March 18 and June 17.[6] Markets, it turns out, are better at profiting off crises than quelling them.

Still, there are those who would argue that it is humans as a whole who have too much freedom, whether to not wear masks or deny the science of climate change—as if a lack of personal responsibility had been the problem all along. Environmentalism's misanthropic streak has been an antidemocratic one, too: if we're so greedy and irresponsible then maybe we really do need to be repressively managed by a set of benevolent technocrats. The eco-pessimists similarly argue that hardwired flaws of human nature mean it's time to come to terms with the troubles ahead. It's too late. Hole up and learn how to die. That is its own kind of climate denial. That we should just resign ourselves to doom ignores the bare fact that every tenth of a degree of additional warming translates to tens of thousands of lives lost. The fight to stop more deaths and degrees, unfortunately, will never end, and comfortable parts of the Global North resigning to close themselves off against some imagined end-time leaves thousands of others to experience catastrophes outside their purview. It's entirely possible that even the best efforts of the world's governments will yield a world that has warmed by more than 2 degrees Celsius by the end of the century. Some level of serious disruption is at this point inevitable, but a great deal can still be prevented. It also stands to reason that the governments that rally to prevent catastrophic outcomes will be better prepared to deal with them than those that didn't try at all. Battling emissions and the forces most responsible for them isn't, indeed, can't be a purely defensive fight.

From carbon taxes to consumption cuts, climate policy has long been framed as an issue of stiff-lipped sacrifice: What will we have to give up to save our skins? On the left this can translate into a kind of low-carbon asceticism and longing for a simpler and quieter life in the woods. The right takes this characterization to extremes, accusing climate hawks of wanting to ban cars and hamburgers and throw civilization back into the Dark

Ages. While its critics like to pretend otherwise, the Green New Deal—thankfully—turns that question on its head, asking instead how to invest society's vast resources to maximize human and planetary well-being. We now have a few decades' worth of evidence that just exposing the public to the horrors of climate change doesn't do much to keep more of them from happening. Neither does bartering behind closed doors with industry. The main barrier to climate action isn't a technological one: the core tools needed to deal with this problem already exist. The problem has been power, and that the people proposing the most workable, reasonable solutions don't have enough of it. Rather than showcasing doom and gloom, trying to guilt us into action, those looking to build power around the climate fight might consider a different approach. Why not start with the freer, happier, and more functional world we can win?

LET'S START WITH work and doing less of it. Economist Juliet Schor has been drawing connections between work hours and climate change for well over a decade. In her 1993 bestseller *The Overworked American*, she delineated how Americans have come to work more and what effect that has on how people spend their dwindling leisure time. Namely, they're doing more shopping—a habit spurred on by copious corporate advertising. "Many potentially satisfying leisure skills are off limits because they take too much time: participating in community theater, seriously taking up a sport or a musical instrument, getting involved with a church or community organization," she wrote. "We have gotten ourselves entrenched in a cycle of work and spend—a cycle of long hours and consumer mentality as a way of life." As Schor's recent work has pointed out more directly, all that manufactured consumption comes at a high carbon cost. Examining data from twenty-nine high-income OECD countries conducted between 1970 and 2007, Schor and the late Eugene Rosa and Kyle Knight found in comparing nations that shorter work hours have a significant impact on ecological (i.e., resource usage) and carbon footprints. The mechanism isn't complicated: shortening work hours and allowing workers to take raises in the form of time instead of money creates opportunities for people to spend more time doing things they actually like rather than buying cheap consumer goods to fill the gap.

Things didn't necessarily need to turn out this way. At the start of the Great Depression, John Maynard Keynes famously predicted that workweeks could shrink down to just fifteen hours as people opted for more leisure time, their material needs being met and then some as living standards rose.[7] The labor militants who helped push for and win the original New Deal also campaigned for shorter workweeks and higher wages, allowing more people to do less work overall while getting more of their basic needs met by a freshly minted welfare state. Combined with rising automation, many expected the trend toward shorter workweeks to continue as Keynes predicted. Yet years later, work hours in the US have ballooned and remain stubbornly high, thanks in no small part, as Schor documents, to the right wing's persistent attacks on unions. And as many workers obtained decent standards of living, companies had to convince them they needed more things through advertising, inventing needs that could only be met with more work. Productivity has skyrocketed as wages have stagnated—a split that widened starkly as neoliberalism and the giddy consumerism it brought with it took hold.[8]

Though the effects of these dynamics are experienced worldwide, they are felt most acutely in the US—an oddly tired man out among wealthy countries. Reiterating Schor's and others' work, economists Anders Fremstad and Mark Paul note how different things are in other high-income countries: "For example, the average German worker toils 23 percent fewer hours than their American counterpart, and the average German emits 46 percent less carbon." None of that happened by accident, of course: in Germany, shorter workweeks have been a perennial demand of the country's labor movement, which has a formal role in the governance of its biggest companies. An analysis by economists David Rosnick and Mark Weisbrot in 2006 found that, had the US "adopted European standards for work hours, US carbon dioxide emissions in 2000 would have been 7 percent lower than its actual 1990 emissions," thereby satisfying the targets outlined in the Kyoto Protocol.[9] Counterintuitively, perhaps, a federal job guarantee could be one of the quicker routes to a more leisurely society. Just as the program could establish a wage floor that the private sector would be forced to compete against, it could also set an hours floor, mandating that all jobs provided as part of direct hiring programs feature thirty-two-hour workweeks with no reduction in wages.

Like most climate policies, this one won't work in a vacuum. Any climate plan worth its salt ultimately needs to do two things: change the amount of energy people consume and the composition of that energy by electrifying the economy and having much more of it run off zero-carbon power. There's mounting evidence that doing one or the other simply won't be enough. A World War II–style mobilization can help achieve the latter, which—at least in the short term—would stimulate carbon-intensive consumer demand in ways that would need to be balanced out to mute its environmental impact. Yet mainstream conversations about carbon footprints have tended to fixate on individual action: Do you use a plastic bag at the store or bring your own? Do you drive a gas-guzzling SUV or a Prius? What focusing on these consumer choices ignores is the fact that government policy structures consumption choices at every turn. Contra Fox News's fearmongering, the sorts of policies proposed as part of a Green New Deal are not premised on creating some draconian rationing system of secret police confiscating hamburgers. And the developed countries with much smaller per capita emissions than the US are hardly dystopias.

In order, Finland, Denmark, Norway, Iceland, and the Netherlands claimed the top 5 slots in 2019's UN World Happiness Report.[10] Researchers measure happiness based on six specific categories: GDP per capita in terms of purchasing power; life expectancy; social support from networks of friends and families; having the "freedom to choose what you do with your life"; generosity; perceptions of corruption; and both positive and negative affect, or how often people reported experiencing positive or negative emotions. Aside from GDP per capita and life expectancy, the data for all these categories are drawn from self-reported answers to the Gallup World Poll.[11] The United States dropped from 11 to 19 between two periods UN researchers compared, 2006–2008 and 2016–2018. We now land between Belgium (18) and the Czech Republic (20).

What makes Americans so unhappy? It turns out that wealth has a small impact. "We found the only one of the six factors that has grown, income per capita, and that's contributed to helping happiness but only by a small amount. It's been offset by a declining sense of freedom and generosity and an increase in perceived levels of corruption," said economist

John Helliwell, an editor on the report. "One thing that we have noticed in psychological experiments," he said, is "that people overestimate the amount of happiness you're going to get from more income or more consumption, and underestimate the happiness they get from more time with family and friends."

Happy countries are doing well on measures outside the report's bounds, too. On average, the carbon footprint of the average American is more than twice that of residents in the world's happiest countries. They also work 330 hours less each year—about forty-one fewer days for those working 8 hours a day, owed at least partially to sky-high levels of union density and the European Union–wide mandate that workers get four paid weeks off per year. These places back fairly ambitious climate policies, at least compared to the United States: pledging carbon neutrality by 2030, pushing the whole European Union to go net-zero by midcentury, investing in renewables-based heating systems—the list goes on.[12]

But less obviously green investments also go a long way toward allowing their residents to live less carbon-intensive lives and creating a built environment to make that possible and pleasurable. State-supported dense and affordable housing, for instance—critical to building a low-carbon world—can encourage people to work and learn closer to where they live.[13] Well-funded public housing in other parts of the world includes things like kindergartens, bars, and restaurants—the kind of "social infrastructure," as NYU sociologist Eric Klinenberg calls it, that enables hyperlocal communities to grow and thrive.[14] Reliable and well-funded public transit, similarly, lets people avoid traffic jams en route to work and play, and the pollution that accompanies them. Just 7 percent of Americans currently use public transit to commute. And much of the carbon costs embedded in things like education and even sports stem from the fact that people in many parts of the country use cars to reach them, lacking any viable alternative or the resources to buy a Tesla. Beyond common-sense reforms like fuel efficiency standards, building out robust transit networks can help to remedy that and chip away at the carbon-heavy car culture, one of the biggest contributors to US emissions.[15]

As sociologist Daniel Aldana Cohen has pointed out, though density does lower emissions, it alone is not enough, particularly in urban areas built for the carbon-hungry rich. "When the people clustered are prosperous professionals, the carbon benefits of density can be cancelled out by the emissions their consumption causes. The smokestacks, of course, are elsewhere," he writes. "It's by expanding collective consumption—in housing, transit, services, and leisure—that we can democratize and decarbonize urban life."[16]

Happy countries' investments in collective consumption and their shorter workweeks allow for a whole host of activities that make us happier—and happen to not destroy the planet.[17] Research by Indiana University's Joseph Kantenbacher has found that the things that bring us the most joy also tend not to spew greenhouse gases into the atmosphere. These generally involve some kind of human connection. "Intimate relations" and "socializing after work" take the top two spots in a 2006 study by economists Daniel Kahneman and Alan B. Krueger, ranking activities that improve self-reported life satisfaction.[18] Shopping ranks relatively low, along with non-work-related time on the computer, commutes, and (of course) work. Volunteering also ranks highly, a category that can include everything from neighborhood trash pickups to community organizing that pushes for more climate-friendly policies. And people who get more sleep are generally more satisfied with life, as well as healthier than their overworked and underslept counterparts.

So provided we're not filling up a gas tank to get to them, our favorite creature comforts—sex, socialization, and sleep—simply aren't carbon intensive. Ensuring that we don't just seek out more polluting, less-gratifying things to do with our leisure time isn't a matter of buying the right car or light bulb. It's about building a society that makes a low-carbon and altogether happier life possible for everyone. For the most part, the world's happiest countries are social democracies, or at least far closer to that model than the United States. That's not to lionize existing social democracies or overstate the extent to which they can be seen as a clean blueprint for building a better and greener America. They're for the most part tiny countries that are leagues less diverse than the US—a fact made painfully obvious by rising anti-immigrant politicians bent on protecting their generous welfare states from mostly nonwhite outsiders. Let's also not forget that Norway is a prolific oil exporter, greening the homefront while continuing to ship its toxic exports overseas. These countries are neither socialist utopias nor carbon-neutral ones.[19]

But aside from investing more money in the kinds of public goods listed above, they also tend to treat things like health care and childcare as basic rights. Relatedly, they're far less unequal than the US, according to the Gini index measuring inequality within countries.

People in these countries spend less of their free time in cars, and public institutions create literal spaces to relax. Iceland's extensive network of public pools (*sundlaugs*), for instance, are geothermically heated, open year-round, and a core part of the island's civic culture. One *New York Times Magazine* reporter observed wistfully that they are "key to Icelandic well-being." "The more local swimming pools I visited," he wrote in 2016, "the more convinced I became that Icelanders' remarkable satisfaction is tied inextricably to the experience of escaping the fierce, freezing air and sinking into warm water among their countrymen."[20]

Again, though, Iceland isn't the United States. We can aim higher. Our *sundlaugs* might crop up around new geothermal wells that provide round-the-clock power to luxurious public housing and daycares one short electric van share ride away. The rigs will be worked by former oil and gas drillers and engineers now enjoying union benefits and a four-day week. The saunas would be flanked by taco trucks and blaring stereo systems to be enjoyed by students attending college debt-free, people freed from jails, and new immigrants who aren't worried about ICE coming to raid their homes because ICE doesn't exist. Instead of importing its leisure and social democracy from abroad, a Green New Deal can create more opportunities for the types of fun and freedom Americans already know.

THOUGH MOST REFERENCES to the New Deal order focus on rapid-fire economic mobilization and jobs programs, it would be a mistake to describe it as all work and no play. One of Franklin Roosevelt's first acts after taking office in March 1933 was repealing prohibition, if only by degree; he legalized and taxed beverages with no more than 3.2 percent alcohol nationwide before the Twenty-first Amendment repealed prohibition altogether that December. Doing so was meant as much to lift the nation's spirits as to stimulate the economy, providing a boost to tavern owners and grain and grape growers.[21] Local governments saved on the cost of enforcing the ban. Because there were still moral misgivings about the role of alcohol in society, the New Deal invested generously in public leisure infrastructure like parks, playhouses, and hunting lodges, giving people some way to spend their time other than at the bar. Rexford Tugwell, a Brain Trust member and undersecretary of the Department of Agriculture, was especially bullish on America's newly unleashed alcohol production, seeing it as central to what he called his fellow New Dealers' "political dedication to the pursuit of happiness."

"Wine and beer," he said in a speech two months after the end of prohibition, "are made from agricultural produce and the consumption of American wine and beer cannot only serve the broader purposes of the New Deal in making for a calmer and happier type of existence, but will help the American farmer to find a better market for his produce," assuring his audience that California vintages were every bit as good as those imported from France. He at one point proposed opening a model winery in Maryland under the auspices of the Department of Agriculture to serve as a state-of-the-art research facility for viticulture and enology, although it was shot down. "I foresee a plethora of small local vintages, some good, some mediocre, some perfectly dreadful, out of which will arise in future some great names and great traditions of American wine," he continued. "I anticipate a calmer and more leisurely type of civilization, in which there will be time for friendly conversation, philosophical speculation, gaiety and substantial happiness. For today we have in our possession all the elements which are necessary to that more abundant life."[22]

The obvious parallel to prohibition today is legalizing marijuana and releasing all the people who have been locked up because of its criminalization. But Green New Dealers can take Tugwell's broader point too. At its best, the New Deal didn't just put people to work for work's sake but to satisfy needs and wants alike. Though talk of green jobs fixates on solar panels and big infrastructure projects, there's much more that can be done to build that more abundant life.

In my hometown of Millville, New Jersey—built on a sandy patch of earth just next to the Pine Barrens—the WPA's Federal Art Project funded the restoration of artisan glassblowing that had once thrived there but which had long been displaced by modern manufacturing methods by the time of the Great Depression. Dozens of glassblowers got back to work making vases, perfume bottles, pitchers, and candle holders, some having lost the calluses built from handling the tools used to shape hot glass. "When one of them breathed into his blowpipe to form the cavity he wanted in a piece of molten glass," Nick Taylor writes, "the others joined in a collective pause. The process required a steady breath and concentration, but sometimes the glass cracked with a loud pop, and then all the blowers joined in a shout of 'Hallelujah!'" The products were distributed throughout the hospitals and libraries for a nominal fee. Corning Glass— feeling threatened—complained to Washington when the reputation of the Millville glassblowers started to spread and succeeded in getting the program shut down and the glassblowers put back on relief.[23]

Other New Deal arts and recreation investments have more enduring legacies: Diego Rivera's frescoes, for instance, and Zora Neal Hurston's oral history interview with the last survivor of the Transatlantic Slave Trade, Cudjo Lewis.[24] There wasn't a profit to be made supporting artists like the glassblowers or Jacob Lawrence and Jackson Pollock through the Depression, but the government did it anyway. The same was true of the hundreds of courthouses, libraries, and schools it built, and the farmland and rural ecosystems that the federal jobs program built and revived. That thousands of people took to social-distancing-friendly state parks and hiking trails amid pandemic lockdowns is a testament to just how beloved New Deal infrastructure remains. And with hundreds of performers having been cast out of work and public arts institutions facing a dire crisis, there's now ample opportunity to provide public jobs not only for the bread of decarbonization—wetland remediation, tree planting, and solar technician trainings —but the roses too, for outdoor stand-up comedy shows or avant-garde theater productions.

Inevitably, the right will call it socialism. In some cases, they might be right. What it's called is mostly beside the point. "It isn't big government that gives you happiness, nor is it small government. It's the right kind of government," Helliwell told me. "You want a government that people believe in. If you're in a high-trust environment, people don't fight" about which policies to implement, at least not to the extent that we're used to here in the US, he added. It's also hard to argue with the fact that ensuring people's basic needs are met—for food, health care, housing, childcare, and more—creates the conditions for a happier life, and far more directly than simply raising per capita income. As it ratchets down workweeks, then, that right kind of government can value the work society actually needs fairly. Nurses, teachers, and caregivers kept the United States alive amid COVID-19, at tremendous risk. Yet they're still valued less in monetary terms than ad tech salesmen and private equity vultures. That work caring for and repairing society is every bit as green and essential to a healthy twenty-first century as building wind turbines and designing new batteries. It may not be a coincidence that the places where this work is valued, and where basic needs are basic rights, are also generally ahead of the curve on climate.

A group of researchers preparing their findings for the UN Intergovernmental Panel on Climate Change's next report have started to sketch out that link, laying out a series of what they called Shared Socioeconomic Pathways (SSPs) that forecast how we do or don't avert planetary catastrophe. SSP 1—a kind of best-case scenario—envisions "more inclusive development that respects perceived environmental boundaries. Management of the global commons slowly improves . . . and the emphasis on economic growth shifts toward a broader emphasis on human well-being," thus reducing inequality across and within countries.[25] Perhaps unsurprisingly, the more unequal economies sketched out in further SSPs are also far less likely to rein in emissions along the time line needed, which may not be too difficult to guess based on what the Donald Trumps and Jair Bolsonaros of the world have planned.

In other words, the Green New Deal components that its critics have decried as wasteful add-ons—social housing, a federal job guarantee, universal health care—are anything but. Federal job guarantee workers on four-day workweeks can work to create sites of low-carbon leisure like the original New Deal did with the Works Projects Administration; we may get our *sundlaugs* yet. Offering decent wages for less work is also a good way to entice people away from poorly paid jobs in carbon-intensive supply chains like Walmart. Eliminating unemployment further means eliminating a major driver of unhappiness, as Helliwell and fellow researchers have found.[26] Because if being overworked makes people unhappy, being unemployed or underemployed may be worse still. At the same time, the kind of robust safety net envisioned by a Green New Deal means that people won't be dependent on work to satisfy their basic needs. If the Green New Deal has an overarching goal, it's to both decarbonize and decommodify survival.

Importantly, a strong social safety net can also help to undermine another one of the twentieth century's more extractive institutions: the nuclear family and the bad sex that flows from the web of economic dependencies bound up in it.[27] Historian Kristen Ghodsee finds that in Soviet East Germany—itself very far from an ideal society—taking such things as housing and education out of the market created a dramatically different relationship to sex than in capitalist societies, including in nearby West Germany. Long after the collapse of the Berlin Wall, several men in their late forties recounted sex under socialism to Dagmar Herzog, another historian:

> it was really annoying that East German women had so much sexual self-confidence and economic independence. Money was useless, they complained. The few extra Eastern Marks that a doctor could make in contrast with, say, someone who worked in the theater, did absolutely no good, they explained, in luring or retaining women the way a doctor's salary could and did in the West. 'You had to be interesting.' What pressure. And as one revealed: 'I have much more power now as a man in unified Germany than I ever did in communist days.[28]

To wit, studies found that East German women were more consistently satisfied by their sexual encounters than their West German counterparts; in one survey, 82 percent of the former reported feeling "happy" after sex, compared to 52 percent of the latter.[29] You don't need to look back to a country with a repressive secret police force to find similar dynamics. As journalist Katie Baker noted in *Dissent,* the pickup artist known as Roosh V was appalled to come to Denmark while writing his series of travel guides, whose formulaic titles include *Bang Iceland* and *Bang Brazil* and have been criticized as guides to sexual assault. His title for a country that guarantees free health care and college, universal childcare, and eighteen weeks of maternity pay was a bit different: *Don't Bang Denmark.* "Fans of the travel writer," Baker wrote, "will be disappointed that 'pussy literally goes into hibernation' in this 'mostly pacifist nanny state,' where the social programs rank among the best in the world. . . . He concludes that the typical fetching Nordic lady doesn't need a man 'because the government will take care of her and her cats, whether she is successful at dating or not.'"[30]

DECARBONIZING THE GLOBAL economy and adapting to the climate-changed century ahead will be the single hardest and most important thing our species has ever done. It's impossible without a big, democratic government and massive state investment, as well as the dismantling of the most powerful industry that has ever existed. That, in turn, seems dangerously far off unless some critical mass of people see the Green New Deal as their path to a better life and manage to overcome the rank and racist divide-and-conquer politics that have been so successful at stopping efforts to turn these United States into a more perfect union, and this planet into a fairer place. A few people might get ideas for how to do that from reading this book, and I'll be thrilled if they do. But as Robin D. G. Kelley has written: "Revolutionary dreams erupt out of political engagement; collective social movements are incubators of new knowledge."[31] Indeed, many of the good ideas now percolating around the climate movement can be traced back to grassroots struggles waged by people whose homes lay in the path of fossil fuel infrastructure and its consequences. For decades, they have demanded what today seems so obvious, while the wonks and politicians wasted time we won't get back. In a real democracy, they might have listened.

A Green New Deal isn't just about subbing out one form of energy for another as all else stays equal. It means rooting out the deep power imbalances that have made the fossil fuel economy possible and that will keep toxic and deadly extraction humming along if they remain in place. It means writing a new social contract in its place that ensures people are as free as money to move around the places where they were born and be welcomed with open arms where they weren't. It's about building a genuinely sustainable United States, if such a thing is even possible. Should all that succeed, the resulting society will also be a happier one, where people are free to enjoy hiking and hedonism in whatever quantities they choose without the fear of being hunted down by police and immigration agents; where no one will have to work a bullshit job just to survive; where the work that keeps society running—that cares for lands and lives—is valued more than work that kills; where democracy is a process, perfected and debated by free people with a say over how their workplaces, schools, homes, and governments are run; where those people can dream about abundant, joyful futures and have the tools to turn those dreams into reality.

"In a world where no one is compelled to work more than four hours a day every person possessed of scientific curiosity will be able to indulge it, and every painter will be able to paint without starving, however excellent his pictures may be," Bertrand Russell wrote in 1932, arguing for shorter workweeks amid a deepening Depression. "Modern methods of production have given us the possibility of ease and security for all; we have chosen instead to have overwork for some and starvation for others. Hitherto we have continued to be as energetic as we were before there were machines. In this we have been foolish, but there is no reason to go on being foolish for ever."[32]

Today's world is dying. Let's have no regrets about the one we build to replace it.

ACKNOWLEDGMENTS

I can't imagine having worked through this book with anyone other than Katy O'Donnell, who patiently stewarded it through its various iterations as the world changed around it, for better and for worse. Aside from sharpening its arguments and prose and picking at drafts in all the right ways, she was leagues more understanding than she needed to be through more delay than I'd like to admit. Thanks as well to my agent, Ian Bonaparte, for shepherding a left-wing book about climate change through the wilds of publishing. I couldn't be happier with where it ended up. The entire team at Bold Type Books and Hachette—Clive Priddle, Kelly Lenkevich, Johanna Dickson, and many others—have made this such a smooth, pleasant process that I worry I'll be in for a shock writing anywhere else. The Lannan Foundation and the Schumann Center for Media and Democracy have generously supported my writing and reporting over the years.

I'm indebted to all the editors who've indulged me in exploring the ideas in these pages, going out to protests and strange conferences, and who improved my writing and thinking enormously. To name just a few: Bryan Farrell and Eric Stoner at *Waging Nonviolence*; Natasha Lewis, Kaavya Asoka, and Colin Kinniburgh at *Dissent*; Jessica Stites, Miles Kampf-Lassen, and Sarah Lazare at *In These Times*; Ari Bloomekatz at *Rethinking Schools*; Ryan Grim at The Intercept; and Heather Souvaine Horn at the *New Republic*. Guido Girgenti, Tobita Chow, Benjamin Rubin, David Stein, Karthik Ganapathy, Nancy Fraser, Kert Davies, Christian Parenti, David Pomerantz, and Rajiv Sicora were all generous enough to read over and provide thoughtful comments to versions of these chapters, which made them considerably better. So did Alyssa Battistoni and Daniel Aldana Cohen, who—along with Thea Riofrancos—have been unfailingly brilliant comrades, coauthors, and collaborators to think through these strange times with.

This book is the product of too many sprawling conversations, sidebars, interviews, and late-night debates to possibly list here, stretching from college divestment campaign meetings to the Paris climate talks to postpanel dive-bar drinks. I'm so grateful to the many people whose work has been an invaluable resource to me and who were kind enough to talk me through their research and organizing. Those include but are by no means limited to: David Adler, Kevin Anderson, Jenny Andersson, Aaron Benanav, Neil Bhatiya, Saqib Bhatti, Ted Boettner, Johanna Bozuwa, Jedediah Britton-Purdy, J. Mijin Cha, Saikat Chakrabarti, Danny Cullenward, Michael Dobson, Peter Erickson, Billy Fleming, Bill Fletcher Jr., Ben Franta, Lili Geismer, Connor Gibson, Pedro Glatz, Tom Goldtooth, Charles Goodman, Jake Grumbach, Rhiana Gunn-Wright, Nicolas Haeringer, Darrick Hamilton, Mary Annaïse Heglar, Ellie Johnston, Sivan Kartha, Stephanie Kelton, Dario Kenner, Mike Konczal, Richard Kozul-Wright, Cathy Kunkel, Deepak Lambda-Nieves, Mathew Lawrence, Laurie MacFarlane, Paasha Mahdavi, Mercedes Martinez, Sergio Marxauch, JW Mason, Lara Merling, Philip Mirowski, Tadzio Mueller, Lidy Nacpil, Raj Patel, Mark Paul, Glen Peters, Ann Pettifor, Jasson Perez, Ryan Pollack, Meena Raman, Eric Rauchway, Janet Redman, Asad Rehman, Carla Santos Skandier, Jake Schlacter, Juliet Schor, Shawn Sebastian, Waleed Shahid, Harjeet Singh, Quinn Slobodian, Doreen Stabinsky, Julia Steinberger, Leah Stokes, Sean Sweeney, Nathan Tankus, Astra Taylor, Pavlina Tcherneva, Nathan Thanki, Katie Thomas, John Treat, Todd Tucker, Joe Uehlein, Liz Veazey, Gernot Wagner, Evan Weber, and Miya Yoshitani. Besides having taught me so much with their own writing, Mark Engler, Sarah Jaffe, and Naomi Klein offered some of the soundest advice I've ever received on how to make a book. Jed Bickman and Alex Colston encouraged me to do just that and eagerly talked through these ideas in their earliest stages.

Kate Mueller did a tremendous job copyediting the manuscript, and Nathan Pensler provided invaluable help with citations in the final week. Like my other housemates in Brooklyn, Nathan kindly let me skip out on chores and house dinners to hole up and write. Too many people to name offered welcome distractions, bottles of wine, couches to sleep on, home-cooked meals, and grounding advice at every stage of this process. The biggest thanks are saved for my parents, Barbara and Ethan, who brought me into this world and taught me to think it could be a better place.

NOTES

Introduction: From Great Acceleration to Great Transformation

1. Jan Zalasiewicz, Colin Nell Waters, C. P. Summerhayes, and Alexander P. Wolfe, "The Working Group on the Anthropocene: Summary of Evidence and Interim Recommendations," *Anthropocene* 19 (2017): 55–60.

2. Abrahm Lustgarten, "The Great Climate Migration," *New York Times Magazine,* July 23, 2020, www.nytimes.com/interactive/2020/07/23/magazine/climate-migration.html.

3. Paul J. Crutzen and Eugene F. Stoermer, "The 'Anthropocene,'" *Global Change Newsletter* 41 (May 2000): 17–18.

4. Simon L. Lewis and Mark A. Maslin, "Defining the Anthropocene," *Nature* 519 (2015): 171–180.

5. Nathaniel Rich, "Losing Earth: The Decade We Almost Stopped Climate Change," *New York Times Magazine,* August 1, 2018, www.nytimes.com/interactive/2018/08/01/magazine/climate-change-losing-earth.html.

6. Richard Heede, "Tracing Anthropogenic Carbon Dioxide and Methane Emissions to Fossil Fuel and Cement Producers, 1854–2010," *Climate Change* 122 (2014): 229–241.

7. Matthew Taylor and Jonathan Watts, "Revealed: The 20 Firms Behind a Third of All Carbon Emissions," *Guardian* (Manchester, UK), October 9, 2020, www.theguardian.com/environment/2019/oct/09/revealed-20-firms-third-carbon-emissions.

8. Yannick Oswald, Anne Owen, and Julia K. Steinberger, "Large Inequality in International and Intranational Energy Footprints Between Income Groups and Across Consumption Categories," *Nature Energy* (2020): 231–239.

9. See Dario Kenner, "Polluter Elite Database," *Why Green Economy?*, June 2019, http://whygreeneconomy.org/the-polluter-elite-database/; and Dario Kenner, *Carbon Inequality: The Role of the Richest in Climate Change* (London: Routledge, 2019).

10. Mark Kaufman, "The Carbon Footprint Sham," *Mashable,* July 2020, https://mashable.com/feature/carbon-footprint-pr-campaign-sham/.

11. "Carbon Reduction," BP, archived from February 12, 2006, https://web.archive.org/web/20060212090704/http:/www.bp.com/sectiongenericarticle.do?categoryId=9005334&contentId=7009881.

12. Margaret Thatcher, interview by Douglas Keay, *Woman's Own,* September 23, 1987, transcript, www.margaretthatcher.org/document/106689.

13. Zhu Liu et al., "Near-Real-Time Monitoring of Global CO_2 Emissions Reveals the Effects of the COVID-19 Pandemic," *Nature Communications* 11 (2020): Article 5172.

14. UN Environment Program, *Emissions Gap Report 2019* (Nairobi: UNEP, 2019), https://wedocs.unep.org/bitstream/handle/20.500.11822/30797/EGR2019.pdf?sequence=1&isAllowed=y.

15. "Climate Change and Health," The World Health Organization, February 1, 2018, https://www.who.int/news-room/fact-sheets/detail/climate-change-and-health.

16. Myles Allen et al., "Summary for Policymakers," in *Global Warming of 1.5°C: An IPCC Special Report [. . .]*, ed. V. Masson-Delmotte et al. (Geneva: Intergovernmental Panel on Climate Change, 2019), www.ipcc.ch/site/assets/uploads/sites/2/2019/05/SR15_SPM_version_report_LR.pdf.

17. "The Biden Plan for a Clean Energy Revolution and Environmental Justice," joebiden.org, accessed October 19, 2020, https://joebiden.com/climate-plan/#.

18. Ruth Wilson Gilmore and Craig Gilmore, "Restating the Obvious," in *Indefensible Space: The Architecture of the National Insecurity State*, ed. Michael Sorkin (New York: Routledge, 2008), 147.

19. UN UNCTAD, "Case Name and Number," Investment Policy Hub, https://investmentpolicy.unctad.org/investment-dispute-settlement.

20. David Coady, Ian Parry, Nighia-Piotr Le, and Baoping Shang, "Global Fossil Fuel Subsidies Remain Large: An Update Based on Country-Level Estimates," International Monetary Fund, IMF Working Papers, May 2, 2019, www.imf.org/en/Publications/WP/Issues/2019/05/02/Global-Fossil-Fuel-Subsidies-Remain-Large-An-Update-Based-on-Country-Level-Estimates-46509; and Friends of the Earth, "New Report: Big Oil's Money Pit to Reap Stimulus Billions," April 15, 2020, https://foe.org/news/new-report-big-oils-money-pit-to-reap-stimulus-billions/.

21. The Sentencing Project, "Fact Sheet: Trends in U.S. Corrections," updated August 2020, https://sentencingproject.org/wp-content/uploads/2016/01/Trends-in-US-Corrections.pdf.

22. The Center for Popular Democracy, "Report Examining 12 City, County Budgets Reveals Heavy Spending on Policing," Campaign Updates, July 5, 2017, https://populardemocracy.org/news-and-publications/report-examining-12-city-county-budgets-reveals-heavy-spending-policing; and Kate Hamaji et al., *Freedom to Thrive: Reimagining Safety & Security in Our Communities*, The Center for Popular Democracy, Law for Black Lives, Black Youth Project, https://populardemocracy.app.box.com/v/FreedomtoThrive.

23. H. Res. 109, "Recognizing the Duty of the Federal Government to Create a Green New Deal," 116th Congress, First Session, February 7, 2019, www.congress.gov/116/bills/hres109/BILLS-116hres109ih.pdf.

24. Noah Smith, "The Green New Deal Would Spend the U.S. into Oblivion," Bloomberg Opinion, February 8, 2019, www.bloomberg.com/opinion/articles/2019-02-08/alexandria-ocasio-cortez-s-green-new-deal-is-unaffordable?sref=tXknHEwo.

25. Nick Estes, *Our History Is the Future: Standing Rock versus the Dakota Access Pipeline, and the Long Tradition of Indigenous Resistance* (London: Verso, 2019), 40.

26. For a detailed account of the #NoDAPL movement and its historical significance, see Estes, *Our History Is the Future*.

27. Jacey Fortin and Lisa Friedman, "Dakota Access Pipeline to Shut Down Pending Review, Judge Rules," *New York Times*, July 6, 2020, www.nytimes.com/2020/07/06/us/dakota-access-pipeline.html.

28. Nick Estes, "A Red Deal," *Jacobin*, August 6, 2019, www.jacobinmag.com/2019/08/red-deal-green-new-deal-ecosocialism-decolonization-indigenous-resistance-environment.

29. Coral Davenport and Campbell Robertson, "Resettling the First American 'Climate Refugees,'" *New York Times*, May 2, 2016, www.nytimes.com/2016/05/03/us/resettling-the-first-american-climate-refugees.html.

30. Movement for Black Lives (M4BL), "Invest-Divest," 2020, https://m4bl.org/policy-platforms/invest-divest/.

31. Kate Aronoff, "What the Green New Deal Could Mean for Iowa," *New Republic*, January 31, 2020, https://newrepublic.com/article/156392/green-new-deal-mean-iowa

32. Shawn Sebastian, interview with author.

33. Lawrence Mishel, Elise Gould, and Josh Bivens, "Wage Stagnation in Nine Charts," Economic Policy Institute, January 6, 2015, www.epi.org/publication/charting-wage-stagnation/; and "The Productivity–Pay Gap," Economic Policy Institute, updated July 2019, www.epi.org/productivity-pay-gap/.

34. Darrick Hamilton, "Neoliberalism and Race," *Democracy: A Journal of Ideas*, 2019, https://democracyjournal.org/magazine/53/neoliberalism-and-race/.

35. Dino Grandoni, "The Energy 202: California's Fires Are Putting a Huge Amount of Carbon Dioxide into the Air," *Washington Post*, September 17, 2020, www.washingtonpost.com/politics/2020/09/17/energy-202-california-fires-are-putting-huge-amount-carbon-dioxide-into-air/.

Chapter 1: Climate Denial Is Dead

1. Alexander C. Kaufman, "Scott Pruitt's First Year Set the EPA Back Anywhere from a Few Years to 3 Decades," HuffPost, January 20, 2018, www.huffpost.com/entry/pruitt-one-year_n_5a610a5ce4b074ce7a06beb4.

2. Naomi Oreskes and Erik M. Conway, *Merchants of Doubt: How a Handful of Scientists Obscured the Truth on Issues from Tobacco Smoke to Climate* (New York: Bloomsbury Press, 2010), 248.

3. Naomi Klein, "The Right Is Right: The Revolutionary Power of Climate Change," in *This Changes Everything: Capitalism vs. the Climate* (New York: Simon & Schuster, 2014), 31–63.

4. Klein, *This Changes Everything*, 73.

5. See Kim Phillips-Fein, *Invisible Hands: The Businessmen's Crusade Against the New Deal* (New York: W. W. Norton, 2009); Nancy MacLean, *Democracy in Chains: The Deep History of the Radical Right's Stealth Plan for America* (New York: Viking, 2017); Quinn Slobodian, *Globalists: The End of Empire and the Birth of Neoliberalism* (Cambridge, MA: Harvard University Press, 2018); Philip Mirowski, *Never Let a Serious Crisis Go to Waste: How Neoliberalism Survived the Financial Meltdown* (London: Verso, 2013); and David Harvey, *A Brief History of Neoliberalism* (New York: Oxford University Press, 2005).

6. MacLean, *Democracy in Chains*, 149.

7. Koch Docs, "1991 CATO Climate Denial Conference Flyer," https://kochdocs.org/2019/08/12/1991-cato-climate-denial-conference-flyer-and-schedule/.

8. Jedediah Britton-Purdy, David Singh Grewal, Amy Kapczynski, and K. Sabeel Rahman, "Building a Law-and-Political-Economy Framework: Beyond the TwentiethCentury Synthesis," *Yale Law Journal* 129, no. 6 (April 2020), www.yalelawjournal.org/feature/building-a-law-and-political-economy-framework.

9. InfluenceMap, "Big Oil's Real Agenda on Climate Change," March 2019, https://influencemap.org/report/how-big-oil-continues-to-oppose-the-paris-agreement-38212275958aa21196dae3b76220bddc.

10. Alternative für Deutschland, *Manifesto for Germany: The Political Progamme of the Alternative for Germany* (Berlin: AfD, 2017), 78, www.afd.de/grundsatzprogramm/#englisch.

11. Alexander C. Kaufman and Chris D'Angelo, "Interior Officials Are Citing Coal Execs and Crank Bloggers to Defend Climate Stances," Politics (blog), HuffPost, March 10, 2018, www.huffpost.com/entry/interior-climate-change-crank-blogs_n_5aa2df63e4b086698a9da922.

12. Patrick Michaels, interview with author, July 25, 2019. Michaels is listed as an academic member in this ESEF 1998 working paper: Robert Nilsson, "Environmental Tobacco Smoke Revisited: The Reliability of the Evidence for Risk of Lung Cancer and Cardiovascular Disease," European Science and Environment Forum, March 1998, RJ Reynolds Records, www.industrydocuments.ucsf.edu/docs/rsmx0078.

13. Fred Palmer, interview with author, July 25, 2019.

14. Brad Johnson, "Oil-Funded Pat Michaels Admits Solving Global Warning Is a Problem of 'Political Acceptability,'" Think Progress, August 15, 2010, https://thinkprogress.org/oil-funded-pat-michaels-admits-solving-global-warming-is-a-problem-of-political-acceptability-bebeea48b4b4/.

15. Gayathri Vaidyanathan, "Think Tank That Cast Doubt on Climate Change Science Morphs into Smaller One," E&E News, December 10, 2015, www.eenews.net/stories/1060029290.

16. $686,500 from Exxon and ExxonMobil from 1997 to 2006, with grants of over $100 per year in 2004–2006. See Heartland Institute, "Funding," Desmog Blog, www.desmogblog.com/heartland-institute#funding.

17. Scott Waldman, "White House: Adviser Who Applauded Rise in CO_2 to Leave Administration," E&E News, September 11, 2019, www.eenews.net/stories/1061113085.

18. *Squishy* was a term attributed to Ebell by Caleb Rossiter in an interview, which Ebell confirmed, both in interviews with author, July 25, 2019.

19. Kate Yoder, "Frank Luntz, the GOP's Message Master, Calls for Climate Action," Grist, July 25, 2019, https://grist.org/article/the-gops-most-famous-messaging-strategist-calls-for-climate-action/.

20. Matt Gaetz, "Congressman Matt Gaetz Unveils the 'Green Real Deal,'" press release, April 3, 2019, https://gaetz.house.gov/media/press-releases/congressman-matt-gaetz-unveils-green-real-deal.

21. Mirowski, *Never Let a Serious Crisis Go to Waste*, 336.

22. AEI resident scholar Benjamin Zycher—who headed up energy work on Reagan's Council of Economic Advisors—is a regular attendee of Heartland Institute gatherings.

23. Jack Gerard, "Jack Gerard Outlines Market-Based Emissions Reduction Model in Press Call," American Petroleum Institute, November 16, 2015, www.api.org/news-policy-and-issues/testimony-and-speeches/2015/11/16/jack-gerard-outlines-market-based-emissi.

24. Legislative Analyst's Office, "Assessing California's Climate Policies—Electricity Generation," LAO Report, January 8, 2020, https://lao.ca.gov/Publications/Report/4131.

25. American Enterprise Institute, "To Tax or Not to Tax: Senators Sheldon Whitehouse and Brian Schatz Present Their American Opportunity Carbon Free Act," June 10, 2015, www.aei.org/events/to-tax-or-not-to-tax-sen-sheldon-whitehouse-presents-his-american-opportunity-carbon-fee-act/.

26. Frank I. Luntz, "The New American Lexicon," 2006, https://docu.tips/documents/70467439-leaked-luntz-republican-playbook-5c1618aa71002.

27. Mike Lillis, "Dem Leaders Embrace Pay-Go," *The Hill*, June 6, 2018, https://thehill.com/homenews/house/390898-dem-leaders-embrace-pay-go.

28. ConservativeHome, "Margaret Thatcher's Greatest Achievement: New Labour," CentreRight (blog), April 11, 2008, https://conservativehome.blogs.com/centreright/2008/04/making-history.html.

29. Zach Boren and Damian Kahya, "German Far Right Targets Greta Thunberg in Anti-climate Push," Unearthed, May 14, 2019, https://unearthed.greenpeace.org/2019/05/14/germany-climate-denial-populist-eike-afd/.

30. ntv, "AfD-Jugend meutert gegen Klimaleugner" [AfD Youth Mutinies Against Climate Deniers], May 28, 2019, www.n-tv.de/politik/AfD-Jugend-meutert-gegen-Klimaleugner-article21054615.html.

31. Aude Mazoue, "Le Pen's National Rally Goes Green in Bid for European Election Votes," France 24, April 20, 2019, www.france24.com/en/20190420-le-pen-national-rally-front-environment-european-elections-france.

32. Benjamin Opratko, "Austria's Green Party Will Pay a High Price for Its Dangerous Alliance with the Right," *Guardian* (Manchester, UK), January 9, 2020, www.theguardian.com/commentisfree/2020/jan/09/austria-greens-right-peoples-party-anti-immigration.

33. Bojan Pancevski, "Austrian Conservatives and Greens Form Coalition Government," *Wall Street Journal*, January 1, 2020, www.wsj.com/articles/austrian-conservatives-and-greens-form-coalition-government-11577919908?emailToken=2706bb405320cdd47eb2e0830d9457a5OKklaa3WrQoAb1RH9EE7r8fxTBOJ29XypFhVQWOlw640+5K9gKzRWHOALUoLwpVsXRZaaFkS1A2mAS9zGkoxSLmGnicuT+uBazcFBdwKt+Rh/hZP6mYmJu1HvbO0a14j&reflink=article_copyURL_share.

34. Naomi O'Leary, "Danish Left Veering Right on Immigration," Politico, September 6, 2018, www.politico.eu/article/danish-copenhagen-left-veers-right-on-immigration-policy-integration/.

35. Mette Frederiksen, "Realistic & Fair Immigration: A Policy to Unite Denmark," Inter Press Service, May 27, 2019, www.ipsnews.net/2019/05/realistic-fair-immigration-policy-unite-denmark/.

Chapter 2: Long Live Climate Denial!

1. Binyamin Appelbaum, "2018 Nobel in Economics Is Awarded to William Nordhaus and Paul Romer," *New York Times*, October 8, 2018, www.nytimes.com/2018/10/08/business/economic-science-nobel-prize.html

2. Of course, these figures change if socioeconomic conditions change: see Daniel Aldana Cohen, "Apocalyptic Climate Reporting Completely Misses the Point," *Nation*, November 2, 2019, www.thenation.com/article/archive/mainstream-media-un-climate-report-analysis/.

3. John D. Sutter, "On 6 Degrees of Climate Change," CNN, May 22, 2015, www.cnn.com/2015/05/21/opinions/sutter-6-degrees-climate/index.html.

4. Frank Ackerman, Elizabeth A. Stanton, and Ramón Bueno, "Fat Tails, Exponents, Extreme Uncertainty: Simulating Catastrophe in DICE," *Ecological Economics* 69, no. 8 (June 15, 2010): 1657–1665, www.sciencedirect.com/science/article/pii/S0921800910001096.

5. In such models, carbon pricing is effectively a proxy for a wide suite of policies, for example, a $5,500 per ton price effectively amounts to a ban on carbon-intensive fuels. See J. Rogelj et al., "Mitigation Pathways Compatible with 1.5°C in the Context of Sustainable Development," chap. 2 in *Global Warming of 1.5°C: An IPCC Special Report [. . .]*, ed. V. Masson-Delmotte et al. (Geneva: Intergovernmental Panel on Climate Change, 2019), www.ipcc.ch/site/assets/uploads/sites/2/2019/02/SR15_Chapter2_Low_Res.pdf.

6. Alexander Kaufman, "'Sobering' New UN Report Challenges Republican Climate Hawks' Free-Market Dogma," HuffPost, October 8, 2018, www.huffpost.com/entry/carbon-tax-intergovernmental-panel-on-climate-change_n_5bbbaf55e4b028e1fe404102?aeh=.

7. Rosemary D. Marcuss and Richard E. Kane, "U.S. National Income and Product Statistics Born of the Great Depression and World War II," *Survey of Current Business*, February 2007, 32–46, https://apps.bea.gov/scb/pdf/2007/02%20February/0207_history_article.pdf.

8. Arthur C. Pigou, "Preface to the Third Edition (1928)," in *The Economics of Welfare*, 3rd ed. (London: Macmillan, 1929), http://files.libertyfund.org/files/1410/0316_Bk.pdf: "The complicated analyses which economists endeavour to carry through are not mere gymnastic. They are instruments for the bettering of human life. The misery and squalor that surround us, the injurious luxury of some wealthy families, the terrible uncertainty overshadowing many families of the poor—these are evils too plain to be ignored. By the knowledge that our science seeks it is possible that they may be restrained. Out of the darkness light!"

9. Irving Fisher, *The Nature of Capital and Income* (New York: Macmillan, 1906): "The national dividend or income consists solely of services as received by ultimate consumers, whether from their material or from their human environment."

10. Pigou, *The Economics of Welfare*, 172.

11. Susie Cagle, "Richmond v. Chevron: The California City Taking on Its Most Powerful Polluter," *Guardian* (Manchester, UK), October 9, 2019, www.theguardian.com/environment/2019/oct/09/richmond-chevron-california-city-polluter-fossil-fuel.

12. Pigou, *The Economics of Welfare*, 184.

13. Pigou, *The Economics of Welfare*, 185–186. Pigou, though, might have gotten the idea from playwright and socialist George Bernard Shaw, who two years earlier had written *The Common Sense of Municipal Trading* (London: Constable, 1904), a Fabian Society treatise on the value of bringing certain citywide utilities under public ownership released in advance of a local election in London in 1904. Sussing out the relative merits of public and private enterprise, Shaw wrote: "Take the most popular branch of a commercial enterprise: the drink traffic. It yields high profits. Take the most obvious and unchallenged branch of public enterprise: the making of roads. It is not commercially profitable at all. But suppose the drink trade were debited with what it costs in disablement, inefficiency, illness and crime, with all their depressing effects on industrial productivity, and with their direct cost on doctors, policemen, prisons, &c. &c. &c.! Suppose at the same time the municipal highways and bridges account were credited with the value of the time and wear and tear saved by them! It would at once appear that the roads and bridges pay for themselves many times over, whilst the pleasures of drunkenness are costly beyond all reason." Pigou cites Shaw's book. On page 4, Shaw

describes why electric utilities should be publicly owned and, on page 5, a protean version of the entrepreneurial state: "The moment public spirit and business capacity meet on a municipality you get an irresistible development of municipal activity."

14. Bryan Cosby and Katie Tubb, "Why the 'Conservative' Carbon Tax Is Still a Non-Starter," Heritage Foundation, July 17, 2018, www.heritage.org/environment/commentary/why -the-conservative-carbon-tax-still-non-starter-0.

15. Milton Friedman, "Population and Economics," YouTube video, 5:51, 1977, excerpt from a Q&A session, www.youtube.com/watch?v=KH0O_JjH06k&t=1s.

16. EPA, "Benefits and Costs of the Clean Air Act Amendments of 1990," https://web.archive .org/web/20120507213317/http://www.epa.gov/air/sect812/feb11/factsheet.pdf.

17. Paul W. Hansen, "Return on Investment Is High for Regulations," *Jackson Hole News & Guide* (WY), June 27, 2018, www.jhnewsandguide.com/opinion/columnists/common_ground/return -on-investment-is-high-for-regulations/article_e7679c85-c461-5d0a-b39d-ad653737ed71.html.

18. See NRDC, "EPA's Endangerment Finding: The Legal and Scientific Foundation for Climate Change," fact sheet, May 2017, www.nrdc.org/sites/default/files/epa-endangerment-finding-fs.pdf; and Devin Watkins, "Support Builds for EPA to Reconsider Endangerment Finding," Blog, Competitive Enterprise Institute, April 12, 2019, https://cei.org/blog/support-builds-epa-reconsider-endanger ment-finding. It was Bob Nordhaus, William Nordhaus's brother, who wrote on the ability of the EPA to regulate as-of-yet unknown pollutants: "after languishing in obscurity for decades, [it] is now the legal rationale for the Obama administration's plan to regulate carbon emissions without a law passed by Congress": see Coral Davenport, "Brothers Battle Climate Change on Two Fronts," *New York Times*, May 10, 2014, www.nytimes.com/2014/05/11/us/brothers-work-different-angles-in-taking -on-climate-change.html?hp.

19. Rachel Carson, *Silent Spring* (New York: Library of America, 2018), 161.

20. Bill McKibben, "Global Warming's Terrifying New Math," *Rolling Stone*, July 19, 2012, www .rollingstone.com/politics/politics-news/global-warmings-terrifying-new-math-188550/.

21. Philip Shabecoff, "Reagan Order on Cost-Benefit Analysis Stirs Economic and Political Debate," *New York Times*, November 7, 1981, www.nytimes.com/1981/11/07/us/reagan-order-on -cost-benefit-analysis-stirs-economic-and-political-debate.html.

22. Adam Meyerson, "One Hundred Conservative Victories," Hoover Institution, April 1, 1989, www.hoover.org/research/one-hundred-conservative-victories.

23. William D. Nordhaus, "Can We Control Carbon Dioxide?," Working Paper, June 1975, https://pure.iiasa.ac.at/id/eprint/365/1/WP-75-063.pdf.

24. See Jerry Taylor, "Nordhaus vs. Stern," Cato at Liberty (blog), Cato Institute, November 28, 2006, www.cato.org/blog/nordhaus-vs-stern.

25. Paul Voosen, "Cool Head on Global Warming," The Chronicle of Higher Education, November 4, 2013, www.chronicle.com/article/Cool-Head-on-Global-Warming/142713.

26. Paul A. Samuelson and William D. Nordhaus, *Economics*, 19th ed. (New York: McGraw-Hill, 2010), http://pombo.free.fr/samunord19.pdf.

27. William D. Nordhaus, "Do Real-Output and Real-Wage Measures Capture Reality? The History of Lighting Suggests Not," Cowles Foundation paper No. 957, 1998, https://lucept.files.word press.com/2014/11/william-nordhaus-the-cost-of-light.pdf.

28. Peter Howard and Derek Sylvan, *Expert Consensus on the Economics of Climate Change*, Institute for Policy Integrity, NYU School of Law, 2015, https://policyintegrity.org/files/publications /ExpertConsensusReport.pdf.

29. Greg Mankiw, "Rogoff Joins the Pigou Club," Greg Mankiw's Blog, September 16, 2006, http://gregmankiw.blogspot.com/2006/09/rogoff-joins-pigou-club.html.

30. Leslie Hook, "Surge in US Economists' Support for Carbon Tax to Tackle Emissions," *Financial Times* (London), February 17, 2019, www.ft.com/content/fa0815fe-3299-11e9-bd3a-8b2a211d90d5.

31. Steve Keen, "The Appallingly Bad Neoclassical Economics of Climate Change," *Globalizations*, September 1, 2020, www.tandfonline.com/doi/full/10.1080/14747731.2020.1807856?scroll =top&needAccess=true.

32. Jason Hickel, "The Nobel Prize for Climate Catastrophe," FP, December 6, 2018, https:// foreignpolicy.com/2018/12/06/the-nobel-prize-for-climate-catastrophe/.

33. Keen, "The Appallingly Bad Neoclassical Economics of Climate Change."

34. Jason Hickel, "Is It Time for a Post-growth Economy?," Aljazeera, July 16, 2018, www .aljazeera.com/opinions/2018/7/16/is-it-time-for-a-post-growth-economy/?gb=true.

35. Gernot Wagner, "Why Oil Giants Figured Out Carbon Costs First," Bloomberg News, January 22, 2020, www.bloomberg.com/news/articles/2020-01-22/why-oil-giants-figured-out-carbon -costs-first-gernot-wagner?sref=tXknHEwo.

36. Michael Greenstone and Cass R. Sunstein, "Donald Trump Should Know: This Is What Climate Change Costs Us," New York Times, December 15, 2016, www.nytimes.com/2016/12/15 /opinion/donald-trump-should-know-this-is-what-climate-change-costs-us.html.

37. Michael Greenstone, interview with author, April 23, 2019.

38. Cribbed from Donald Rumsfeld; David Wallace-Wells, The Uninhabitable Earth: Life After Warming (New York: Duggan Books, 2019), 153.

39. Carbon Pricing Leadership Coalition, Report of the High-Level Commission on Carbon Prices, May 29, 2017; and "Interview with Nobel Laureate Joseph Stiglitz and Lord Nicholas Stern Co-Chairs of the High-Level Commission on Carbon Prices," YouTube video, 9:17, May 31, 2017, www.carbon pricingleadership.org/report-of-the-highlevel-commission-on-carbon-prices.

40. Anders Fremstad and Mark Paul, Disrupting the Dirty Economy: A Progressive Case for a Carbon Dividend, People's Policy Project, September 2018, www.peoplespolicyproject.org/wp -content/uploads/2018/09/CarbonTax.pdf.

41. Gernot Wagner, interview with author, February 21, 2020.

42. OECD, "Few Countries Are Pricing Carbon High Enough to Meet Climate Targets," September 18, 2018, www.oecd.org/environment/few-countries-are-pricing-carbon-high-enough-to-meet -climate-targets.htm.

43. See Kate Aronoff, "California Gov. Jerry Brown Was a Climate Leader, but His Vision Had a Fatal Flaw," The Intercept, November 28, 2018, https://theintercept.com/2018/11/28/california -jerry-brown-climate-legacy/.

44. Brad Plumer, "New U.N. Climate Report Says Put a High Price on Carbon," New York Times, October 8, 2018, www.nytimes.com/2018/10/08/climate/carbon-tax-united-nations-report -nordhaus.html.

45. William D. Nordhaus, The Climate Casino: Risk, Uncertainty, and Economics for a Warming World (New Haven, CT: Yale University Press, 2013), 233.

46. Mike Cummings, "Cheers and Roses from Undergrads for Yale's Latest Nobel Laureate," Yale News, October 8, 2018, https://news.yale.edu/2018/10/08/cheers-and-roses-undergrads-yales -latest-nobel-laureate.

Chapter 3: First as Tragedy

1. Julian Borger, "Bush Kills Global Warming Treaty," Guardian (Manchester, UK), March 29, 2001, www.theguardian.com/environment/2001/mar/29/globalwarming.usnews.

2. For more on the Copenhagen process and broader North–South dynamics at the UNFCCC, see Martin Khor, "A Clash of Paradigms—UN Climate Negotiations at a Crossroads," Development Dialogue 3, no. 61 (September 2012): 76–105.

3. Roy Scranton, "A Wicked Problem," chap. 2 in Learning to Die in the Anthropocene: Reflections on the End of a Civilization (San Francisco: City Lights Books, 2015).

4. Ellen Meiksins Wood, The Origin of Capitalism: A Longer View (London: Verso, 2017), 1.

5. Solveclimate staff, "Oil Giants BP, ConocoPhillips Drop Out of US Climate Action Partnership," InsideClimate News, February 16, 2010, https://insideclimatenews.org/news/20100216/oil -giants-bp-conocophillips-drop-out-us-climate-action-partnership.

6. Anne C. Mulkern, "Grass-Roots Organizer Jumps from Nature Conservancy to API," New York Times, February 26, 2010, https://archive.nytimes.com/www.nytimes.com/gwire/2010/02/26 /26greenwire-grass-roots-organizer-jumps-from-nature-conser-65511.html?pagewanted=2.

7. Canadian Association of Petroleum Producers (CAPP), "Harnessing Passion Through Grassroots," slide 20, April 2015, www.slideshare.net/OilGasCanada/capp-presentation-draft-2.

8. Eric Pooley, "Rumors of War," chap. 2 in *The Climate War: True Believers, Power Brokers, and the Fight to Save the Earth* (New York: Hyperion Books, 2010).

9. Lily Geismer, interview with author, August 16, 2019.

10. Pooley, *The Climate War.*

11. Frederic D. Krupp, "New Environmentalism Factors in Economic Needs," *Wall Street Journal,* November 20, 1986, www.wsj.com/articles/SB117269353475022375.

12. Charles Peters, "A Neo-Liberal's Manifestor," *Washington Post,* September 5, 1982, www.washingtonpost.com/archive/opinions/1982/09/05/a-neo-liberals-manifesto/21cf41ca-e60e-404e-9a66-124592c9f70d/.

13. Lily Geismer, "Democrats and Neoliberalism," Vox, June 11, 2019, www.vox.com/polyarchy/2019/6/11/18660240/democrats-neoliberalism.

14. Jonas Meckling, *Carbon Coalitions: Business, Climate Politics, and the Rise of Emissions Trading* (Cambridge, MA: MIT Press, 2011), 85–86.

15. Steven F. Bernstein, *The Compromise of Liberal Environmentalism* (New York: Columbia University Press, 2001), 205.

16. Naomi Oreskes and Erik M. Conway, *Merchants of Doubt: How a Handful of Scientists Obscured the Truth on Issues from Tobacco Smoke to Climate* (New York: Bloomsbury Press, 2010), 84.

17. Oreskes and Conway, *Merchants of Doubt,* 88.

18. Oreskes and Conway, *Merchants of Doubt,* 103.

19. Dallas Burtraw and Sarah Jo Szambelan, *U.S. Emissions Trading Markets for SO2 and NOx* (Washington, DC: Resources for the Future, October 2009), https://media.rff.org/documents/RFF-DP-09-40.pdf.

20. DG Environment, "Case Study 1: Comparison of the EU and US Approaches Towards Acidification, Eutrophication and Ground Level Ozone," Assessment of the Effectiveness of European Air Quality Policies and Measures, Milieu Ltd., Danish National Environmental Research Institute, and Center for Clean Air Policy, October 4, 2004, https://ec.europa.eu/environment/archives/cafe/activities/pdf/case_study1.pdf.

21. Gene E. Likens, "The Role of Science in Decision Making: Does Evidence-based Science Drive Environmental Policy?," *Frontiers in Ecology and the Environment* 8, no. 6 (August 1, 2010), https://esajournals.onlinelibrary.wiley.com/doi/full/10.1890/090132.

22. Bill Clinton, *Preface to the Presidency: Selected Speeches of Bill Clinton 1974–1992,* ed. Stephen A. Smith (Fayetteville: University of Arkansas Press, 1996), 142.

23. John Shanahan, "How to Help the Environment Without Destroying Jobs," Heritage Foundation, January 19, 1993, www.heritage.org/environment/report/how-help-the-environment-without-destroying-jobs.

24. Dawn Erlandson, "The BTU Tax Experience: What Happened and Why It Happened," *Pace Environmental Law Review* 12, no. 1 (Fall 1994), https://core.ac.uk/download/pdf/46711423.pdf.

25. Erlandson, "The BTU Tax Experience."

26. David S. Hilzenrath, "Miscalculations, Lobby Effort Doomed BTU Tax Plan," *Washington Post,* June 11, 1993, www.washingtonpost.com/archive/business/1993/06/11/miscalculations-lobby-effort-doomed-btu-tax-plan/d756dac3-b2d0-46a4-8693-79f6f8f881d2/.

27. See John C. Hughes, *Slade Gordon: A Half Century in Politics* (Olympia: Washington State Legacy Project, 2011), www.sos.wa.gov/legacy/stories/slade-gorton/pdf/complete.pdf; and Branko Marcetic, "Joe Biden and the Disastrous History of Bipartisanship," *In These Times,* August 22, 2019, http://inthesetimes.com/features/joe-biden-bipartisanship-nostalgia-centrism-2020.html.

28. Nathaniel Loewentheil, "Of Stasis and Movements: Climate Legislation in the 111th Congress," ISPS Working Paper, January 14, 2013, https://papers.ssrn.com/sol3/papers.cfm?abstract_id=2202979.

29. John Cook, "Ten Years On: How Al Gore's An Inconvenient Truth Made Its Mark," The Conversation, May 30, 2016, https://theconversation.com/ten-years-on-how-al-gores-an-inconvenient-truth-made-its-mark-59387.

30. Pooley, *Climate Wars,* 156–157.

31. James Verini, "The Devil's Advocate," *New Republic,* September 24, 2007, https://newrepublic.com/article/62836/the-devils-advocate.

32. Reed Hundt, *A Crisis Wasted: Barack Obama's Defining Decisions* (New York: RosettaBooks, 2019), 289.

33. Meckling, *Carbon Coalitions*, 85: "BP bridged two divides: the one between business and the environmental community, on the one hand, and that between U.S. and European business, on the other hand. While the firm did not pursue its new political strategy by itself, it was extraordinarily influential in organizing and shaping the protrading NGO-business coalition. Other business groups did not actually belong to the NGO-business coalition but instead quietly supported emissions trading."

34. Matto Mildenberger, *Carbon Captured: How Business and Labor Control Climate Politics* (Cambridge, MA: MIT Press, 2020), 130.

35. Christopher Leonard, *Kochland: The Secret History of Koch Industries and Corporate Power in America* (New York: Simon & Schuster, 2019), 402.

36. David Victor, "The Problem with Cap and Trade," *MIT Technology Review*, June 23, 2009, www.technologyreview.com/s/414025/the-problem-with-cap-and-trade/.

37. Full quote: "—because, in fact, the cap and trade bills debated in Congress were all about inter-organizational deals among corporations, unions, advocates, and industrial sectors, not about specific benefits that would be directly delivered to individual citizens. The best ad writers would be able to do would be to personally dramatize threats from climate change, and they rarely even did that" in Theda Skocpol, "Naming the Problem: What It Will Take to Counter Extremism and Engage Americans in the Fight Against Global Warming," page 52, paper presented at "The Politics of America's Fight Against Global Warming" symposium, Harvard University, January 2013, https://scholars.org/sites/scholars/files/skocpol_captrade_report_january_2013_0.pdf.

38. FRED Economic Data, "Unemployment Rate," Federal Reserve Bank of St. Louis, September 2020, https://fred.stlouisfed.org/series/UNRATE; and IER, "Winners and Losers in the Waxman-Markey Stealth Tax," Institute for Energy Research, May 22, 2009, www.instituteforenergyresearch.org/uncategorized/winners-and-losers-in-the-waxman-markey-stealth-tax/.

39. Rachel Morris, "Could Cap and Trade Cause Another Market Meltdown?," *Mother Jones*, June 8, 2009, www.motherjones.com/politics/2009/06/could-cap-and-trade-cause-another-market-meltdown/.

40. Jacob M. Grumbach, "Polluting Industries as Climate Protagonists: Cap and Trade and the Problem of Business Preferences," *Business and Politics* 17, no. 4 (2015): 633–659. doi:10.1515/bap-2015-0012.

41. Clifford Krauss and Jad Mouawad, "Oil Industry Backs Protests of Emissions Bill," *New York Times*, August 28, 2009, www.nytimes.com/2009/08/19/business/energy-environment/19climate.html.

42. Kevin Grandia, "Leaked Memo: Oil Lobby Launches Fake 'Grassroots' Campaign," Desmog Blog, August 13, 2009, www.desmogblog.com/oil-lobbys-%E2%80%98energy-citizens%E2%80%99-astroturf-campaign-exposed-launch.

43. Robert J. Brulle, "The Climate Lobby: A Sectoral Analysis of Lobbying Spending on Climate Change in the USA, 2000 to 2016," *Climatic Change* 149 (2018): 289–303, https://link.springer.com/article/10.1007%2Fs10584-018-2241-z.

44. Leonard, *Kochland*, 406.

45. Leonard, *Kochland*, 433.

46. Barack Obama, "Remarks by the President in State of the Union Address," The White House, Office of the Press Secretary, January 27, 2010, https://obamawhitehouse.archives.gov/the-press-office/remarks-president-state-union-address.

47. Amanda Reilly and Kevin Borgardus, "7 Years Later, Failed Waxman-Markey Bill Still Makes Waves," E&E News, June 27, 2016, www.eenews.net/stories/1060039422.

48. Paul Harris and Ewen MacAskill, "US Midterm Election Results Herald New Political Era as Republicans Take House," *Guardian* (Manchester, UK), November 3, 2019, www.theguardian.com/world/2010/nov/03/us-midterm-election-results-tea-party.

49. "How the Tea Party Fared," *New York Times*, November 4, 2010, http://archive.nytimes.com/www.nytimes.com/interactive/2010/11/04/us/politics/tea-party-results.html.

50. Skocpol, "Naming the Problem."

51. Nick Schulz, "Government Planning Is Back," AEIdeas (blog), American Enterprise Institute, June 25, 2009, www.aei.org/publication/government-planning-is-back/.

52. Nick Confessore, "Meet the Press: How James Glassman Reinvented Journalism—as Lobbying," *Washington Monthly*, December 2003, https://washingtonmonthly.com/magazine/december-2003/meet-the-press/.

53. "About Us," Tech Central Station, http://techcentralstation.com/about.html.

54. This model isn't uncommon in conservative media outlets. See Callum Borchers, "Charity Doubles as a Profit Stream at the Daily Caller News Foundation, *Washington Post*, June 2, 2017, www.washingtonpost.com/news/the-fix/wp/2017/06/02/charity-doubles-as-a-profit-stream-at-the-daily-caller-news-foundation/?noredirect=on.

55. Nick Schulz, "Researchers Question Key Global-Warming Study," *USA Today*, October 28, 2003, https://usatoday30.usatoday.com/news/opinion/editorials/2003-10-28-schulz_x.htm.

56. Nick Schulz, "Vanity Scare," TCS Daily, April 14, 2006, https://web.archive.org/web/20070118045935/http://www.tcsdaily.com/article.aspx?id=041406F.

57. ExxonMobil, "2019 Worldwide Giving Report," August 14, 2020, https://corporate.exxonmobil.com/Community-engagement/Worldwide-giving/Worldwide-Giving-Report#Civicandcommunity.

58. Union of Concerned Scientists, "ExxonMobil Foundation & Corporate Giving to Climate Change Denier & Obstructionist Organizations, 1998–2017," https://www.ucsusa.org/sites/default/files/attach/2019/ExxonMobil-Worldwide-Giving-1998-2017.pdf.

59. Nick Schulz, "The Merit of a Carbon Tax," *The Hill*, July 31, 2007, https://thehill.com/opinion/op-ed/7377-the-merit-of-a-carbon-tax.

60. Nick Schulz, "The Greening of Capitalism" *New Atlantis*, Spring 2007, www.thenewatlantis.com/publications/the-greening-of-capitalism.

61. Business Climate Leaders (BCL), "Panel 2, Carbon Tax Forum: The Business Case for Carbon Pricing," YouTube video, 43:08, September 23, 2018, www.youtube.com/watch?v=MgLOjzm30i8.

62. ExxonMobil, *2007 Corporate Citzenship Report*, www.globalhand.org/system/assets/a25f2fc01db618e4965a9618b5cd66bc49dfe471/original/2007_Coporate_Citizenship_Report.pdf.

63. Steve Horn, "Oklahoma, Colorado and Arizona Push ALEC Bill to Require Teaching Climate Change Denial in Schools," ThinkProgress, January 31, 2013, https://thinkprogress.org/oklahoma-colorado-and-arizona-push-alec-bill-to-require-teaching-climate-change-denial-in-schools-abfa67acb2ad/.

64. ExxonMobil, "2019 Worldwide Contributions and Community Investments," https://corporate.exxonmobil.com/-/media/Global/Files/worldwide-giving/2019-Worldwide-Giving-Report.pdf.

65. David Hasemyer and Bob Simison, "Exxon's Support of a Tax on Carbon: Rhetoric or Reality?," InsideClimate News, December 21, 2015, https://insideclimatenews.org/news/18122015/exxon-mobil-carbon-tax-rhetoric-or-reality-climate-change-rex-tillerson.

66. Ed King, "ExxonMobil Boss Warns Against Tougher Climate Regulations," Climate Home News, July 10, 2015, www.climatechangenews.com/2015/10/07/exxon-boss-warns-against-tougher-climate-regulations/.

67. David Hasemyer, "With Bare Knuckles and Big Dollars, Exxon Fights Climate Probe to a Legal Stalemate, InsideClimate News, June 5, 2017, https://insideclimatenews.org/news/05062017/exxon-climate-change-fraud-investigation-eric-schneiderman-rex-tillerson-exxonmobil.

68. Amy Harder, "Carbon Tax Campaign Unveils New Details and Backers," Axios, September 11, 2019, www.axios.com/carbon-tax-campaign-unveils-new-details-and-backers-37a1f955-8231-4022-9372-be624aef86ae.html.

69. Email from Alan Jeffers, December 19, 2016.

70. Steve Horn, "California Dems Give Up on New Oil Safety Regulations, "The Daily Poster, September 1, 2020, www.dailyposter.com/p/california-dems-give-up-on-new-oil.

71. Alexander C. Kaufman, "Fossil Fuel Industries Outspend Clean Energy Advocates on Climate Lobbying by 10 to 1," HuffPost, July 18, 2018, www.huffpost.com/entry/fossil-fuel-industries-climate-lobbying_n_5b4f8fdee4b0de86f4894831.

72. Alexander Hertel-Fernandez, Matto Mildenberger, and Leah C. Stokes, "Legislative Staff and Repesentation in Congress." *American Political Science Review* 113, no. 1 (2019): 1–18.

73. Carbon Tracker, "Paying with Fire: How Oil and Gas Executives Are Rewarded for Chasing Growth and Why Shareholders Could Get Burned," February 14, 2019, www.carbontracker.org/reports/paying-with-fire/.

74. Carbon Tracker, "Breaking the Habit: Why None of the Large Oil Companies Are 'Paris-Aligned,' and What They Need to Do to Get There," September 13, 2019, https://carbontransfer.wpengine.com/reports/breaking-the-habit/.

75. Donald Trump, "Remarks by President Trump at 9th Annual Shale Insight Conference," White House, October 23, 2019, www.whitehouse.gov/briefings-statements/remarks-president-trump-9th-annual-shale-insight-conference-pittsburgh-pa/.

Chapter 4: Parallel Worlds

1. We Are Still In, "We Showed the World That America Is Still In," www.wearestillin.com/COP23.

2. Natalie Sauer, "Twelve Activists Denied Entry to Poland for UN Climate Summit, Says Campaign Group," Climate Home News, August 12, 2018, www.climatechangenews.com/2018/12/08/twelve-activists-denied-entry-poland-un-climate-summit-says-campaign-group/.

3. See Pierre Briançon, "Macron's 'Jupiter' Model Unlikely to Stand Test of Time," Politico, June 16, 2017, www.politico.eu/article/emmanuel-macron-jupiter-model-unlikely-to-stand-test-of-time-leadership-parliamentary-majority/.

4. Editorial Board, "The Global Carbon Tax Revolt," Wall Street Journal, December 3, 2018, www.wsj.com/articles/the-global-carbon-tax-revolt-1543880507.

5. Reuters, "Brazil's Leader Falsely Blames Leonardo DiCaprio for Amazon Fires," New York Times, November 30, 2019, www.nytimes.com/2019/11/30/world/americas/amazon-fires-dicaprio-bolsonaro.html.

6. Nancy MacLean, Democracy in Chains: The Deep History of the Radical Right's Stealth Plan for America (New York: Viking, 2017), 158–159.

7. Pascale Bonnefoy, "'An End to the Chapter of Dictatorship': Chileans Vote to Draft a New Constitution," New York Times, October 25, 2020, www.nytimes.com/2020/10/25/world/americas/chile-constitution-plebiscite.html.

8. Paolo Tamma and Kalina Oroshakoff, "France's Fuel Tax Retreat Dismays COP24 Climate Talks," Politico, December 4, 2018, www.politico.eu/article/france-fuel-tax-retreat-dismays-cop24-climate-change-summit/.

9. Patrick Galey, "Un Summit 'Parallel Universe' to Climate Emergency: NGOs," Yahoo! News, December 12, 2019, https://news.yahoo.com/un-summit-parallel-universe-climate-emergency-ngos-172846024.html.

10. Quinn Slobodian, Globalists: The End of Empire and the Birth of Neoliberalism (Cambridge, MA: Harvard University Press, 2018), 2.

11. Carl Schmitt, The Nomos of the Earth in the International Law of the Jus Publicum Europaeum (New York: Telos Press, 2003), 235.

12. Slobodian, Globalists, 95.

13. Vijay Prishad, The Darker Nations: The People's History of the Third World (New York: New Press, 2007), 68.

14. Timothy Mitchell, "McJihad: Islam in the U.S. Global Order," Social Text 20, no. 4 (2002): 10.

15. Timothy Mitchell, Carbon Democracy: Political Power in the Age of Oil (London: Verso, 2011), 113.

16. Mitchell, Carbon Democracy, 114.

17. Mitchell, Carbon Democracy, 123.

18. Giuliano Garavini, The Rise and Fall of OPEC in the Twentieth Century (Oxford, UK: Oxford University Press, 2019), 128.

19. Timothy Mitchell, "The Resources of Economics," Journal of Cultural Economy 3, no. 2 (2010): 189–204.

20. Barry Weisberg, *Our Lives Are at Stake: Workers Fight for Health and Safety; Shell Strike 1973* (San Francisco: United Front Press, 1973), www.marxists.org/history/usa/workers/ocaw/1973 /shell-strike.pdf.

21. Adom Getachew, *Worldmaking After Empire: The Rise and Fall of Self-Determination* (Princeton, NJ: Princeton University Press, 2019), 72.

22. Prashad, *Darker Nations*, 13.

23. Getachew, *Worldmaking After Empire*, 74. For detailed account of contests over the UN, see pages 71–106.

24. Declaration on the Establishment of a New International Economic Order, General Assembly Resolution 3201 (S–VI), United Nations, 2010, https://legal.un.org/avl/pdf/ha/ga_3201/ga_3201_ph _e.pdf.

25. Karl Brunner, *Economic Analysis and Political Ideology: The Selected Essays of Karl Brunner*, vol. 1, ed. Thomas Lys (Cheltenham, UK: Edward Elgar, 1996), 152.

26. Victor McFarland, "The New International Economic Order, Interdependence, and Globalization," *Humanity* 6, no. 1 (Spring 2015): 217–233, http://humanityjournal.org/wp-content/uploads /2014/06/HUM-6.1-final-text-McFARLAND.pdf.

27. Daniel P. Moynihan, "The United States in Opposition," *Commentary*, March 1975, www .commentarymagazine.com/articles/daniel-moynihan/the-united-states-in-opposition/.

28. "Share of Oil Reserves, Oil Production and Oil Upstream Investment by Company Type, 2018," International Energy Agency, January 17, 2020, https://www.iea.org/data-and-statistics /charts/share-of-oil-reserves-oil-production-and-oil-upstream-investment-by-company-type-2018.

29. Nils Gilman, "The New International Economic Order: A Reintroduction," *Humanity* 6, no. 1 (March 19, 2015), http://humanityjournal.org/issue6-1/the-new-international-economic -order-a-reintroduction/.

30. Richard T. Ely, "The Economics of Resources or the Resources of Economics," *American Economic Review* 64, no. 2 (May 1974): 1–14, http://pombo.free.fr/solow1974.pdf.

31. John Wihbey, "Jimmy Carter's Solar Panels: A Lost History That Haunts Today," Climate Connections, November 11, 2008, www.yaleclimateconnections.org/2008/11/jimmy-carters-solar -panels/.

32. Jimmy Carter, "Crisis of Confidence," The Carter Center, July 14, 1979, www.cartercenter.org /news/editorials_speeches/crisis_of_confidence.html.

33. See Rick Perlstein, *Reaganland: America's Right Turn 1976–1980* (New York: Simon & Schuster, 2020).

34. See Tim Barker, "Other People's Blood," *n + 1* 34 (Spring 2019), https://nplusonemag.com /issue-34/reviews/other-peoples-blood-2/.

35. Yakov Feygin and Dominik Leusder, "The Class Politics of the Dollar System," Phenomenal World, May 1, 2020, https://phenomenalworld.org/analysis/the-class-politics-of-the-dollar-system.

36. Nancy Pelosi, "Pelosi, Bicameral Delegation Conclude Visit to COP25 Madrid," press release, Speaker of the House newsroom, December 3, 2019, www.speaker.gov/newsroom/12319-1.

37. Megan Cassella, "'We Ate Their Lunch': How Pelosi Got to 'Yes' on Trump's Trade Deal," Politico, December 10, 2010, www.politico.com/news/2019/12/10/democrats-trumps-trade-deal-080719.

38. Scott Wong and Cristina Marcos, "Pelosi's Whiplash Moment Brings Praise and Criticism," *The Hill*, December 10, 2019, https://thehill.com/homenews/house/473970-pelosis-whiplash -moment-brings-praise-and-criticism.

39. Kate Aronoff, "The New U.S. Trade Deal Is Climate Sabotage," *New Republic*, January 17, 2020, https://newrepublic.com/article/156240/new-us-trade-deal-climate-sabotage.

40. Sierra Club, "Trump's NAFTA 2.0: An Environmental Failure," www.sierraclub.org/sites /www.sierraclub.org/files/Trump-NAFTA-Environment-Failure.pdf.

41. Nancy Pelosi, "Transcript of Speaker Pelosi Press Conference Announcing New USMCA Agreement," Speaker of the House, December 10, 2019, www.speaker.gov/newsroom/121019-2.

42. API, "API Commends House Passage of USMCA," American Petroleum Institute, www.api .org/news-policy-and-issues/news/2019/12/19/api-commends-house-passage-of-usmca.

43. UN Environment Programme, "Emission Gap Report," www.unenvironment.org/explore -topics/climate-change/what-we-do/mitigation/emissions-gap-report.

44. CSO Equity Review, *Can Climate Change Fuelled Loss and Damage Ever Be Fair?* (Manila, London, Cape Town, Washington et al.: CSO Equity Review Coalition, 2019), 11, http://civilsociety review.org/report2019/.

45. CSO Equity Review, *Can Climate Change Fuelled Loss and Damage Ever Be Fair?*

46. Damian Carrington, "Wikileaks Cables Reveal How US Manipulated Climate Accord," *Guardian* (Manchester, UK), December 3, 2010, www.theguardian.com/environment/2010/dec/03 /wikileaks-us-manipulated-climate-accord.

47. Suzanne Goldenberg, "US Denies Climate Aid to Countries Opposing Copenhagen Accord," *Guardian* (Manchester, UK), April 9, 2010, www.theguardian.com/environment/2010/apr/09 /us-climate-aid.

48. WHO, *WHO Framework Convention on Tobacco Control* (Geneva: World Health Organization, 2005), https://apps.who.int/iris/bitstream/handle/10665/42811/9241591013.pdf;jsessionid=4C 4C7A4AC591D646AB52F83E0863D6BD?sequence=1.

49. Jan-Philipp Scholz, "Gas Flaring in Niger Delta Ruins Lives, Business," Deutsche Welle, March 11, 2017, www.dw.com/en/gas-flaring-in-the-niger-delta-ruins-lives-business/a-41221653.

50. For a detailed look at the oil industry's use of carbon credits around flaring in Nigeria, see Isaac 'Asume' Osuoka, Gibson Ikanone, Vivian Bellonwu-Okafor, and Orike Didi, *Up in Smoke: Gas Flaring, Communities and Carbon Trading in Nigeria* (Port Harcourt, Nigeria: Social Development Integrated Centre, 2016), https://saction.org/wp-content/uploads/publications/Up_in_Smoke.pdf.

51. Philip Jakpor, interview with author, December 7, 2019.

52. Carbon Brief, "Mapped: The World's Largest CO_2 Importers and Exporters," July 5, 2017, www.carbonbrief.org/mapped-worlds-largest-co2-importers-exporters; and Enno Schröder and Servaas Storm, "Economic Growth and Carbon Emissions: The Road to 'Hothouse Earth' Is Paved with Good Intentions," Working Paper No. 84, Institute for New Economic Thinking, November 2018, www.ineteconomics.org/uploads/papers/WP_84.pdf.

53. Glen P. Peters, Jan C. Minx, Christopher L. Weber, and Ottmar Edenhofer, "Growth in Emission Transfers via International Trade from 1990 to 2008," *PNAS* 108, no. 21 (May 24, 2011): 8903–8908, www.pnas.org/content/108/21/8903.

Chapter 5: New Scenarios

1. Climate Files, "1998 Shell Internal TINA Group Scenarios 1998–2020 Report," Shell, www .climatefiles.com/shell/1998-shell-internal-tina-group-scenarios-1998-2020-report/.

2. Thomas J. Chernak, *Foundations of Scenario Planning: The Story of Pierre Wack* (New York: Routledge, 2017), 99.

3. Chernak, *Foundations of Scenario Planning*, 135.

4. Betty Sue Flowers, interview with author, September 3, 2019.

5. Betty Sue Flowers, interview with author, September 3, 2019.

6. Phia Steyn, "Oil Exploration in Colonial Nigeria, c. 1903–1958," University of Stirling, Scotland, https://dspace.stir.ac.uk/bitstream/1893/2735/1/Oil%20exploration%20in%20colonial%20Nigeria.pdf.

7. Art Kleiner, *The Age of Heretics: A History of the Radical Thinkers Who Reinvented Corporate Management* (San Francisco: Jossey-Bass, 2008), 125.

8. Ed Pilkington, "Shell Pays Out $15.5m over Saro-Wiwa Killing," *Guardian* (Manchester, UK), June 8, 2009, www.theguardian.com/world/2009/jun/08/nigeria-usa.

9. Jenny Andersson, interview with author, January 10, 2020.

10. Royal Dutch/Shell spokeswoman Barbara Calvert, 1986: "'Our policy is that by being there we are doing the right thing, helping to bring about change through our social and employment policies, and by speaking out against the apartheid system," quoted in Robin Toner, "Shell Oil Boycott Urged; Pretoria Policy at Issue," *New York Times*, January 10, 1986, www.nytimes.com/1986/01/10/world /shell-oil-boycott-urged-pretoria-policy-at-issue.html.

11. Shell, *Listening and Responding: The Profits & Principles Advertising Campaign* (London: Shell International Petroleum, 1999), www.documentcloud.org/documents/4425677-Shell-Documents-Trove-2-10.html.

12. T Brand Studio for Shell, "Moving Forward: Net-Zero Emissions by 2070," *New York Times*, www.nytimes.com/paidpost/shell/net-zero-emissions-by-2070.html; and Ben Van Beurden, "What's Needed to Implement the Paris Agreement? Shell's CEO Shares His Views," Politico, August 26, 2019, www.politico.eu/sponsored-content/whats-needed-to-implement-the-paris-agreement-shells-ceo-shares-his-views/.

13. Shell, *Global Supply Model Oil & Gas: A View to 2100*, Shell International BV, 2017, www.shell.com/energy-and-innovation/the-energy-future/scenarios/shell-scenarios-energy-models/global-supply-model/_jcr_content/par/textimage.stream/1500439104411/88d3c54a304eb12000e6f71bf27c322456ca66d5/shell-global-supply-model.pdf. Shell isn't alone in this; a similar document from ExxonMobil, its 2019 outlook for energy, projects no reduction in the company's carbon emissions through 2040 and extolls the company's support for the Paris Agreement. See ExxonMobil, *2019 Outlook for Energy: A Perspective to 2040* (Irving, TX: Exxon Mobil Corporation), https://corporate.exxonmobil.com/-/media/Global/Files/outlook-for-energy/2019-Outlook-for-Energy_v4.pdf.

14. Shell, "Moving with the Times," October 9, 2018, www.shell.com/media/speeches-and-articles/2018/moving-with-the-times.html.

15. Kelly Ilblom and Annmarie Hordern, "Shell CEO Says Blaming Oil Suppliers Won't Solve Climate Change," Bloomberg, October 9, 2019, www.bloomberg.com/news/articles/2019-10-09/shell-ceo-says-blaming-oil-suppliers-won-t-solve-climate-change?sref=tXknHEwo.

16. Naomi Klein, "Climate Rage," *Rolling Stone*, November 12, 2009, www.rollingstone.com/politics/politics-news/climate-rage-193377/.

17. Saikat Chakrabarti, interview with author, November 23, 2018.

18. Bryan Farrell, "Meet the Activists with a Plan to Make Climate Change Matter in Elections," Waging Nonviolence, June 20, 2017, https://wagingnonviolence.org/2017/06/sunrise-movement-climate-change-elections/.

19. Becky Bond and Zack Exley, *Rules for Revolutionaries: How Big Organizing Can Change Everything* (White River Junction, VT: Chelsea Green, 2016).

20. Benjamin Wallace-Wells, "A Night for Pragmatism, and Nancy Pelosi," *New Yorker*, November 7, 2018, www.newyorker.com/news/our-columnists/a-night-for-pragmatism-and-nancy-pelosi.

21. Ryan Grim and Briahna Gray, "Podcast Special: Alexandria Ocasio-Cortez on Her First Weeks in Washington," The Intercept, January 28, 2019, https://theintercept.com/2019/01/28/alexandria-ocasio-cortez-podcast/.

22. Draft Text for Proposed Addendum to House Rules for 116th Congress of the United States, https://web.archive.org/web/20181220022447/https://docs.google.com/document/d/1jxUzp9SZ6-VB-4wSm8sselVMsqWZrSrYpYC9slHKLzo/edit.

23. See Noah Smith, "The Green New Deal Would Spend the U.S. into Oblivion," Bloomberg, February 8, 2019, www.bloomberg.com/opinion/articles/2019-02-08/alexandria-ocasio-cortez-s-green-new-deal-is-unaffordable.

24. Domenico Montanaro, "Poll: Climate Becomes Top Priority for Democrats; Trump Struggles on Race, COVID-19," NPR, September 19, 2020, www.npr.org/2020/09/19/914233038/poll-climate-becomes-top-priority-for-democrats-trump-struggles-on-race-covid-19.

25. Susan Clayton, Christie Manning, Kirra Krygsman, and Meighen Speiser, *Mental Health and Our Changing Climate: Impacts, Implications, and Guidance* (Washington, DC: American Psychological Association, Climate for Health, ecoAmerica, March 2017), www.apa.org/news/press/releases/2017/03/mental-health-climate.pdf.

26. Jason Plautz, "The Environmental Burden of Generation Z," *Washington Post*, February 3, 2020, www.washingtonpost.com/magazine/2020/02/03/eco-anxiety-is-overwhelming-kids-wheres-line-between-education-alarmism/?arc404=true.

27. Mark Fisher, *Capitalist Realism: Is There No Alternative?* (Ropley, Hampshire, UK: O Books, 2009), 5.

28. Luntz Global Partners, "Findings and Insights on GOP Climate Strategy," memo, June 10, 2019, www.clcouncil.org/media/Luntz-Memo-on-GOP-Climate-Strategy.pdf.

Chapter 6: Pick Good! Be Smart!

1. Ali Vingiano, "Meet the College Dropout Who Started an Elite Credit Card That You're Probably Not Cool Enough to Own," BuzzFeed, July 7, 20114, www.buzzfeednews.com/article/alisonvingiano/magnises-credit-card.

2. John Carreyrou, "Theranos Whistleblower Shook the Company—and His Family," *Wall Street Journal*, November 18, 2016, www.wsj.com/articles/theranos-whistleblower-shook-the-companyand-his-family-1479335963.

3. Bethany McLean, *Saudi America: The Truth About Fracking and How It's Changing the World* (New York: Columbia Global Reports, 2019).

4. McLean, *Saudi America*.

5. McLean, *Saudi America*, 34.

6. See Helen Thompson, *Oil and the Western Economic Crisis* (London: Palgrave MacMillan, 2017): 14; and Robert L. Hirsch, Roger Bezdek, and Robert Wendling, "Peaking of World Oil Production: Impacts, Mitigation, & Risk Management," US Department of Energy, Office of Scientific and Technical Information, February 2005, www.osti.gov/servlets/purl/939271.

7. See Paul Mutter, "The Not-About-Iraqi-Oil Iraqi Oil Map," Foreign Policy In Focus (FPIF), January 10, 2010, https://fpif.org/the_not-about-iraqi-oil_iraqi_oil_map/.

8. Atif Mian and Amir Sufi, "Household Debt and Defaults from 2000 to 2010: The Credit Supply View," chap. 11 in *Evidence and Innovation in Housing Law and Policy*, ed. Lee Anne Fennell and Benjamin J. Keys (Cambridge, UK: Cambridge University Press, 2017), www.cambridge.org/core/books/evidence-and-innovation-in-housing-law-and-policy/household-debt-and-defaults-from-2000-to-2010-the-credit-supply-view/422C07E7A01A81DC454A083E9E4911DB/core-reader.

9. L. Randall Wray et al., *Reforming the Fed's Policy Response in the Era of Shadow Banking* (Ford Foundation and Levy Economics Institute of Bard College, April 2015), www.levyinstitute.org/pubs/rpr_4_15.pdf.

10. James Andrew Felkerson, *A Detailed Look at the Fed's Crisis Response by Funding Facility and Recipient*, Public Policy Brief, No. 123 (Annandale-on-Hudson, NY: Levy Economics Institute of Bard College, 2012), www.econstor.eu/bitstream/10419/121982/1/689983247.pdf.

11. Thompson, *Oil and the Western Economic Crisis*, 50.

12. Jennifer A. Dlouhy, "Industry Veteran to Head Energy Department's Oil and Gas Office," *Houston Chronicle*, August 28, 2013, www.houstonchronicle.com/business/article/Industry-veteran-to-head-Energy-Department-s-oil-4769845.php.

13. Kimberly Amadeo, "US Shale Oil Boom and Bust," The Balance, May 1, 2020, www.thebalance.com/us-shale-oil-boom-and-bust-3305553.

14. Robert Grattan, "Abundant Light Crude Creates a Sweet Situation for Refiners," *Houston Chronicle*, August 12, 2014, www.houstonchronicle.com/business/energy/article/Abundant-light-crude-creates-a-sweet-situation-5684930.php.

15. Ahiza Garcia, "The Bay Area's Tech Boom Is Hurting Businesses," CNN, October 15, 2018, https://www.cnn.com/2018/10/15/tech/san-francisco-workers-tech-boom/index.html.

16. Kevin Roose, "The Entire Economy Is MoviePass Now. Enjoy It While You Can," *New York Times*, May 16, 2018, www.nytimes.com/2018/05/16/technology/moviepass-economy-startups.html.

17. McLean, *Saudi America*, 92.

18. Will Tucker, "Crude Export Issue Saw a Surge of Lobbying Ahead of Omnibus," Center for Responsive Politics, December 17, 2015, www.opensecrets.org/news/2015/12/crude-export-issue-saw-a-surge-of-lobbying-ahead-of-omnibus/.

19. Tucker, "Crude Export Issue Saw a Surge of Lobbying Ahead of Omnibus."

20. McLean, *Saudi America*, 79.

21. Producers for American Crude Oil Exports (PACE), "Mission," https://oilexports.com/mission/.

22. Columbia University, "International Launch of the Center on Global Energy Policy," press release, June 6, 2013, www.energypolicy.columbia.edu/international-launch-center-global-energy-policy; and School of International and Public Affairs, *SIPA Annual Report 2012–2013* (New York: Columbia University, 2014), 39, https://web.archive.org/web/20140831192744/https://sipa.columbia.edu/system/files/annual_report_2013.pdf.

23. School of International and Public Affairs, *SIPA Annual Report 2013–2014* (New York: Columbia University, 2015), 56–57, https://issuu.com/columbiauniversitysipa/docs/annual_report _2013-14.

24. School of International and Public Affairs, *SIPA Annual Report 2014–2015* (New York: Columbia University, 2016), 56, https://issuu.com/columbiauniversitysipa/docs/sipa_annual_report _2014_2015.

25. Matt Lee-Ashley and Alison Cassady, "The Environmenal Impacts of Exporting More American Crude Oil," Center for American Progress, August 21, 2015, www.americanprogress.org/issues/green /news/2015/08/21/119756/the-environmental-impacts-of-exporting-more-american-crude-oil/.

26. Jason Furman and Jim Stock, "New Report: The All-of-the-Above Energy Strategy as a Path to Sustainable Economic Growth," The White House, President Barack Obama, May 29, 2014, https://obamawhitehouse.archives.gov/blog/2014/05/29/new-report-all-above-energy-strategy -path-sustainable-economic-growth.

27. Tyson Slocum, interview with author, February 14, 2020.

28. Bill McKibben, "Oil Export Ban Hypocrisy," *The Hill*, December 15, 2015, https://thehill.com /opinion/op-ed/263297-oil-export-ban-hypocrisy.

29. Tim Donaghy, John Noël, and Lorne Stockman, "Policy Briefing: Carbon Impacts of Reinstat-ing the U.S. Crude Export Ban," Greenpeace and Oil Change International, January 28, 2020, https:// www.greenpeace.org/usa/research/crude-export-ban-carbon/.

30. Kathy Hipple, "Bankruptcies in Fracking Sector Mount in 2019," Institute for Energy Eco-nomic and Financial Analysis (IEEFA), January 2020, https://ieefa.org/wp-content/uploads/2020/01 /Bankruptcies-in-Fracking-Sector-Mount-in-2019_January-2020.pdf.

31. Bethany McLean, "The Next Financial Crisis Lurks Underground," *New York Times*, September 1, 2018, www.nytimes.com/2018/09/01/opinion/the-next-financial-crisis-lurks-underground.html.

32. David Roberts, "Rick Perry Tells the Truth About Energy Subsidies, Contradicting His Boss," Vox, August 15, 2018, www.vox.com/energy-and-environment/2018/8/15/17691822/trump -administration-hypocrisy-energy-subsidies-rick-perry.

33. Shakuntala Makhijani, "Cashing in on All of the Above: U.S. Fossil Fuel Production Subsidies under Obama," Oil Change International, July 2014, http://priceofoil.org/content/uploads/2014/07 /OCI_US_FF_Subsidies_Final_Screen.pdf.

34. Kate DeAngelis and Bronwen Tucker, *Adding Fuel to the Fire: Export Credit Agencies and Fossil Fuel Finance* (Friends of the Earth United States and Oil Change International, January 2020), https://1bps6437gg8c169i0y1drtgz-wpengine.netdna-ssl.com/wp-content/uploads/2020/01 /FoE_ECAs_R6_JM.pdf.

35. Mariah Blake, "How Hillary Clinton's State Department Sold Fracking to the World, *Mother Jones*, September/October 2014, www.motherjones.com/politics/2014/09/hillary-clinton -fracking-shale-state-department-chevron/.

36. American Oversight, "DOE Records Regarding U.S. Delegation to Ukraine Led by Former Secretary Perry," DOE-19-1001, DOE-19-1219, DOE-19-1239, US Department of Energy, Ukraine visit May 19–21, 2019, records published April 15, 2020, www.americanoversight.org/document /doe-records-regarding-u-s-delegation-to-ukraine-led-by-former-secretary-perry.

37. Nick Taylor, *American Made: The Enduring Legacy of the WPA; When FDR Put the Nation to Work* (New York: Bantam Books, 2009), 167.

38. Andrew Carnegie, "Wealth," *North American Review*, no. 391 (June 1889), www.swarthmore .edu/SocSci/rbannis1/AIH19th/Carnegie.html.

39. David Jackson, "Obama Vows 'Hands-Off' Approach in GM Stake," ABC News, June 1, 2009, https://abcnews.go.com/Politics/story?id=7731222&page=1.

40. Carter Dougherty, "Sweden's Fix for Banks: Nationalize Them," *New York Times*, January 22, 2009, www.nytimes.com/2009/01/23/business/worldbusiness/23sweden.html.

41. Rainforest Action Network, BankTrack, Indigenous Environmental Network, Oil Change In-ternational, Reclaim Finance, and Sierra Club, *Banking on Climate Change: Fossil Fuel Finance Report 2020*, March 2020, www.ran.org/bankingonclimatechange2019/.

42. Peter Erickson, Adrian Down, Michael Lazarus, and Doug Koplow, "Effect of Subsidies to Fossil Fuel Companies on United States Crude Oil Production," *Nature Energy* 2 (October 2, 2017): 891–898, www.nature.com/articles/s41560-017-0009-8.

43. Brad Plumer, "A Closer Look at Obama's '$90 Billion for Green Jobs,'" *Washington Post*, October 4, 2012, www.washingtonpost.com/news/wonk/wp/2012/10/04/a-closer-look-at-obamas -90-billion-for-clean-energy/.

44. Jeff Brady, "After Solyndra Loss, U.S. Energy Loan Program Turning a Profit," NPR, November 13, 2014, www.npr.org/2014/11/13/363572151/after-solyndra-loss-u-s-energy-loan-program -turning-a-profit. For more on Solyndra, see Mariana Mazzucatto, *The Entrepreneurial State: Debunking Public vs. Private Sector Myths* (New York: Public Affairs, 2015): 138–177.

45. Associated Press, "Softbank Posts $6.4 Billion Loss Due to WeWork as CEO Admits 'Mistaken Investment Moves,'" MarketWatch, November 6, 2019, www.marketwatch.com/story/softbank -posts-64-billion-loss-due-to-wework-as-ceo-admits-mistaken-investment-moves-2019-11-06.

46. Associated Press, "Softbank Posts $6.4 Billion Loss."

47. Ruth DeFries et al., "The Missing Economic Risks in Assessments of Climate Change Impacts," Grantham Research Institute on Climate Change and the Environment, LSE, September 20, 2019, www.lse.ac.uk/GranthamInstitute/publication/the-missing-economic-risks-in-assessments-of -climate-change-impacts/.

48. Chris Lafakis, Laura Ratz, Emily Fazio, and Maria Cosma, "The Economic Implications of Climate Change," Moody's Analytics, June 2019, www.moodysanalytics.com/-/media/article/2019 /economic-implications-of-climate-change.pdf.

Chapter 7: Planning for a Good Crisis

1. "Remarks by President Trump, Vice President Pence, and Members of the Coronavirus Task Force in Press Conference," James S. Brady Press Briefing Room, White House, February 26, 2020, www.whitehouse.gov/briefings-statements/remarks-president-trump-vice-president-pence -members-coronavirus-task-force-press-conference/.

2. Ihab Mikati et al., "Disparities in Distribution of Particulate Matter Emission Sources by Race and Poverty Status," *American Journal of Public Health* 108 (2018): 480–485.

3. Ryan Grim, "Rikers Island Prisoners Are Being Offered PPE and $6 an Hour to Dig Mass Graves," The Intercept, March 31, 2020, https://theintercept.com/2020/03/31/rikers-island-corona virus-mass-graves/.

4. Rachel Abrams and Jessica Silver-Green, "'Terrified' Package Delivery Employees Going to Work Sick," *New York Times*, March 21, 2020, www.nytimes.com/2020/03/21/business/corona virus-ups-fedex-xpo-workers.html.

5. Avie Schneider, "Staggering: Record 10 Million File for Unemployment in 2 Weeks," NPR, April 2, 2020, www.npr.org/sections/coronavirus-live-updates/2020/04/02/825383525/6-6-million -file-for-unemployment-another-dismal-record#:~:text=Updated%20at%2010%3A38%20 a.m.,million%20initial%20claims%20were%20filed.

6. Amanda Holpuch, "Millionaires to Reap 80% of Benefit from Tax Change in US Coronavirus Stimulus," *Guardian* (Manchester, UK), April 15, 2020, www.theguardian.com/world/2020/apr/15 /tax-change-coronavirus-stimulus-act-millionaires-billionaires.

7. David Lawder, "Conservative Groups Advising White House Push Fast Reopening, Not Testing," Reuters, May 1, 2020, www.reuters.com/article/us-health-coronavirus-usa-plans-analysis /conservative-groups-advising-white-house-push-fast-reopening-not-testing-idUSKBN22D6BD.

8. Vaishnavi Chandrashekhar, "How a Communist Physics Teacher Flattened the COVID-19 Curve in Southern India," *Science*, November 9, 2020, www.sciencemag.org/news/2020/11/how -communist-physics-teacher-flattened-covid-19-curve-southern-india.

9. Cara Anna, "As US Struggles, Africa's COVID-19 Response Is Praised," Associated Press, September 22, 2020, https://apnews.com/article/virus-outbreak-ghana-africa-pandemics-donald -trump-0a31db50d816a463a6a29bf86463aaa9.

10. Derek Thompson, "What's Behind South Korea's COVID-19 Exceptionalism?," *Atlantic*, May 6, 2020, www.theatlantic.com/ideas/archive/2020/05/whats-south-koreas-secret/611215/.

11. Laurie Garrett, "Grim Reapers: How Trump and Xi Set the Stage for the Coronavirus Pandemic," *New Republic*, April 2, 2020, https://newrepublic.com/article/157118/trump-xi-jinping-america-china-blame-coronavirus-pandemic.

12. Nina Feldman, "Hahnemann Building's Owner Stymies City's Bid to Lease It for COVID-19 Cases," March 24, 2020, WHYY, https://whyy.org/articles/hahnemann-buildings-owner-stymies-citys-bid-to-lease-it-for-covid-19-cases/.

13. Sarah Owermohle, "The Risk in 'Cutting Red Tape' for Coronavirus Treatments," Politico, March 20, 2020, www.politico.com/newsletters/prescription-pulse/2020/03/20/the-risk-in-cutting-red-tape-for-coronavirus-treatments-488641.

14. Caroline Kelly, "Rep. Katie Porter Gets CDC Chief to Agree to Pay for Coronavirus Testing," CNN, March 12, 2020, www.cnn.com/2020/03/12/politics/katie-porter-cdc-coronavirus-testing-white-board/index.html.

15. Sharon Lerner, "Big Pharma Prepares to Profit from the Coronavirus," The Intercept, March 13, 2020, https://theintercept.com/2020/03/13/big-pharma-drug-pricing-coronavirus-profits/.

16. Kim Willsher, Oliver Holmes, Bethan McKernan, and Lorenzo Tondo, "US Hijacking Mask Shipments in Rush for Coronavirus Protection," *Guardian* (Manchester, UK), April 3, 2020, www.theguardian.com/world/2020/apr/02/global-battle-coronavirus-equipment-masks-tests.

17. EPA, "EPA Announces Enforcement Discretion Policy for COVID-19 Pandemic," press release, March 26, 2020, www.epa.gov/newsreleases/epa-announces-enforcement-discretion-policy-covid-19-pandemic.

18. David Roberts, "Gutting Fuel Economy Standards During a Pandemic Is Peak Trump," Vox, April 2, 2020, www.vox.com/energy-and-environment/2020/4/2/21202509/trump-climate-change-fuel-economy-standards-coronavirus-pandemic-peak.

19. Drilled News kept an impressive tally of climate and environmental rollbacks during the pandemic: Amy Westervelt and Emily Gertz, "The Climate Rules Being Rolled Back During the COVID-19 Pandemic," Drilled News, April 6, 2020, www.drillednews.com/post/the-climate-covid-19-policy-tracker.

20. Michelle Goldberg, "Red States Are Exploiting Coronavirus to Ban Abortion," *New York Times*, April 6, 2020, www.nytimes.com/2020/04/06/opinion/abortion-covid.html.

21. Dave Lawler, "G7 Statement Scrapped After U.S. Insisted Coronavirus Be Called 'Wuhan Virus,'" Axios, March 25, 2020, www.axios.com/wuhan-virus-pompeo-g7-statement-coronavirus-b416d6df-f902-4685-961f-4744b3d19d06.html.

22. Peter Nicholas, "There Are No Libertarians in an Epidemic," *Atlantic*, March 10, 2020, www.theatlantic.com/politics/archive/2020/03/trump-socialism-and-coronavirus-epidemic/607681/.

23. Editorial Board, "Virus Lays Bare the Frailty of the Social Contract," *Financial Times* (London), April 3, 2020, www.ft.com/content/7eff769a-74dd-11ea-95fe-fcd274e920ca.

24. Erin Duffin, "Value of COVID-19 Fiscal Stimulus Packages in G20 Countries as of October 2020, as a Share of GDP," Statista, October 12, 2020, www.statista.com/statistics/1107572/covid-19-value-g20-stimulus-packages-share-gdp/.

25. Nick Taylor, *American Made: The Enduring Legacy of the WPA; When FDR Put the Nation to Work* (New York: Bantam Books, 2009), 21.

26. Herbert Hoover, "Annual Message to the Congress on the State of the Union," December 2, 1930, The American Presidency Project, www.presidency.ucsb.edu/documents/annual-message-the-congress-the-state-the-union-22.

27. Taylor, *American Made*, 25.

28. Herbert Hoover, "Against the Proposed New Deal," Madison Square Garden speech, New York City, October 31, 1932, www.columbia.edu/~gjw10/hoover.newdeal.html.

29. J. J. McEntee, "Federal Security Agency Final Report of the Director of the Civilian Conservation Corps," April 1933 through June 30, 1942, https://livingnewdeal.org/wp-content/uploads/2018/06/CCC-Final-Report-of-Director-1942-.pdf.

30. Michael Hiltzik, *The New Deal: A Modern History* (New York: Free Press, 2011), 67–68.

31. Ira Katznelson, *Fear Itself: The New Deal and the Origins of Our Time* (New York: Liveright, 2014), 36.

32. William E. Leuchtenberg, *Franklin D. Roosevelt and the New Deal: 1932–1940* (New York: Harper Torchbooks, 1963), 121.

33. Paul Krugman, "Not the New Deal," *New York Times*, September 16, 2005, www.nytimes.com/2005/09/16/opinion/not-the-new-deal.html.

34. Tennessee Valley Authority, "Design for the Public Good," www.tva.com/about-tva/our-history/built-for-the-people/design-for-the-public-good.

35. Daniel Aldana Cohen, "A Green New Deal for Housing," *Jacobin*, February 8, 2019, www.jacobinmag.com/2019/02/green-new-deal-housing-ocasio-cortez-climate.

36. Reed Hundt, *A Crisis Wasted: Barack Obama's Defining Decisions* (New York: RosettaBooks, 2019), 161–162.

37. Michael Grunwald, *The New Deal: The Hidden Story of Change in the Obama Era* (New York: Simon & Schuster, 2012), 383.

38. Barack Obama, "Remarks by the President in State of Union Address," Office of the Press Secretary, The White House, January 27, 2010, https://obamawhitehouse.archives.gov/the-press-office/remarks-president-state-union-address.

39. Gabriel Debenedetti, "Biden Is Planning an FDR-Size Presidency," *New York*, May 11, 2020, https://nymag.com/intelligencer/2020/05/joe-biden-presidential-plans.html.

40. Katznelson, *Fear Itself*, 477.

41. David Segal, "Housekeepers Face a Disaster Generations in the Making," *New York Times*, September 24, 2020, www.nytimes.com/2020/09/18/business/housekeepers-covid.html.

42. Katznelson, *Fear Itself*, 485.

43. Rexford G. Tugwell, *The Battle for Democracy* (New York: Columbia University Press, 1935), 203–204.

44. "The Cabinet: Tugwell Upped," *Times*, June 25, 1934, http://content.time.com/time/magazine/0,9263,7601340625,00.html.

45. Chip Gibbons, "The Trial(s) of Harry Bridges," *Jacobin*, September 15, 2016, www.jacobinmag.com/2016/09/harry-bridges-longshore-strike-deportation-communist-party.

46. Raj Patel and Jim Goodman, "The Long New Deal," *Journal of Peasant Studies* 47, no. 3 (April 8, 2020): 440, www.tandfonline.com/doi/abs/10.1080/03066150.2020.1741551.

47. Eric Schickler, "New Deal Liberalism and Racial Liberalism in the Mass Public, 1937–1952," APSA 2010 Annual Meeting Paper, July 19, 2010, https://ssrn.com/abstract=1644391.

48. Hiltzik, *The New Deal*, 53.

49. Christine Berry, "The COVID-19 Pandemic Will Change Everything—for Better or Worse," Blog, Verso, March 24, 2020, www.versobooks.com/blogs/4613-the-covid-19-pandemic-will-change-everything-for-better-or-worse.

Chapter 8: Power to the People

1. Ivan Penn, "PG&E Gives Wildfire Victims More Stock in Bankruptcy Plan," *New York Times*, June 12, 2020, www.nytimes.com/2020/06/12/business/energy-environment/pge-wildfire-victims-stock.html.

2. Brandon Rittiman, "PG&E Disasters Killed 117 People Last Decade," ABC10 News, January 2, 2020, www.abc10.com/article/news/local/wildfire/pge-disasters-killed-117-people-last-decade/103-3ca212b6-c502-4b7f-948e-ad6e73bf55a3.

3. Katherine Blunt, "PG&E Equipment Might Have Ignited Northern California Wildfire," *Wall Street Journal*, October 9, 2020, www.wsj.com/articles/pg-e-equipment-might-have-ignited-northern-california-wildfire-11602284242.

4. Marisa Lagos, "Californians Are Angry at PG&E over Blackouts—And They're not Sparing Newsom," KQED, October 23, 2019, http://kqed.org/news/11782081/poll-shows-anger-at-pge-gavin-newsom-over-blackouts.

5. Taryn Luna, "Californians Want to End PG&E's Operations as They Exist Now, New Poll Says," *Los Angeles Times*, December 10, 2019, www.latimes.com/california/story/2019-12-10/poll -californians-pge-operations.

6. Sharon Beder, *Power Play: The Fight to Control the World's Electricity* (New York: New Press, 2003), 24–28.

7. GE—one of NELA's most active members—founded the Radio Corporation of America and was NBC's longtime parent company, which became a regular gag on the network's long-running comedy *30 Rock*.

8. Beder, *Power Play*, 39.

9. David Morris, "Defending the Public Good: FDR's Portland Speech," Institute for Local Self-Reliance, September 11, 2013, https://ilsr.org/defending-public-good-fdrs-portland-speech/.

10. US Public Works Administration, Mississippi Valley Committee, and M. Llewellyn Cooke, *Report of the Mississippi Valley Committee of the Public Works Administration*, (Washington, DC: GPO, 1934), 77, http://babel.hathitrust.org/cgi/pt?id=mdp.39015031947065;view=1up;seq=77.

11. Michael Hiltzik, *The New Deal: A Modern History* (New York: Free Press, 2011), 73.

12. PEC, "Happy Birthday, LBJ," PEC Cooperative News, August 27, 2019, www.pec.coop /news/2019/happy-birthday-lbj/.

13. Appalachian Voices, "A People's History of the Tennessee Valley Authority," www.tiki-toki .com/timeline/entry/1330290/A-Peoples-History-of-the-Tennessee-Valley-Authority/.

14. Tennesse Valley Energy Democracy Movement, *The People's Vision for a Democratic, Just and Green TVA*, 2020, http://prod.energydemocracyyall.org/wp-content/uploads/2020/04/peoples visionTVA.pdf.

15. U.S. Energy Information Administration (EIA), "Profile Overview," Nebraska State Profile and Energy Estimates, www.eia.gov/state/?sid=NE#tabs-1.

16. Kavya Balaraman, "'Backstop' Bill to Turn PG&E into a Non-profit Headed to California Governor's Desk," Utility Dive, June 29, 2020, www.utilitydive.com/news/backstop-bill-pge-non -profit-california-gov/580773/.

17. Michael McDonald and Mark Chediak, "Hedge Fund Collects $3 Billion in Bet on Wild-life Insurance Claims," Insurance Journal, August 24, 2020, www.insurancejournal.com/news/na-tional/2020/08/24/580129.htm.

18. Marie J. French, "Pressure Mounts on National Grid for Alternatives to Moratorium," Politico, November 19, 2019, www.politico.com/states/new-york/city-hall/story/2019/11/18/pressure -mounts-on-national-grid-for-alternatives-to-moratorium-1227708.

19. IBEW 1245, "A Message from IBEW Workers: State Takeover of PG&E Is Expensive and Dangerous," IBEW Local 1245 fact sheet, www.documentcloud.org/documents/6774097-IBEW -Local-1245-response.html.

20. For a detailed account of state level utility lobbying, see Leah Cardamore Stokes, *Short Cir-cuiting Policy: Interest Groups and the Battle over Clean Energy and Climate Policy in the American States* (New York: Oxford University Press, 2019).

21. David Pomerantz, interview with author, July 31, 2019.

22. There have been some inroads made to fight back against this practice. Challenges brought to the CPUC by the Utility Reform Network managed to stop California IOUs from giving ratepayer cash over to EEI in a 2019 rate case. The Missouri Public Service Commission has taken similar steps, but such payments continue freely through most of the country. In addition to EEI, utility customers variously fund the American Gas Association and US Chamber of Commerce, which have actively battled both renewables and emissions reductions measures at the local, state, and national level.

23. Joby Warrack, "Utilities Wage Campaign Against Rooftop Solar," *Washington Post*, March 7, 2015, www.washingtonpost.com/national/health-science/utilities-sensing-threat-put-squeeze -on-booming-solar-roof-industry/2015/03/07/2d916f88-c1c9-11e4-ad5c-3b8ce89f1b89_story.html ?utm_term=.eea1c5cbab32.

24. David Pomerantz, "EEI Used Anti-clean Energy Campaigns as Role Models in Political Boot Camp for Utility Execs," Energy and Policy Institute, August 27, 2020, www.energyandpolicy.org /eei-campaign-institute/.

25. Ryan Randazzo, "APS Parent Company Spent $37.9M Fighting Clean-Energy Measure," January 17, 2019, azcentral.com, *Arizona Republic* (Phoenix), www.azcentral.com/story/news /politics/arizona/2019/01/17/pinnacle-west-spent-38-million-fight-arizonas-prop-127-clean-energy -measure/2595711002/.

26. See Clean Virginia, "Breaking: Virginia Candidates Refusing Dominion Money Win 50 General Assembly Seats," press release, November 6, 2019, https://web.archive.org/web/201911 08192449/https://www.cleanvirginia.org/breaking-virginia-candidates-refusing-dominion-money -win-nearly-50-general-assembly-seats/; and Mel Leonor, "Democratic Party of Virginia Rejects Political Dollars from Dominion," *Richmond Times-Dispatch*, September 19, 2019, https://richmond .com/news/virginia/democratic-party-of-virginia-rejects-political-dollars-from-dominion/article _d47bd616-46f1-5186-836d-b2af1f1dd091.html.

27. Lee Carter, interview with author, July 24, 2020.

28. Ryan Grim and Akela Lacy, "Ohio Republicans Balked at a Nuclear Bailout, so the Industry Elected New Republicans—and Walked Away with $1.1 Billion," The Intercept, July 26, 2019, https://theintercept.com/2019/07/26/ohio-billion-dollar-nuclear-bailout-firstenergy/?comments =1&menu=1.

29. See Gavin Benke, *Risk and Ruin: Enron and the Culture of American Capitalism* (Philadelphia: University of Pennsylvania Press, 2018).

30. Public Citizen, "The Best Energy Bill Corporations Could Buy," Public Citizen's Analysis of the Domenici-Barton Energy Policy Act of 2005, www.citizen.org/wp-content/uploads/aug2005 ebsum.pdf.

31. Joseph Daniel, "The Coal Bailout Nobody Is Talking About," Blog, Union of Concerned Scientists, September 24, 2018, https://blog.ucsusa.org/joseph-daniel/the-coal-bailout-nobody -is-talking-about.

32. Rocky Mountain Institute, "Case Study: Transition to Clean Energy for Cooperative Utilities; Tri-State's Responsible Energy Plan," January 2020, https://rmi.org/wp-content/uploads/2020/05/tri _states_responsible_energy_program_case_study.pdf.

33. Erik Hatlestad, Katie Rock, and Liz Veazey, *Rural Electrification 2.0: The Transition to a Clean Energy Economy*, CURE, Center for Rural Affairs, We Own It, 2019, www.cureriver.org/wp -content/uploads/2019/06/Rural-Electrification-2.0-report_CURE-1.pdf.

34. C2ES, "U.S. State Electricity Portfolio Standards," Center for Climate and Energy Solutions, www.c2es.org/document/renewable-and-alternate-energy-portfolio-standards/.

35. Gretchen Bakke, *The Grid: The Fraying Wires Between Americans and Our Energy Future* (New York: Bloomsbury, 2016), 26.

36. Amazon Web Services (AWS), "Pacific Gas & Electric Company Case Study," https://aws .amazon.com/solutions/case-studies/pacific-gas-electric/.

Chapter 9: A Postcarbon Democracy

1. Cathy Kunkel, "Saving West Virginia," *Jacobin*, February 27, 2018, www.jacobinmag.com /2018/02/west-virginia-teachers-strike-energy-industry.

2. Luc Cohen and Joshua Schneyer, "When the Oil Boom Went Bust, Oklahoma Protected Drillers and Squeezed Schools," Reuters, May 17, 2016, www.reuters.com/investigates/special-report/usa -oklahoma-bust/.

3. Eric Blanc, "Educators Versus Exxon: An Interview with Angela Reams-Brown and Tia Mills," *Jacobin*, October 26, 2018, www.jacobinmag.com/2018/10/louisiana-exxonmobil-walkout-public -education-funding?fbclid=IwAR1e21vmH3dLEhoVayobanWvw5FZYVE0VV6EzfeQNfkmdibfa LKNVkZGJT8.

4. Michael Mitchell, Michael Leachman, and Kathleen Masterson, *A Lost Decade in Higher Education* (Washington DC: Center on Budget Priorities, 2017).

5. Bill Estep and Liz Mooney, "Trump Promised to Revive KY Coal. He Didn't, but He's Still Expected to Win 'Huge,'" *Lexington Herald Leader*, September 17, 2020, www.kentucky.com/news /politics-government/article245657425.html.

6. Basav Sen, "Labor Unions Must Lead a Just Transition from Fossil Fuels to Clean Energy," *American Prospect*, April 15, 2019, https://prospect.org/labor/labor-unions-must-lead-just-transition-fossil-fuels-clean-energy/.

7. David Biello, "Fact or Fiction?: Natural Gas Will Reduce Global Warming Pollution," *Scientific American*, August 5, 2015, www.scientificamerican.com/article/fact-or-fiction-natural-gas-will-reduce-global-warming-pollution/.

8. Kathy Hipple, "Bankruptcies in Fracking Sector Mount in 2019," Institute for Energy Economics and Financial Analysis (IEEFA), January 2020, https://ieefa.org/wp-content/uploads/2020/01/Bankruptcies-in-Fracking-Sector-Mount-in-2019_January-2020.pdf.

9. Gregory Meyer, "Chevron to Take $10BN Write Down on Shale Gas Glut," *Financial Times* (London), www.ft.com/content/1efa7bfe-1b97-11ea-9186-7348c2f183af.

10. David Roberts, "Fracking May Be a Bigger Climate Problem Than We Thought," Vox, August 29, 2019, www.vox.com/energy-and-environment/2019/8/15/20805136/climate-change-fracking-methane-emissions.

11. EPA, "Understanding Global Warming Potentials," Greenhouse Gas Emissions, US Environmental Protection Agency, www.epa.gov/ghgemissions/understanding-global-warming-potentials.

12. Un Environment Programe, "World's Governments Plan to Produce 120% More Fossil Fuels by 2030 Than Can Be Burned Under 1.5°C Warming," press release, November 20, 2019, www.unenvironment.org/news-and-stories/press-release/worlds-governments-plan-produce-120-more-fossil-fuels-2030-can-be.

13. Ramón A. Alvarez et al., "Assessment of Methane Emissions from the U.S. Oil and Gas Supply Chain," *Science* 361, no. 6398 (July 13, 2018): 186–188; Yuzhong Zhang et al., "Quantifying Methane Emissions from the Largest Oil-Producing Basin in the United States from Space," *Science Advances* 6, no. 17 (April 22, 2020), https://advances.sciencemag.org/content/6/17/eaaz5120; and Nicholas Kusnetz, "Is Natural Gas Really Helping the U.S. Cut Emissions?," InsideClimate News, January 30, 2020, insideclimatenews.org/news/30012020/natural-gas-methane-carbon-emissions.

14. Adam Vaughan, "Fracking Wells in the US Are Leaking Loads of Planet-Warming Methane," *New Scientist*, April 22, 2020, www.newscientist.com/article/2241347-fracking-wells-in-the-us-are-leaking-loads-of-planet-warming-methane/.

15. Sudhanshu Pandey et al., "Satellite Observations Reveal Extreme Methane Leakage from a Natural Gas Well Blowout," *PNAS* 116, no. 52 (December 26, 2019): 26376–26381.

16. David Roberts, "A Beginner's Guide to the Debate over 100% Renewable Energy," Vox, February 6, 2018, www.vox.com/energy-and-environment/2017/4/4/14942764/100-renewable-energy-debate.

17. Ted Boettner, "The Fracking Boom in Appalachia: Big GDP Growth, Small Amount of Jobs and Local Income," Ohio River Valley Institute, September 22, 2020, https://ohiorivervalleyinstitute.org/the-fracking-boom-in-appalachia-big-gdp-growth-small-amount-of-jobs-and-local-income/.

18. "I think for anybody who has spent any time in Alaska, there is an awareness that we all have that we are seeing the impacts of climate change perhaps more readily than in other parts of the country because of our Arctic environment," Alaska senator Lisa Murkowski (Republican) said in a *New York Times* interview. "But we're also a place where we recognize that in order to stay warm, we have a resource that can keep us warm, and oil has been a mainstay for us. We've provided it to the country and that has allowed for jobs and revenues, it has allowed for schools and roads and institutions that everybody else around the country enjoys." See Lisa Friedman, "An Alaska Senator Wants to Fight Climate Change and Drill for Oil, Too," *New York Times*, November 1, 2017, www.nytimes.com/2017/11/01/climate/murkowski-alaska-anwr.html.

19. Jo Michell and Rob Calvert Jump, "Labour, the 'Red Wall,' and the Vicissitudes of Britain's Voting System," openDemocracy, August 20, 2020, www.opendemocracy.net/en/oureconomy/labour-red-wall-and-vicissitudes-britains-voting-system/.

20. Timothy Mitchell, "Fixing the Economy," *Cultural Studies* 12, no. 1 (1998): 82–101, www.tandfonline.com/doi/abs/10.1080/095023898335627.

21. IDMC, "Briefing Paper: Seizing the Momentum: Displacement on the Global Climate Change Agenda," Internal Displacement Monitoring Centre, Geneva, November 2, 2016, www.internal-displacement.org/sites/default/files/publications/documents/20161102-climate-change.pdf.

22. Christian Parenti, *Tropic of Chaos* (New York: Public Affairs, 2011), 7.

23. Will Evans, "Inside Tesla's Factory, a Medical Clinic Designed to Ignore Injured Workers," Reveal, November 5, 2018, www.revealnews.org/article/inside-teslas-factory-a-medical-clinic -designed-to-ignore-injured-workers/; and Julia Carrie Wong, "Tesla Factory Workers Reveal Pain, Injury and Stress: 'Everything Feels Like the Future but Us,'" *Guardian* (Manchester, UK), May 18, 2017, www.theguardian.com/technology/2017/may/18/tesla-workers-factory-conditions-elon-musk.

24. See Myles Lennon and Douglas Rogers, "Decentralizing Energy," *American Anthropologist*, www.americananthropologist.org/2018/07/03/decentralizing-energy/.

25. Kate Aronoff, "Green Tea Party," *Harper's Magazine*, May 30, 2017, https://harpers.org /2017/05/green-tea-party/.

26. R. B. Jackson et al., "Persistent Fossil Fuel Growth Threatens Paris Agreement and Planetary Health," *Environmental Research Letters* 14, no. 12 (December 4, 2019), https://iopscience.iop.org /article/10.1088/1748-9326/ab57b3.

27. See Carbon Tracker, "Decline and Fall: The Size & Vulnerability of the Fossil Fuel System," June 4, 2020, https://carbontracker.org/reports/decline-and-fall/. The same report notes that $4 trillion per year was spent on demand infrastructure, with the largest amounts by far invested in highways.

28. Kevin Crowley and Ashkat Rathi, "Exxon's Plan for Urging Carbon Emissions Revealed in Leaked Documents," Bloomberg, October 5, 2020, www.bloomberg.com/news/articles/2020-10-05 /exxon-carbon-emissions-and-climate-leaked-plans-reveal-rising-co2-output?sref=tXknHEwo.

29. Alec Tyson and Brian Kennedy, "Two-Thirds of Americans Think Government Should Do More on Climate," Pew Research Center, June 23, 2020, www.pewresearch.org/science/2020 /06/23/two-thirds-of-americans-think-government-should-do-more-on-climate/.

30. David Roberts, "Friendly Policies Keep US Oil and Coal Afloat Far More Than We Thought," Vox, July 26, 2018, www.vox.com/energy-and-environment/2017/10/6/16428458/us -energy-coal-oil-subsidies.

31. Jessica Resnick-Ault and Arathy S. Nair, "U.S. Oil Producers on Pace for Most Bankruptcies Since Last Oil Downturn," Reuters, October 1, 2020, www.reuters.com/article/us-usa -shale-bankruptcy-idUSKBN26M7EM.

32. Paul Davidson, "ExxonMobil Posts Biggest Quarterly Profit Ever, $14.8B," ABC News, October 30, 2008, https://abcnews.go.com/Business/story?id=6151102&page=1; and Pippa Stevens, "Exxon Announces U.S. Job Cuts, Global Workforce Could See 15% Reduction," CNBC, October 29, 2020, www.cnbc.com/2020/10/29/exxon-announces-additional-job-cuts-amid-ongoing-covid -19-hit-to-oil-demand.html.

33. Carolyn Davis, "Oxy Taking 'Contrarian Approach' to Net-Zero Emissions by Developing Oil Resources, Reusing CO_2," Natural Gas Intelligence, November 13, 2020, www.naturalgasintel .com/oxy-taking-contrarian-approach-to-net-zero-emissions-by-developing-oil-resources-reusing -co2/.

34. Ben German, "BP Boss: Peak Oil Demand May Have Just Happened," Axios, May 12, 2020, www.axios.com/bp-boss-peak-oil-demand-may-have-just-happened-6bd668ec-b325-472d-af06 -e0747ea4f342.html; Christopher Helman, "BP Set to Slash 10,000 Jobs: 'I Can't Make Your Worries Disappear,' says CEO," *Forbes*, June 8, 2020, www.forbes.com/sites/christopherhelman/2020/06 /08/bp-set-to-slash-10000-jobs-i-cant-make-your-worries-disappear-says-ceo/#3e5b690053bc; and Benji Jones, "One of the World's Biggest Oil Companies Plans to Cut Up to 10% of Its Workforce. Here Are the Jobs Most at Risk," Business Insider, October 1, 2020, www.businessinsider.com /shell-ceo-job-cuts-layoffs-future-of-oil-and-gas-2020-9.

35. Carla Skandier, "Quantitative Easing for the Planet," Democracy Collaborative, August 30, 2018, https://democracycollaborative.org/learn/publication/quantitative-easing-planet.

36. See Steve Coll, *Private Empire: ExxonMobil and American Power* (New York: Penguin Books, 2013).

37. Robert McNally, *Crude Volatility: The History and Future of Boom-Bust Oil Prices* (New York: Columbia University Press, 2017), 77.

38. Gabriel Mathy, "Hot Oil," Phenomenal World, September 5, 2020, www.phenomenalworld .org/analysis/hot-oil.

39. Giuliano Garavini, *The Rise and Fall of OPEC in the Twentieth Century* (Oxford, UK: Oxford University Press, 2019), 128.

40. Bracken Hendricks, Rhiana Gunn-Wright, and Sam Ricketts, "The Greatest Mobilization Since WWII," *Democracy* 56 (Spring 2020), https://democracyjournal.org/magazine/56/the-greatest -mobilization-since-wwii/; and Andrew Bossie and J. W. Mason, "The Public Role in Economic Transformation: Lessons from World War II," Roosevelt Institute, March 26, 2020, https://roosevelt institute.org/publications/the-public-role-in-economic-transformation-lessons-from-world-war-ii/.

41. John H. Ohly, *Industrialists in Olive Drab: The Emergency Operation of Private Industry During World War II* (Washington, DC: Center of Military History, 2000), 4.

42. Cornell Law School, "Presidential Action in the Domain of Congress: The Steel Seizure Case," Legal Information Institute, www.law.cornell.edu/constitution-conan/article-2/section-3 /presidential-action-in-the-domain-of-congress-the-steel-seizure-case.

43. United States Senate, "Roll Call Vote 107th Congress, 1st Session," On Passage of the Bill S.1447 (Aviation and Transportation Security Act), October 11, 2001, www.senate.gov/legislative /LIS/roll_call_lists/roll_call_vote_cfm.cfm?congress=107&session=1&vote=00295.

44. Ian Katz, "Geithner in Book Says U.S. Considered Nationalizing Banks," Bloomberg, May 8, 2014, www.bloomberg.com/news/articles/2014-05-08/geithner-in-book-says-obama-aides -weighed-nationalizing-banks?sref=tXknHEwo; and David Dayen, "He Was the Resistance Inside the Obama Administration," *New Republic*, September 11, 2018, https://newrepublic.com/article /151159/tim-geithner-resistance-inside-obama-administration.

45. Sarah Bloom Raskin, "Why Is the Fed Spending So Much Money on a Dying Industry?," *New York Times*, May 28, 2020, www.nytimes.com/2020/05/28/opinion/fed-fossil-fuels.html.

46. Analysis by Accountable US of U.S. Small Business Administration (SBA), "Paycheck Pro- tection Program (PPP) Report," www.sba.gov/sites/default/files/2020-06/PPP_Report_200530.pdf.

47. Lukas Ross, Alan Zibel, Dan Wagner, and Chris Kuveke, *Big Oil's $100 Billion Bender: How the U.S. Government Provided a Safety Net for the Flagging Fossil Fuel Industry*, Bailout Watch, Friends of the Earth, Public Citizen, September 2020, https://prismic-io.s3.amazonaws.com/bailout/1b1e1458 -bbff-49bc-a636-f6cbd47a88af_Big+Oils+Billion+Dollar+Bender.pdf?utm_campaign=Hot%20 News&utm_medium=email&_hsmi=96461767&_hsenc=p2ANqtz-9lkS8M1whibNfOIgBIeYw7xE _content=96461767&utm_source=hs_email.

48. Gregory Meyer, "Trump Campaign for US Coal Undermined by Industry Slump," *Financial Times* (London), May 18, 2020, www.ft.com/content/39909dd4-7c6c-425e-8f05-4f6e63b60c43.

49. Baker Hughes, "Rig Count Overview & Summary Count," https://rigcount.bakerhughes .com/rig-count-overview.

50. Lisa Friedman, "A Coal Baron Funded Climate Denial as His Company Spiraled into Bank- ruptcy," *New York Times*, December 17, 2019, www.nytimes.com/2019/12/17/climate/murray-energy -climate-denial-coal.html.

51. Joshua Macey and Jackson Salovaara, "Bankruptcy as Bailout: Coal Company Insolvency and the Erosion of Federal Law," *Stanford Law Review* 71, no. 4 (2019): 879–962, www.stanfordlaw review.org/print/volume-71/issue-4/.

52. Jennifer A. Dlouhy, "'Stealth Bailout' Shovels Millions of Dollars to Oil Companies," Bloomberg, May 15, 2020, www.bloomberg.com/news/articles/2020-05-15/-stealth-bailout-shovels -millions-of-dollars-to-oil-companies.

53. U.S. Department of Labor, "Oil, Gas Industy Workers in 9 States Owed More Than $1.6M in Back Wages Ongoing Labor Depatment Enforcement Initiative Finds," press release, March 14, 2016, www.dol.gov/newsroom/releases/whd/whd20160314-1.

54. There's been an ongoing debate as to whether or not the Fed can perform buyouts under its existing authorization section 13(3), which was used extensively during the financial crisis to lend to the banking sector without congressional authorization. Thomas Hanna, of the Democracy Collabo- rative, told me: "Ironically, the Dodd-Frank Act limited the Fed's ability to use 13(3) by imposing lan- guage around 'broad-based eligibility' and restricting use for a single company. So, the debate comes down to the question of what constitutes broad-based eligibility, with some of the people we have

spoken to contending that if a buyout was for a whole sector (i.e., all fossil fuel companies) rather than targeted at just one or two, then it might still qualify under the Fed's existing 13(3) authority. Beyond this, though, as we have seen during COVID, the Fed has wide latitude during a crisis and often makes up the rules as it goes along." See also: Arthur S. Long, "Revised Section 13(3) of the Federal Reserve Act," American Bar Association, March 22, 2019, https://businesslawtoday.org/2019/03/revised-section-133-federal-reserve-act/.

55. Green New Deal Coalition, "Supporting Groups," www.greennewdealforall.org/supporting-groups.

56. Trade Unions for Energy Democracy, website home page, http://unionsforenergydemocracy.org/.

57. Jared Odessky, "Bargaining Green New Deals at Work," Data for Progress, August 1, 2019, www.dataforprogress.org/memos/gnd-unions-contract.

58. Terry O'Sullivan, "To All Members of the LIUNA General Executive Board, District Councils, and Local Unions in the United States," letter, October 26, 2016, www.liunaactionnetwork.org/site/DocServer/2016-10-26_DAPL_Letter_to_Membership_Oct_2016_FINAL.pdf?docID=2721.

59. Kate Aronoff, "LIUNA's Rank-and-File Is Challenging Union Leadership on Standing Rock—and Beyond," *In These Times*, November 7, 2016, http://inthesetimes.com/working/entry/19605/liunas_rank_and_file_is_challenging_union_leadership_on_standing_rockand_be.

60. The Electrical Worker online, "Mass IBEW Fights to Keep Green Jobs Good Jobs," September 2013, www.ibew.org/articles/13ElectricalWorker/EW1309/Mass.IBEWSolar.0913.html.

61. Texas AFL-CIO, "Federal Environmental Policy Must Have Strong Labor and Community," 2019, www.texasaflcio.org/2019-adopted-resolutions/federal-environmental-policy-must-have-strong-labor-and-community. Ryan Pollock's account of working to pass the resolution can be found here: Ryan Pollock, "A Green New Deal Can Win, Even Among Building Trades Unions," *Jacobin*, December 2019, https://jacobinmag.com/2019/12/a-green-new-deal-can-win-even-among-building-trades-unions.

62. Ryan Pollock, interview with author, January 24, 2020.

63. U.S. Bureau of Labor Statistics, "Fastest Growing Occupations," Occupational Outlook Handbook, www.bls.gov/ooh/fastest-growing.htm.

64. J. Mijin Cha, "Climate Jobs Scorecard," Data for Progress, http://filesforprogress.org/memos/Climate_Jobs_Scorecard.pdf.

65. Arthur Neslen, "Spain to Close Most Coalmines in €250M Transition Deal," *Guardian* (Manchester, UK), October 26, 2018, www.theguardian.com/environment/2018/oct/26/spain-to-close-most-coal-mines-after-striking-250m-deal?fbclid=IwAR0iaBEf-2POHIPn17jt7MWRTl8b4RX-JVcBp0gcvVs5X-dDM5hw9iYZgmy8; and European Trade Union Confederation (ETUC), "Spain Guarantees a Just Transition for Miners," www.etuc.org/en/spain-guarantees-just-transition-miners.

66. J. Mijin Cha, *A Roadmap to an Equitable Low-Carbon Future: Four Pillars for a Just Transition*, Climate Equity Network, April 2019, https://dornsife.usc.edu/assets/sites/242/docs/JUST_TRANSITION_Report_FINAL_12-19.pdf.

67. Greg Muttitt, Anna Markova, and Matthew Crighton, *Sea Change: Climate Emergency, Jobs and Managing the Phase-Out of UK Oil and Gas Extraction*, Platform, Oil Change International, and Friends of the Earth Scotland, May 2019, https://platformlondon-org.exactdn.com/wp-content/uploads/2019/05/SeaChange-final-r1.pdf.

68. E. Allison and B. Mandler, "Abandoned Wells," American Geosciences Institute (AGI), 2018, www.americangeosciences.org/geoscience-currents/abandoned-wells.

69. ECONorthwest, *Reclaiming Oil and Gas Wells on Federal Lands: Estimate of Costs*, Center for Western Priorities, February 2018, http://westernpriorities.org/wp-content/uploads/2018/02/Bonding-Report.pdf.

70. Frank Rusco, "Federal Energy Development: Challenges to Ensuring a Fair Return for Federal Energy Resources," Testimony Before the Subcommittee on Energy and Mineral Resources, Committee on Natural Resources, House of Representatives, US Government Accountability Office, September 24, 2019, www.gao.gov/assets/710/701616.pdf.

71. Jason Bordoff, Danile Raimi, and Neelesh Nerurkar, "Green Stimulus for Oil and Gas Workers: Considering a Major Federal Effort to Plug Orphaned and Abandoned Wells," Columbia SIPA

Center on Global Energy Policy, July 20, 2020, www.energypolicy.columbia.edu/research/report/green
-stimulus-oil-and-gas-workers-considering-major-federal-effort-plug-orphaned-and-abandoned.

72. US Environmental Protection Agency (EPA), *Opportunites for Petroleum Brownfields*, March
2014, www.epa.gov/sites/production/files/2014-03/documents/pbfopportunities.pdf.

73. Caroline Spivack, "NYC Passes Its Own 'Green New Deal' in Landmark Vote," Curbed,
Vox Media, April 22, 2019, https://ny.curbed.com/2019/4/18/18484996/nyc-council-passes-climate
-mobilization-act-green-new-deal.

74. Drew Anderson, "$1.7B to Clean Up Orphaned and Abandoned Wells Could Cre-
ate Thousands of Jobs," CBC News, April 17, 2020, www.cbc.ca/news/canada/calgary/
federal-oil-and-gas-orphan-wells-program-1.5535943.

75. Nichola Groom, "States Ask Trump Administration to Pay Laid Off Workers to Plug
Abandoned Wells," Reuters, May 6, 2020, www.reuters.com/article/us-global-oil-usa-wells/states
-ask-trump-administration-to-pay-laid-off-oil-workers-to-plug-abandoned-wells-idUSKBN22I2KA.

76. See Michael Dobson, "Revisiting OPEC's Democratic Roots in the Age of Emergency," *E-International Relations*, January 17, 2020, https://www.e-ir.info/2020/01/17/revisiting
-opecs-democratic-roots-in-the-age-of-climate-emergency/.

77. See Paasha Mahdavi, *Power Grab: Political Surivival through Extractive Resource Nationalization* (Cambridge: Cambridge University Press, 2020).

Chapter 10: Toward a Nonviolent Economy

1. C. L. R. James, *A History of Negro Revolt* (London: Fact, 1938).

2. Edward E. Baptist, *The Half Has Never Been Told: Slavery and the Making of American Capitalism* (New York: Basic Books, 2014).

3. Baptist, *The Half Has Never Been Told*, 322.

4. Hannah Spahn, *Thomas Jefferson, Time, and History* (Charlottesville: University of Virginia
Press, 2011), 15.

5. Baptist, *The Half Has Never Been Told*, 245.

6. Tera W. Hunter, "When Slaveowners Got Reparations," *New York Times*, April 16, 2019,
www.nytimes.com/2019/04/16/opinion/when-slaveowners-got-reparations.html.

7. J.-F. Mercure et al., "Macroeconomic Impact of Stranded Fossil Fuel Asssets," *Nature Climate
Change* 8 (2018): 588–593, www.nature.com/articles/s41558-018-0182-1.

8. W. E. Burghardt Du Bois, "The Freemen's Bureau," *Atlantic*, March 1901, www.theatlantic
.com/magazine/archive/1901/03/the-freedmens-bureau/308772/.

9. Eric Foner, *Reconstruction: America's Unfinished Revolution, 1863–1877* (New York: Harper
Perennial, 2014), 219.

10. Eric Foner, "There Have Been 10 Black Senators Since Emancipation," *New York Times*, February 14, 2020, www.nytimes.com/2020/02/14/opinion/sunday/hiram-revels-reconstruction-150
.html.

11. This was the first film to ever be screened at the White House by Woodrow Wilson.

12. Foner, *Reconstruction*, 596.

13. Foner, *Reconstruction*, 588.

14. Foner, *Reconstruction*, 596.

15. W. E. B. Du Bois, *Black Reconstruction in America* (New York: Free Press, 1998), 30.

16. Philadelphia Energy Authority, "Residential Energy Consumption Survey: Implications for
Philadelphia," Energy News, March 12, 2018, www.philaenergy.org/residential-energy-consump
tion-survey-implications-for-philadelphia/.

17. Jeremy S. Hoffman, Vivek Shandas, and Nicholas Pendleton, "The Effect of Historical Housing Policies on Resident Exposure to Intra-Urban Heat: A Study of 108 US Urban Areas," *Climate* 8,
no. 1 (2020), www.eenews.net/assets/2020/04/23/document_cw_01.pdf.

18. See Robert D. Bullard, "Sacrifice Zones: The Front Lines of Toxic Chemcial Exposure in the United States," *Environmental Health Perspective* 119, no. 6 (June 2011): A266, www.ncbi.nlm.nih .gov/pmc/articles/PMC3114843/.

19. Robin D. G. Kelley, *Hammer and Hoe* (Chapel Hill: University of North Carolina Press, 1990), 45.

20. Kelley, *Hammer and Hoe*, 47.

21. Kelley, *Hammer and Hoe*, 59.

22. Dylan Brown, "Coal: Mining Union Faces 'Life-and-Death' Test," E&E News, April 11, 2017, www.eenews.net/stories/1060052929.

23. For an extended discussion of Black mayors in this era and their relationship to the Black freedom movement, see "Black Faces in High Places," chap. 3 in Keeanga-Yamahtta Taylor, *From #BlackLivesMatter to Black Liberation* (Chicago: Haymarket Books, 2016).

24. See Bill Fletcher, "Governing Socialism," in *We Own the Future: Democratic Socialism— American Style*, ed. Kate Aronoff, Peter Dreier, and Michale Kazin (New York: New Press, 2020); see also Edward Greer, *Big Steel: Black Politics and Corporate Power in Gary, Indiana* (New York: Monthly Review Press, 1979).

25. Ron Grossman, "Flashback: Chicago's Council Wars Pitted Defiant White Aldermen Against a Reform-minded Harold Washington," *Chicago Tribune*, May 9, 2019, www.chicagotribune.com /opinion/commentary/ct-perspec-flashback-chicago-council-wars-washington-vrdolyak-051219 -story.html.

26. Ravi K. Perry, *Black Mayors, White Majorities* (Lincoln: University of Nebraska Press, 2014), https://digitalcommons.unl.edu/cgi/viewcontent.cgi?article=1236&context=unpresssamples.

27. Dena Holland-Neal, "Richard Hatcher's Gary," Belt Magazine, January 10, 2020, https:// beltmag.com/richard-hatcher-gary/.

28. Tejasvi Nagaraja, "On Foreign Policy," in *We Own the Future.*

29. Richard Rothstein, "A 'Forgotten History' of How the U.S. Government Segregated America," interview by Terry Gross, *Fresh Air*, NPR, transcript of 35 min. interview, May 3, 2017, www.npr.org /transcripts/526655831.

30. James Boggs, *The American Revolution: Pages from a Negro Worker's Notebook* (New York: Monthly Review Press, 1963), www.historyisaweapon.com/defcon1/amreboggs.html.

31. See Elizabeth Hinton, *From the War on Poverty to the War on Crime: The Making of Mass Incarceration in America* (Cambridge, MA: Harvard University Press, 2016).

32. See Ruth Wilson Gilmore, *Golden Gulag: Prisons, Surplus, Crisis, and Opposition in Globalizing California* (Berkeley: University of California Press, 2007).

33. Ruth Wilson Gilmore and Craig Gilmore, "Restating the Obvious," in *Indefensible Space: The Architecture of the National Insecurity State*, ed. Michael Sorkin (New York: Routledge, 2013), www.taylorfrancis.com/books/e/9780203939482/chapters/10.4324/9780203939482-8.

34. Franklin Delano Roosevelt, "The Economic Bill of Rights," Historic Documents, January 11, 1944, www.ushistory.org/documents/economic_bill_of_rights.htm.

35. US Supreme Court, *Board of Regents of State Colleges v. Roth, 408 U.S. 564 (1972)*, volume 408, Justia, https://supreme.justia.com/cases/federal/us/408/564/case.html#587.

36. David Stein, "Containing Keynesianism in an Age of Civil Rights: Jim Crow Monetary Policy and the Struggle for Guaranteed Jobs, 1956–1979," chap. 7 in *Beyond the New Deal Order: U.S. Politics from the Great Depression to the Great Recession*, ed. Gary Gerstle, Nelson Lichtenstein, and Alice O'Connor (Philadelphia: University of Pennsylvania Press, 2019), 124.

37. Steven Attewell, *People Must Live by Work: Direct Job Creation in America, from FDR to Reagan* (Philadelphia: University of Pennsylvania Press, 2018), 202.

38. Attewell, *People Must Live by Work,* 207.

39. Stein, "Containing Keynesianism," 129.

40. "Lincoln Memorial Program," March on Washington for Jobs and Freedom, August 28, 1963, www.crmvet.org/docs/mowprog.pdf.

41. Stein, "Containing Keynesianism," 132.

42. Bayard Rustin, "From Protest to Politics: The Future of Civil Rights," *Commentary* 39, no. 2 (February 1965): 6, http://digital.library.pitt.edu/islandora/object/pitt%3A3173506622783 /viewer#page/6/mode/2up.

43. A. Philip Randolph and Bayard Rustin, *A "Freedom Budget" for All Americans: A Summary* (New York: A. Philip Randolph Institute, January 1967), www.crmvet.org/docs/6701_freedom budget.pdf.

44. Stein, "Containing Keynesianism," 129.

45. David P. Stein, "'This Nation Has Never Honestly Dealt with the Question of a Peacetime Economy': Coretta Scott King and the Struggle for a Nonviolent Economy in the 1970s," *Souls* 18, no. 1 (March 2016): 94.

46. Stein, "'This Nation Has Never Honestly Dealt,'" 85–86.

47. Attewell, *People Must Live by Work,* 236.

48. Attewell, *People Must Live by Work,* 237.

49. Attewell, *People Must Live by Work,* 240.

50. Attewell, *People Must Live by Work,* 249.

51. For more on Carter's turn to the right, see Rick Perlstein, *Reaganland: America's Right Turn 1976–1980* (New York: Simon & Schuster, 2020); Jefferson Cowie, *Stayin' Alive: The 1970s and the Last Days of the Working Class* (New York: New Press, 2012); and Judith Stein, *Pivotal Decade: How the United States Traded Factories for Finance in the Seventies* (New Haven, CT: Yale University Press, 2010).

52. The Sentencing Project, "Criminal Justice Facts," The Facts, www.sentencingproject.org /criminal-justice-facts/.

53. Josh Gerstein, "Obama: Government 'Can't Create Jobs,'" Under the Radar (blog), Politico, September 28, 2010, www.politico.com/blogs/under-the-radar/2010/09/obama-government -cant-create-jobs-029576.

54. See Mark Paul, William Darity Jr., and Darrick Hamilton, "The Federal Job Guarantee: A Policy to Achieve Permanent Full Employment," Center on Budget and Policy Priorities, March 9, 2018, www.cbpp.org/research/full-employment/the-federal-job-guarantee-a-policy-to -achieve-permanent-full-employment.

55. Darrick Hamilton, interview with author, March 24, 2018.

56. Archival document tweeted by David Stein on a private account.

57. Samantha Chapman, "Mangroves Protect Coastlines, Store Carbon—and Are Expanding with Climate Change," The Conversation, February 9, 2018, http://theconversation.com /mangroves-protect-coastlines-store-carbon-and-are-expanding-with-climate-change-81445.

58. Kate Aronoff, "The Coronavirus's Lesson for Climate Change," *New Republic*, February 20, 2020, https://newrepublic.com/article/156626/coronaviruss-lesson-climate-change.

59. Randall Wray et al., "Guaranteed Jobs Through a Public Service Employment Program," Policy Note, Levy Economic Institute of Bard College, 2018, www.levyinstitute.org/pubs/pn_18 _2.pdf.

60. See Pavlina R. Tcherneva, *The Case for a Job Guarantee* (Cambridge, UK: Polity Press, 2020); and Stephanie Kelton, *The Deficit Myth: Modern Monetary Theory and the Birth of the People's Economy* (New York: PublicAffairs, 2020).

61. Kate Hamaji et al., *Freedom to Thrive: Reimagining Safety & Security in Our Communities,* The Center for Popular Democracy, Law for Black Lives, Black Youth Project, https://popular democracy.org/sites/default/files/Freedom%20To%20Thrive%2C%20Higher%20Res%20Version.pdf.

62. David E. Rosenbaum, "A Closer Look at Cheney and Halliburton," *New York Times*, September 28, 2004, www.nytimes.com/2004/09/28/us/a-closer-look-at-cheney-and-halliburton.html; and Jonathan Blitzer, "How Climate Change Is Fuelling the U.S. Border Crisis," *New Yorker*, April 3, 2019, www.newyorker.com/news/dispatch/how-climate-change-is-fuelling-the-us-border-crisis.

63. Alexander C. Kaufman, "States Have Put 54 New Restrictions on Peaceful Protest Since Ferguson," HuffPost, June 5, 2020, www.huffpost.com/entry/anti-protest-laws-free-speech_n_5 ed958c3c5b6a11919831bb3; and Julia Carrie Wong and Sam Levin, "Standing Rock Protesters Hold Out Against Extraordinary Police Violence," *Guardian* (Manchester, UK), November 29, 2016, www.theguardian.com/us-news/2016/nov/29/standing-rock-protest-north-dakota-shutdown -evacuation.

64. Rose Braz and Craig Gilmore, "Joining Forces: Prisons and Environmental Justice in Recent California Organizing," *Radical History Review* 96 (Fall 2006): 95–111, http://realcostofprisons

.org/materials/Joining_Forces_Braz.pdf; also see Brett Story and Seth J. Prins, "A Green New Deal for Decarceration," *Jacobin*, August 28, 2019, https://jacobinmag.com/2019/08/green-new-deal-decarceration-environment-prison-incarceration.

65. Movement for Black Lives (M4BL), "Invest-Divest," 2020, https://m4bl.org/policy-platforms/invest-divest/.

66. People's Budget, "Defund the Police. Reimagine Public Safety," https://peoplesbudgetla.com/.

67. Eric Heinz, "Garcetti Halting LAPD Budget Increase, Prohibiting Adding Names to Gang Database," KNBC Channel 4 News, Los Angeles, June 4, 2020, www.nbclosangeles.com/news/local/garcetti-halting-lapd-budget-increase-prohibiting-adding-names-to-gang-database/2374361/.

68. Melissa Gira Grant, "The Pandemic Is the Right Time to Defund the Police," *New Republic*, May 28, 2020, https://newrepublic.com/article/157875/pandemic-right-time-defund-police.

69. Rev. William J. Barber II, "A Third Reconstruction for Our Common Home," in *Winning the Green New Deal: Why We Must, How We Can*, ed. Varshini Prakash and Guido Girgenti (New York: Simon & Schuster, 2020), 204.

Chapter 11: Managing Eco-Apartheid

1. Matt Ford, "Dismantle the Department of Homeland Security," *New Republic*, February 21, 2018, https://newrepublic.com/article/147099/dismantle-department-homeland-security.

2. US Department of Homeland Security, *National Incident Management System*, 3rd ed. (Washington, DC: FEMA, October 2017), www.fema.gov/media-library-data/1508151197225-ced8c603 78c3936adb92c1a3ee6f6564/FINAL_NIMS_2017.pdf.

3. Erik Loomis, "This Day in Labor History: October 19, 1935," Lawyers, Guns & Money (blog), October 19, 2014, www.lawyersgunsmoneyblog.com/2014/10/day-labor-history-october-19-1935.

4. American Public Media, "Fannie Lou Hamer: Testimony Before the Credentials Committee, Democratic National Convention; Atlantic City, New Jersey, August 22, 1964," Say It Plain: A Century of Great American Speeches, http://americanradioworks.publicradio.org/features/sayit plain/flhamer.html.

5. National Oceanic and Atmospheric Administration (NOAA), "Tides & Currents," https://tidesandcurrents.noaa.gov/sltrends/sltrends_station.shtml?id=8534720; www.ft.com/content/f95aa4e2-b3e6-11e7-aa26-bb002965bce8.

6. Richard D. Hylton, "Talking Deals: How Trump Got a Second Chance," *New York Times*, November 22, 1990, www.nytimes.com/1990/11/22/business/talking-deals-how-trump-got-a-second-chance.html?searchResultPosition=15.

7. "Revel Casino Timeline," *Press of Atlantic City*, June 9, 2014, https://pressofatlanticcity.com/revel-casino-timeline/article_544302ae-f037-11e3-8ebf-001a4bcf887a.html.

8. "A Look at Key Moments in the History of Revel Casino Hotel," *Press of Atlantic City*, January 9, 2019, www.pressofatlanticcity.com/a-look-at-key-moments-in-the-history-of-revel-casino-hotel/article_7a8abe14-e84f-5968-a5f8-1b1c74e087e9.html.

9. Nicholas Huba, "City Tax Revenue Down 70 Percent from 2012," *Press of Atlantic City*, July, 28, 2017, https://pressofatlanticcity.com/news/casinos_tourism/city-tax-levy-down-70-percent-from-2012/article_3b4e62c7-1de7-577d-a869-50d818b9ce97.html.

10. See Kenneth Squire, "Carl Ichan, Who's Made a Large Bulk of His Fortune in Energy, Has a New Play in the Industry," CNBC News, May 2, 2020, www.cnbc.com/2020/05/02/carl-icahn-whos-made-a-large-bulk-of-his-fortune-in-energy-has-a-new-play-in-the-industry.html. More than a decade on from negotiating Trump's sweetheart bankruptcy deal, Icahn would lead the charge to oust Aubrey McClendon from Chesapeake Energy.

11. Bloomberg Businessweek, "Everything Is Private Equity Now," October 8, 2019, www.bloomberg.com/news/features/2019-10-03/how-private-equity-works-and-took-over-everything.

12. Haynes and Boone, LLP, *Oil Patch Bankruptcy Monitor*, October 31, 2020, https://www.haynesboone.com/-/media/Files/Energy_Bankruptcy_Reports/Oil_Patch_Bankruptcy_Monitor.

13. Beth Fitzgerald, "Borgata Wins Major Victory on Tax Assessment in Ruling That Could Be Trouble for Atlantic City," NJBIZ, October 21, 2013, https://njbiz.com/borgata-wins -major-victory-on-tax-assessment-in-ruling-that-could-be-trouble-for-atlantic-city/.

14. CBS New York, "Christie Vetoes Bills That Would Help Struggling Atlantic City, Casino Industry," November 9, 2015, https://newyork.cbslocal.com/2015/11/09/christie-vetoes-bills-atlantic-city/.

15. Christian Hetrick, "Atlantic City Bars Offer $1 Shots to Signers of Anti-takeover Petition," Press of Atlantic City, January 26, 2016, https://pressofatlanticcity.com/news/atlantic-city-bars-offer -1-shots-to-signers-of-anti-takeover-petition/article_4aa95a32-c442-11e5-9301-4ba44c5be20c.html.

16. Simon Davis-Cohen, "Meet the Legal Theorists Behind the Financial Takeover of Puerto Rico," Nation, October 30, 2017, www.thenation.com/article/archive/meet-the-legal-theorists -behind-the-financial-takeover-of-puerto-rico/.

17. Clayton P. Gillette and David A. Skeel Jr., eds., "Governance Reform and the Judicial Role in Municipal Bankruptcy," Yale Law Journal 125 (2016): 1150–1237, www.yalelawjournal.org /pdf/a.1150.Gillette-Skeel.1237_pjjg35cu.pdf.

18. See Kim Phillips-Fein, Fear City: New York's Fiscal Crisis and the Rise of Austerity Politics (New York: Metropolitan Books, 2017).

19. Jonathan Oosting, "Michigan Proposal 1: Voters Reject Measure, Repeal Controversial Emergency Manager Law," MLive, November 7, 2012, www.mlive.com/politics/2012/11/election _results_michigan_vote.html.

20. Shea Howell, "Resisting Emergency Management," Infrastructure in America, e-flux Architecture, April 25, 2019, https://power.buellcenter.columbia.edu/essays/resisting-emergency -management.

21. Christian Hetrick, "State Privatizes Trash Collection Without City Council," Press of Atlantic City, May 16, 2017, https://pressofatlanticcity.com/news/state-privatizes-atlantic-city-trash-collec tion-without-city-council/article_0c98e906-4e16-5a7d-999f-70487f2ab6e3.html.

22. Nelson Johnson, Boardwalk Empire: The Birth, High Times, and Corruption of Atlantic City (Medford, NJ: Medford Press Book/Plexus, 2010). For a detailed historical account of Black life in Atlantic City and South Jersey, see Nelson Johnson, The Northside: African Americans and the Creation of Atlantic City (Medford, NY: Plexus, 2010).

23. Zac Spencer, "60 Years Ago, These South Jersey Black Leaders Fought for Change. They See the Same Fight Taking Place Now," Press of Atlantic City, July 11, 2020, www.pressofatlanticcity.com /news/local/60-years-ago-these-south-jersey-black-leaders-fought-for-change-they-see-the-same /article_9579272e-2157-54bb-b854-82faa5de3ab9.html.

24. Leana Hosea and Sharon Lerner, "From Pittsburgh to Flint, the Dire Consequences of Giving Private Companies Responsibility for Ailing Public Water Systems," The Intercept, May 20, 2018, https://theintercept.com/2018/05/20/pittsburgh-flint-veolia-privatization-public-water-systems -lead/.

25. Simon Davis-Cohen, "Meet the Legal Theorists Behind the Financial Takeover of Puerto Rico," Nation, October 30, 2017, www.thenation.com/article/archive/meet-the-legal-theorists -behind-the-financial-takeover-of-puerto-rico/.

26. Carlos Garcia and Jose Ramon Gonzalez, "Update: Pirates of the Caribbean: How Santander's Revolving Door with Puerto Rico's Development Bank Exacerbated a Fiscal Catastrophe for the Puerto Rican People," Hedge Clippers, December 13, 2016, http://hedgeclippers.org/pirates -of-the-caribbean-how-santanders-revolving-door-with-puerto-ricos-development-bank-exacer bated-a-fiscal-catastrophe-for-the-puerto-rican-people/. It's worth noting that Puerto Rico is one of many Caribbean islands to be visited by vultures: Peter James Hudson, "How Wall Street Colonized the Caribbean," Boston Review, June 19, 2019, http://bostonreview.net/race/peter-james-hudson -how-wall-street-colonized-caribbean.

27. Lehman College, "Operation Bootstrap," Department of Latin America, Latino and Puerto Rican Studies, http://lcw.lehman.edu/lehman/depts/latinampuertorican/latinoweb/PuertoRico /Bootstrap.htm.

28. P. Pakdeenurit, N. Suthikarnnarunai, and W. Rattanawong, "Special Economic Zone: Facts, Roles, and Opportunities of Investment," Proceedings of the International MultiConference of Engineers and Computer Scientists, vol. 2, International Association of Engineers, March 12–14, 2014, www.iaeng.org/publication/IMECS2014/IMECS2014_pp1047-1051.pdf.

29. Bonnie Mass, "Puerto Rico: A Case Study of Population Control," *Latin American Perspectives* 4, no. 4 (Autumn 1977): 66–81, http://forzadas.pe/wp-content/uploads/2016/05/Mass_Latin-American-Perspectives-1977.pdf.

30. Margaret Sanger, "The Civilizing Power of Civilization," Margaret Sanger Papers, Sophia Smith Collection, May 12, 1955, www.nyu.edu/projects/sanger/webedition/app/documents/show.php?sangerDoc=236042.xml.

31. Bonnie Mass, "Puerto Rico: A Case Study of Population Control." *Latin American Perspectives* 4, no. 4 (1977): 74.

32. Saqib Bhatti and Carrie Sloan, "Beware of Bankers Bearing Gifts," Refund America Project, February 28, 2017, www.scribd.com/document/338702218/Beware-of-Bankers-Bearing-Gifts.

33. Saqib Bhatti and Carrie Sloan, "Scooping and Tossing Puerto Rico's Future," Refund America Project, August 31, 2016, https://acrecampaigns.org/wp-content/uploads/2020/04/ScoopingandTossingPuertoRicosFuture-Aug2016.pdf.

34. Lara Merling, interview with author, September 21, 2018.

35. Ingrid Vila Biaggi, "No Oversight of $1.5 Billion Electric Project Raises Alarm over Privatization of Puerto Rico's Power," interview by Amy Goodman, Democracy Now!, June 17, 2020, www.democracynow.org/2020/6/17/puerto_rico_power_project_privatization.

36. Julia Keleher, Twitter post, October 26, 2017, 9:33 p.m., https://twitter.com/JuliaBKeleher/status/923724280885661696.

37. A. J. Vicens, "Former Puerto Rico Education Secretary Arrested on Federal Corruption Charges, *Mother Jones*, July 10, 2019, www.motherjones.com/politics/2019/07/julia-keheher-puerto-rico-education-secretary-corruption/.

38. Andre Ujifusa, "Why Julia Keleher Doesn't Want Her Fraud Trial to Be in Puerto Rico," Education Week (blog), December 12, 2019, http://blogs.edweek.org/edweek/campaign-k-12/2019/12/julia-keleher-trial-venue-change-puerto-rico.html.

39. Sarah Jaffe, "Bargaining for the Common Good: A Conversation with Maurice Weeks and Saqib Bhatti," The Baffler, May 5, 2017, https://thebaffler.com/interviews-for-resistance/bargaining-for-the-common-good; see also Tony Vachon, Gerry Hudson, Judith LeBlanc, and Saket Soni, "How Workers Can Demand Climate Justice," *American Prospect*, September 2, 2019, https://prospect.org/labor/workers-can-demand-climate-justice/.

40. Luis J. Valentín Ortiz and Carla Minet, "Las 889 Páginas de Telegram Entre Rosselló Nevaresy Sus Allegados" [The 889 Pages of Telegram Between Rosselló Nevaresy and His Relatives], Centro de Periodism Investigativo, July 13, 2019, http://periodismoinvestigativo.com/2019/07/las-889-paginas-de-telegram-entre-rossello-nevares-y-sus-allegados/.

41. Editorial Board, "Puerto Ricans Are Demanding Better. It's About Time Congress and San Juan Deliver," *Washington Post*, July 19, 2019, www.washingtonpost.com/opinions/puerto-ricans-are-demanding-better-its-about-time-congress-and-san-juan-deliver/2019/07/19/c89eca92-a997-11e9-9214-246e594de5d5_story.html?utm_term=.8a4930c93bbb.

42. Colin McAuliffe, "Memo: The Senate Is an Irredeemable Institution," Data for Progress, December 17, 2019, www.dataforprogress.org/memos/the-senate-is-an-irredeemable-institution.

43. The Seasteading Institute, "Patri Friedman," www.seasteading.org/staff/patri-friedman/.

44. Arizona: Steve Horn, "As Climate Change Burns Arizona, State Has More Imprisoned Firefighters Than Employees," Drilled News, July 24, 2020, www.drillednews.com/post/as-climate-change-burns-arizona-state-has-more-inmate-firefighters-than-employees; and California: Mihir Zaveri, "As Inmates, They Fight California's Fires. As Ex-Convicts, Their Firefighting Prospects Wilt," *New York Times*, November 15, 2018, www.nytimes.com/2018/11/15/us/california-paying-inmates-fight-fires.html.

45. Cecilia Rasmussen, "LAPD Blocked Dust Bowl Migrants at State Borders," *Los Angeles Times*, March 9, 2003, http://articles.latimes.com/2003/mar/09/local/me-then9.

46. ACLU, "The Constitution in the 100-Mile Border Zone," June 21, 2018, www.aclu.org/other/constitution-100-mile-border-zone.

47. Elise Foley, "ICE Director to All Undocumented Immigrants: 'You Need to Be Worried,'" HuffPost, June 13, 2017, www.huffingtonpost.com/entry/ice-arrests-undocumented_us_594027c0e4b0e84514eebfbe.

Chapter 12: Emergency Internationalism

1. Karl Marx, "The Eighteenth Brumaire of Louis Bonaparte," chap. 1 in *Die Revolution*, 1852, www.marxists.org/archive/marx/works/1852/18th-brumaire/ch01.htm.

2. Garrett Hardin, "Population and Immigration: Compassion or Responsibility?" *The Ecologist*, 1977.

3. Garrett Hardin, "The Survival of Nations and Civilization," *Science* 172, no. 3990 (June 25, 1971): 1297, www.sciencemag.org/site/feature/misc/webfeat/sotp/pdfs/172-3990-1297.pdf.

4. Hardin likened the all-seeing power of price mechanisms, neoliberals' passion, to Darwin's process of selection. He corresponded with right-wing economic thinkers, including Friedrich Hayek and Gordon Tullock, and shared plenty of intellectual common ground with both. See, on Hayek finding inspiration in Hardin, Gabriel Oliva, "The Road to Servomechanisms: The Influence of Cybernetics on Hayek from the Sensory Order to the Social Order," Center for the History of Political Economy, Working Paper Series, October 8, 2015, https://papers.ssrn.com/sol3/papers.cfm?abstract_id=2670064; and on Tullock: Jason Oakes, "Rent-seeking and the Tragedy of the Commons: Two Approaches to Problems of Collective Action in Biology and Economics," *Journal of Bioeconomics* 18, no. 2 (2016): 137–151, https://oakesj.files.wordpress.com/2016/06/oakes2016rent.pdf.

5. Garrett Hardin, "Living on a Lifeboat," *BioScience* 24, no. 10 (1974): 561–568, www.garretthardinsociety.org/articles_pdf/living_on_a_lifeboat.pdf.

6. Garrett Hardin, "An Ecolate View of the Human Predicament," The Garrett Hardin Society, talk given in 1981, www.garretthardinsociety.org/articles/art_ecolate_view_human_predicament.html.

7. Nicholas Kulish and Mike McIntire, "Why an Heiress Spent Her Fortune Trying to Keep Immigrants Out," *New York Times*, August 14, 2019, www.nytimes.com/2019/08/14/us/anti-immigration-cordelia-scaife-may.html.

8. Laura Parker, "143 Million People May Soon Become Climate Migrants," *National Geographic*, March 19, 2018, https://news.nationalgeographic.com/2018/03/climate-migrants-report-world-bank-spd/.

9. Amy Harder, "John Kerry, Chuck Hagel Testify on National Security, Climate Change," Axios, April 3, 2019, www.axios.com/john-kerry-chuck-hagel-testify-national-security-climate-change-bc89a8f6-1d9b-4de7-a261-d0e2b3fe7f09.html.

10. John Kerry, "Remarks on Climate Change and National Security," U.S. Department of State, November 10, 2015, https://2009-2017.state.gov/secretary/remarks/2015/11/249393.htm.

11. The Center for Climate & Security, "Defense," Climate and Security Resources: U.S. Government, Defense, http://climateandsecurity.org/resources/u-s-government/defense/.

12. US Department of Defense, "DoD Releases 2014 Climate Change Adaptation Roadmap," press release, October 13, 2014, www.defense.gov/Newsroom/Releases/Release/Article/605221/.

13. Eric Wolff, "James Mattis: Trump's Lone Green Hero?," Politico, December 19, 2016, www.politico.com/story/2016/12/james-mattis-climate-change-trump-defense-232833.

14. Kate Aronoff, "Elizabeth Warren Presses Pentagon on Its Planning for Climate Change," The Intercept, April 15, 2019, https://theintercept.com/2019/04/15/climate-change-military-elizabeth-warren/.

15. US House Select Committee on the Climate Crisis, "Solving the Climate Crisis: The Congressional Action Plan for a Clean Energy Future," https://climatecrisis.house.gov/report.

16. E. Tendayi Achiume, "The Postcolonial Case for Rethinking Borders," *Dissent*, 2019, www.dissentmagazine.org/article/the-postcolonial-case-for-rethinking-borders.

17. Stephanie Leutert, "Climate Change: Induced Migration from Central America," Lawfare (blog), June 21, 2017, www.lawfareblog.com/climate-change-induced-migration-central-america.

18. North American Congress on Latin America (NACLA), "Maquiladoras," https://nacla.org/tags/maquiladoras.

19. Christian Parenti, *Tropic of Chaos: Climate Change and the New Geography of Violence* (New York: Bold Type Books, 2012), 226.

20. Daniel Denvir, *All American Nativism: How the Bipartisan War on Immigrants Explains Politics as We Know It* (New York: Verso, 2020).

21. Greg Grandin, *The End of the Myth: From the Frontier to the Border Wall in the Mind of America* (New York: Metropolitan Books, 2019), 277.

22. Abby Smith, "Majority of Young Conservatives Say Climate Policy Will Sway Their Votes: Poll," *Washington Examiner*, February 10, 2020, www.washingtonexaminer.com/policy/energy/majority-of-young-conservatives-say-climate-policy-will-sway-their-votes-poll.

23. Eric L. Ray, "Mexican Repatriation and the Possibility for a Federal Cause of Action: A Comparative Analysis on Reparations," *University of Miami Inter-American Law Review* 37, no. 1 (2005): 176.

24. Ryan Devereaux, "Border Patrol Launches Militarized Raid of Borderlands Humanitarian Aid Camp," The Intercept, August 2, 2020, https://theintercept.com/2020/08/02/border-patrol-raid-arizona-no-more-deaths/.

25. Adam Tooze, "How Coronavirus Almost Brought Down the Global Financial System" *Guardian* (Manchester, UK), April 14, 2020, www.theguardian.com/business/2020/apr/14/how-coronavirus-almost-brought-down-the-global-financial-system; and Trevor Jackson, "The Sovereign Fed," *Dissent*, April 16, 2020, www.dissentmagazine.org/online_articles/the-sovereign-fed.

26. InfluenceMap, "Necessary Intervention or Excessive Risk?," June 23, 2020, https://influencemap.org/report/Necessary-Intervention-or-Moral-Hazard-5e42adc35b315cc44a75c94af4ead29c.

27. U.S. Bureau of Statistics, "Unemployment Rates for Metropolitan Areas," October 28, 2020, www.bls.gov/web/metro/laummtrk.htm.

28. Yanis Varoufakis, *The Global Minotaur: America, Europe and the Future of the Global Economy* (London: Zed Books, 2013).

29. Nathan Tankus, "The Federal Reserve's Coronavirus Crisis Actions, Explained (Part 3)," MR Online, March 30, 2020, https://mronline.org/2020/03/30/the-federal-reserves-coronavirus-crisis-actions-explained-part-3/.

30. David Adler and Andres Arauz, "It's Time to End the Fed's 'Monetary Triage,'" *Nation*, March 23, 2020, www.thenation.com/article/economy/economy-fed-imf/.

31. Elliot Smith, "Emerging Market Currencies Have Been Hit by the Coronavirus, but Analysts Say It's Not All Bad News," CNBC News, April 14, 2020, www.cnbc.com/2020/04/14/emerging-market-currencies-have-been-hammered-by-covid-19.html.

32. Jonathan Báez, "La inversión en salud se redjó un 36% en 2019" [Investment in Health Fell by 36% in 2019], Ecuador Today, https://ecuadortoday.media/2020/03/31/la-inversion-en-salud-se-redujo-un-36-en-2019/.

33. Ana Lucía Badillo Salgado and Andrew M. Fischer, "Ecuador, COVID-19 and the IMF: How Austerity Exacerbated the Crisis," Bliss (blog), International Institute of Social Studies (ISS), April 9, 2020, https://issblog.nl/2020/04/09/covid-19-ecuador-covid-19-and-the-imf-how-austerity-exacerbated-the-crisis-by-ana-lucia-badillo-salgado-and-andrew-m-fischer/.

34. Sydney Maki, Ben Bartenstein, and Stephan Kueffner, "Wall Street Hopes More Time to Pay Can Save Ecuador from Default," Bloomberg, April 16, 2020, www.bloomberg.com/news/articles/2020-04-16/wall-street-hopes-more-time-to-pay-can-save-ecuador-from-default; Gideon Long and Colby Smith, "Ecuador Reaches Deal to Postpone Debt Repayment until August," *Financial Times*, April 17, 2020, www.ft.com/content/e1622284-102c-48f0-b45d-dadb81579d9d.

35. Nancy Lee, "Trillions for SDGs? Time for a Rethink," Center for Global Development, January 22, 2019, www.cgdev.org/blog/trillions-sdgs-time-rethink.

36. bnamericas, "Spotlight: Alberto Fernández's Vaca Muerta Plan," November 18, 2019, www.bnamericas.com/en/features/spotlight-alberto-fernandezs-vaca-muerta-plan.

37. See Vincent Bevins, *The Jakarta Method: Washington's Anticommunist Crusade and the Mass Murder Program That Shaped Our World* (New York: PublicAffairs, 2020).

38. Olúfẹ́mi O. Táíwò and Beba Cibralic, "The Case for Climate Reparations," *Foreign Policy*, October 10, 2020, https://foreignpolicy.com/2020/10/10/case-for-climate-reparations-crisis-migration-refugees-inequality/.

39. Kevin P. Gallagher and Richard Kozul-Wright, *A New Multilateralism for Shared Prosperity: Geneva Principles for a Global Green New Deal*, Global Development Policy Center, UNCTAD, https://unctad.org/en/PublicationsLibrary/gp_ggnd_2019_en.pdf.

40. Todd Tucker, "There's a Big New Headache for the Green New Deal," *Washington Post*, June 28, 2019, www.washingtonpost.com/politics/2019/06/28/theres-big-new-headache-green -new-deal/.

41. Just a handful of law firms worldwide have the expertise to navigate arbitration proceedings. Teams of lawyers often make $3,000 per head per hour in cases that can stretch on for years, meaning that even countries that win ISDS disputes can run up legal bills in the tens of millions of dollars.

42. Investment Policy Hub, Investment Dispute Settlement Navigator, UNCTAD, https:// investmentpolicy.unctad.org/investment-dispute-settlement.

43. Thomas Dauphin, *Beyond Repair? The Energy Charter Treaty* (Brussels, Belgium: Friends of the Earth Europe, December 2019), www.foeeurope.org/sites/default/files/eu-us_trade_deal /2019/media_briefing_-_beyond_repair_-_the_energy_charter_treaty.pdf.

44. Todd Tucker, "How to Fix the Most Controversial Element of Trade Deals," Politico, September 21, 2016, www.politico.com/agenda/story/2016/09/fix-isds-trade-deals-000204.

45. Sean McElwee, "Memo: Voters See Glaring Environmental Problems in the USMCA," Data for Progress, December 17, 2019, www.dataforprogress.org/memos/voters-see-glaring-environ mental-problems-in-the-usmca.

46. On the uncertainty of appellate bodies, see: Todd N. Tucker, "RIP, World Trade Organization?," *Nation*, December 9, 2019, www.thenation.com/article/archive/wto-trade-tariff-trump/.

47. Todd Tucker, "The Green New Deal Has an International Problem," Progressive International, May 28, 2020, https://progressive.international/blueprint/fff9daea-4a91-4de6-9865-b960 7912de53-todd-tucker-the-green-new-deal-has-an-international-law-problem/en.

48. D2D, Data to Decisions, "Federal Fleet Open Data Visualization," GSA (General Services Administration), https://d2d.gsa.gov/report/federal-fleet-open-data-visualization.

49. FDIC, *How America Banks: Household Use of Banking and Financial Services: 2019 FDIC Survey*, Federal Deposit Insurance Corporation, www.fdic.gov/householdsurvey/.

50. Destin Jenkins, "The Fed Could Undo Decades of Damage to Cities. Here's How," *Washington Post*, April 27, 2020, www.washingtonpost.com/outlook/2020/04/27/fed-could-undo -decades-damage-cities-heres-how/.

51. Jasson Perez, "The Federal Reserve Can Help Workers in a Tie of Crisis, *Jacobin*, May 6, 2020, www.jacobinmag.com/2020/05/federal-reserve-coronavirus; and for more see: Mike Konczal and J. W. Mason, "A New Direction for the Federal Reserve: Expanding the Monetary Policy Toolkit," Roosevelt Institute, November 30, 2017, https://rooseveltinstitute.org/expanding -monetary-policy-toolkit/.

52. David Adler and Daniel Bessner, "To End Forever War, End the Dollar's Global Dominance," *New Republic*, January 28, 2020, https://newrepublic.com/article/156325/end-forever-war -end-dollars-global-dominance.

53. See Andrew Bossie and J. W. Mason, "The Public Role in Economic Transformation: Lessons from World War II," Roosevelt Institute, March 26, 2020, https://rooseveltinstitute.org/publications /the-public-role-in-economic-transformation-lessons-from-world-war-ii/.

54. Adam Hochschild, "How Texaco Helped Franco Win the Spanish Civil War," *Mother Jones*, March 29, 2016, www.motherjones.com/politics/2016/03/texaco-franco-spanish-civil-war-rieber.

55. John Helveston and Jonas Nahm, "China's Key Role in Scaling Low-Carbon Energy Technologies," *Science* 366, no. 6467 (2019): 794–796, https://science.sciencemag.org/content /366/6467/794.

56. Tobita Chow and Jake Werner, *The US-China Trade War: A Progressive Internationalist Alternative* (New York: Rosa Luxemburg Stiftung, January 2020), www.rosalux-nyc.org/wp-content /files_mf/uschinatradewarfinalengweb31.pdf.

57. Adam Tooze, "Whose Century?," *London Review of Books*, July 30, 2020, www.lrb.co.uk /the-paper/v42/n15/adam-tooze/whose-century.

58. Anton Jäger and Noam Maggor, "A Popular History of the Fed," Phenomenal World, October 1, 2020, https://phenomenalworld.org/analysis/a-popular-history-of-the-fed.

Conclusion: We Can Have Nice Things

1. Josh White, vocalist "Freedom Road," by Langston Hughes (lyrics) and Emerson Harper (music), originally recorded 1944, on *That's Why We're Marching: World War II and the American Folksong Movement*, Smithsonian Folkways Recordings—SFW40021, released 1996.

2. Susan Steiner, "Top Five Regrets of the Dying," *Guardian* (Manchester, UK), February 1, 2012, www.theguardian.com/lifeandstyle/2012/feb/01/top-five-regrets-of-the-dying.

3. Greg Grandin, *The End of the Myth: From the Frontier to the Border Wall in the Mind of America* (New York: Metropolitan Books, 2019), 138.

4. Ruth Wilson Gilmore, *Golden Gulag: Prisons, Surplus, Crisis, and Opposition in Globalizing California* (Berkeley: University of California Press, 2007).

5. Will Feuer, "Gilead's Coronavirus Treatment Remdesivir to Cost $3,120 per U.S. Patient with Private Insurance," CNBC News, June 29, 2020, www.cnbc.com/2020/06/29/gileads-coronavirus -treatment-remdesivir-to-cost-3120-for-us-insured-patients.html.

6. Chuck Collins, "US Billionaire Wealth Surges to $584 Billion, or 20 Percent, Since the Beginning of the Pandemic," Institute for Policy Studies, June 18, 2020, https://ips-dc.org/us-billionaire -wealth-584-billion-20-percent-pandemic/.

7. Larry Elliott, "Economics: Whatever Happened to Keynes' 15-Hour Working Week?," *Guardian* (Manchester, UK), August 31, 2008, www.theguardian.com/business/2008/sep/01/economics.

8. Economic Policy Institute, "The Productivity–Pay Gap," July 2019, www.epi.org/produc tivity-pay-gap/.

9. David Rosnick and Mark Weisbrot, *Are Shorter Work Hours Good for the Environment? A Comparison of U.S. and European Energy Consumption* (Washington, DC: Center for Economic and Policy Research, Decemer 2006), https://cepr.net/documents/publications/energy_2006_12.pdf.

10. John F. Helliwell, Richard Layard, Jeffrey D. Sachs, and Jan-Emmanuel De Deve, eds., *World Happiness Report* (New York: Sustainable Development Solutions Network, March 20, 2020), http:// worldhappiness.report/.

11. Gallup, "How Does the Gallup World Poll Work?," www.gallup.com/178667/gallup-world -poll-work.aspx.

12. Arthur Neslen, "Norway Pledges to Become Climate Neutral by 2030," *Guardian* (Manchester, UK), June 15, 2016, www.theguardian.com/environment/2016/jun/15/norway-pledges-to-become -climate-neutral-by-2030; Reuters Staff, "Finland Wants EU to Agree Plan for Net-Zero Carbon Footprint," March 4, 2019, Reuters, www.reuters.com/article/us-climate-change-eu-finland/finland-wants -eu-to-agree-plan-for-net-zero-carbon-footprint-idUSKCN1QL11A; and Helle Gronli, "Renewable Thermal Heating: Lessons from Scandinavia," Yale Center for Business and the Environment, November 18, 2016, https://cbey.yale.edu/our-stories/renewable-thermal-heating-lessons-from-scandinavia.

13. Daniel Aldana Cohen, "A Green New Deal for Housing," *Jacobin*, February 8, 2019, https:// jacobinmag.com/2019/02/green-new-deal-housing-ocasio-cortez-climate.

14. Richard Florida, "How 'Social Infrastructure' Can Knit America Together," Bloomberg CityLab, September 11, 2018, www.citylab.com/life/2018/09/how-social-infrastructure-can-knit -america-together/569854/.

15. US Environmental Protection Agency, "Fast Facts on Transportation Greenhouse Gas Emissions," Green Vehicle Guide, 2018, www.epa.gov/greenvehicles/fast-facts-transportation-green house-gas-emissions.

16. Daniel Aldana Cohen, "Seize the Hamptons," *Jacobin*, October 3, 2014, www.jacobinmag .com/2014/10/seize-the-hamptons/.

17. Chris Mooney, "The Surprising Link Between Things That Make Us Happy and Things That Save Energy," *Washington Post*, December 19, 2014, www.washingtonpost.com/news/wonk /wp/2014/12/19/good-news-for-couch-potatoes-staying-home-and-sleeping-in-are-good-for-the -environment/?utm_term=.a928a4e6524c.

18. Daniel Kahneman and Alan B. Krueger, "Developments in the Measurement of Subjective Well-Being," *Journal of Economic Perspectives* 20, no. 1 (Winter 2006): 3–24, https:// international.ucla.edu/media/files/Kahneman.pdf?AspxAutoDetectCookieSupport=1&Aspx AutoDetectCookieSupport=1.

19. Michael A. McCarthy, "Democratic Socialism Isn't Social Democracy," *Jacobin*, August 7, 2018, https://jacobinmag.com/2018/08/democratic-socialism-social-democracy-nordic-countries.

20. Dan Kois, "Iceland's Water Cure," *New York Times Magazine*, April 19, 2016, www.nytimes.com/2016/04/24/magazine/icelands-water-cure.html?_r=1.

21. Michael Hiltzik, *The New Deal: A Modern History* (New York: Free Press, 2011), 65.

22. Rexford G. Tugwell, *The Battle for Democracy* (New York: Columbia University Press, 1935), 184.

23. Nick Taylor, *American-Made: The Enduring Legacy of the WPA; When FDR Put the Nation to Work* (New York: Bantam Books, 2009), 276.

24. Amie Wright, "Zora Neale Hurston and the Depression-Era Federal Writers' Project," NYPL Blogs, New York Public Library, January 8, 2014, www.nypl.org/blog/2014/01/08/zora-neale-hurston-federal-writers-project.

25. Zeke Hausfather, "Explainer: How 'Shared Socioeconomic Pathways' Explore Future Climate Change," Carbon Brief, April 19, 2018, www.carbonbrief.org/explainer-how-shared-socio economic-pathways-explore-future-climate-change.

26. John F. Helliwell and Haifang Huang, "New Measures of the Costs of Unemployment: Evidence from the Subjective Well-being of 3.3 Million Americans," National Bureau of Economic Research, Working Paper 16829, February 2011, www.nber.org/papers/w16829.

27. See Melinda Cooper, *Family Values: Between Neoliberalism and the New Social Conservatism* (New York: Zone Books, 2017).

28. Kristen R. Ghodsee, *Why Women Have Better Sex Under Socialism and Other Arguments for Economic Independence* (New York: Bold Type Books, 2019), 10.

29. Ghodsee, *Why Women Have Better Sex*, 136.

30. Katie J. M. Baker, "Cockblocked by Redistribution: A Pick-up Artist in Denmark," *Dissent*, Fall 2013, www.dissentmagazine.org/article/cockblocked-by-redistribution.

31. Robin D. G. Kelley, *Freedom Dreams: The Black Radical Imagination* (Boston: Beacon Press, 2002), 8.

32. Bertrand Russell, "In Praise of Idleness," *Harper's Magazine*, October 1932, https://harpers.org/archive/1932/10/in-praise-of-idleness/3/.

INDEX

Kate Aronoff is a staff writer at the *New Republic* covering climate and energy issues. A frequent contributor to *The Intercept,* her work has appeared in the *New York Times,* the *Nation, Dissent, Rolling Stone,* the *Guardian,* and *Harper's,* among other outlets. She was previously a fellow at Type Media Center and a writing fellow at *In These Times,* and is a contributing editor at *Waging Nonviolence.* Aronoff is the coeditor of *We Own the Future: Democratic Socialism, American Style* and the coauthor of *A Planet to Win: Why We Need a Green New Deal.* She lives in Brooklyn.